A-LEVEL AND AS-LEVEL   MATHEMATICS

LONGMAN A-LEVEL AND AS-LEVEL REVISE GUIDES

Series editors
Geoff Black and Stuart Wall

Titles available
Art and Design
Biology
Business Studies
Chemistry
Computer Science
Economics
English
French
Geography
Mathematics
Modern History
Physics
Sociology

A-LEVEL
AND AS-LEVEL

LONGMAN
REVISE
GUIDES

MATHEMATICS

Michael Kenwood
Cyril Moss

Longman

Longman Group UK Limited,
Longman House, Burnt Mill, Harlow,
Essex CM20 2JE, England
and Associated Companies throughout the world.

© Longman Group UK Limited 1990

First published 1990
Fifth impression 1992

British Library Cataloguing in Publication Data

Kenwood, H.M.
  Mathematics
  1. Mathematics
  I. Title II. Moss, Cyril
  510

  ISBN 0-582-05165-7

Set in 10/12pt Century Old Style

Produced by Longman Singapore Publishers Pte Ltd
Printed in Singapore

EDITORS' PREFACE

Longman A Level Revise Guides, written by experienced examiners and teachers, aim to give you the best possible foundation for success in your course. Each book in the series encourages thorough study and a full understanding of the concepts involved, and is designed as a subject companion and study aid to be used throughout the course.

Many candidates at A Level fail to achieve the grades which their ability deserves, owing to such problems as the lack of a structured revision strategy, or unsound examination technique. This series aims to remedy such deficiencies, by encouraging a realistic and disciplined approach in preparing for and taking exams.

The largely self-contained nature of the chapters gives the book a flexibility which you can use to your advantage. After starting with the background to the A, AS Level and Scottish Higher courses and details of the syllabus coverage, you can read all other chapters selectively, in any order appropriate to the stage you have reached in your course.

Geoff Black and Stuart Wall

ACKNOWLEDGEMENTS

We would like to thank Geoff Black and Stuart Wall for their kind help and understanding in the production of this book, and for guiding us at each stage of production. We would like in particular to thank Mrs Peggy Sabine for reading the manuscript and making many extremely helpful amendments and contributions.

We are also most indebted to the following GCE A Level Examination Boards for permission to quote questions from past examination papers:

Associated Examining Board
University of Cambridge Local Examinations Syndicate
Joint Matriculation Board
University of London Schools Examinations Board
University of Oxford, Delegacy of Local Examinations
Northern Ireland Schools Examinations Council.

The above Boards do not accept responsibility for the answers we have given to their questions, and consequently we accept full responsibility and apologise for any errors we may have made.

CONTENTS

Circular Motion

NAMES AND ADDRESSES OF THE EXAM BOARDS

Associated Examining Board (AEB)
Stag Hill House
Guildford
Surrey GU2 5XJ

University of Cambridge Local Examinations Syndicate (UCLES)
Syndicate Buildings
1 Hills Road
Cambridge CB1 1YB

Joint Matriculation Board (JMB)
Devas St
Manchester M15 6EU

University of London Schools Examination Board (ULSEB)
Stewart House
32 Russell Square
London WC1B 5DN

Northern Ireland Schools Examination Council (NISEC)
Beechill House
42 Beechill Road
Belfast BT8 4RS

Oxford and Cambridge Schools Examination Board (OCSEB)
10 Trumpington Street
Cambridge CB2 1QB

Oxford Delegacy of Local Examinations (ODLE)
Ewert Place
Summertown
Oxford OX2 7BX

Scottish Examination Board (SEB)
Ironmills Road
Dalkeith
Midlothian EH22 1BR

Welsh Joint Education Committee (WJEC)
245 Western Avenue
Cardiff CF5 2YX

CHAPTER 1

INTRODUCTION AND SYLLABUS CONTENT

GETTING STARTED

The majority of students studying mathematics at GCE A-level follow a course which includes approximately 50% pure mathematics and 50% applied mathematics – the applied mathematics being normally either mechanics or statistics or, in the case of a few Examining Boards, a mixture of both.

A few of the topics studied at GCSE will be extended at A-level, but on the whole most topics will be completely new to you. They will also be developed to a greater level of understanding than in your earlier mathematical education. It is quite likely that you will be required to depend more on your own initiative, both with regard to organising your study and reading time and seeking out new and additional information to supplement your school/college course. This will require a considerable amount of organisation, self-discipline, and a determination not to put off to tomorrow what can easily be done today. Above all, do not fall into the trap of thinking you can consolidate the first-year work in the second year and the second-year work in the revision period leading up to the examinations. That is the road to disaster. The road to success is the development of a steady working routine throughout the whole of your course.

A-level and Scottish Higher examinations

In this guideline book we have tried to cover the work required in most GCE A-level single mathematics syllabuses and in Scottish Higher Examinations. There will be sections which you will not require, so look carefully at the syllabus of the Board whose examination you will be taking and pick out the relevant parts. To help you in this task we have tried to compile a summary of the syllabuses as relevant to each chapter (Table 1.1), but you are advised not to depend entirely on it – simply to compare it with that issued by your Board.

To all students taking AS mathematics examinations

There are many syllabuses and options available at AS-level and it is crucial that you identify your own course and its syllabus content. Write to your Examination Board to obtain the correct syllabus or obtain a copy for the correct year from your teachers. Once you have the syllabus draw up a grid identifying *all* the topics contained in the syllabus.

For those taking *AS Pure Mathematics*, you will find that Chapters 3–13 inclusive, contain all you need *but that some topics may be omitted* (for AS) that are included there.

For those taking other options, you should read through the total list in the book identifying those included in your syllabus. This is particularly important for those taking examinations at AS with titles such as *Mathematics Contrasting* and *Mathematics Complementary*. These will contain some *pure maths*, some *mechanics* and some *statistics* in many cases.

ESSENTIAL PRINCIPLES

SOLVING PROBLEMS

A considerable amount of your time will be spent on solving mathematical problems. This is really the only way to learn mathematics. Within each chapter we have therefore included a considerable number of worked examples and exercises. You should try to work through the exercises as you study the chapter, calling upon the worked examples for help as and when necessary. At the end of each pure mathematics chapter we include examples from recent previous Board examination papers. In mechanics and statistics, Board examination questions tend to include work involving a combination of chapters. We therefore include Board questions at the end of those *sections* rather than at the end of each chapter.

MARK SCHEMES

> Remember the three types of marks: M, A and B.

When an examination paper is marked, three types of marks are normally considered: method marks M; accuracy marks A, dependent on M; B marks which are really accuracy marks not dependent on an M mark.

The method marks M are given for knowing and carrying out a correct method for solving a section of the problem. For example, in pure mathematics the solution of the trigonometric equation $5\sin^2\theta + 2\cos\theta = 2$, $0 < \theta < 180°$, could attract an M1 for using the substitution $\sin^2\theta = 1 - \cos^2\theta$.

Accuracy marks are then awarded for carrying out the method correctly. In the above example use of the substitution and rearrangement of the equation to:
$5\cos^2\theta - 2\cos\theta - 3 = 0$ could attract an A1;
factorisation to $(5\cos\theta + 3)(\cos\theta - 1) = 0$ an M1;
and the two solutions $\theta = 90°$, A1, $\theta = 126·9°$, A1.

Sometimes the work involved in a part of a question is too small for it to rate more than one mark. In such cases where an M mark would not be particularly appropriate we use a B mark. For example, if you were asked to state the range of the function $f : x \mapsto e^{x-2}$, $x > 2$, the answer, range $= (1, \infty)$ could attract B1. Any other answer B0.

In longer questions where it is quite likely that a candidate may make a mistake this does not mean he/she would necessarily lose all further A marks. Frequently the examiner decides that an incorrect part of a question should be followed through for accuracy marks. In this case, A marks are usually denoted by AF1 or A1$\sqrt{}$. Examples of marking are shown in the pure mathematics chapters.

SYLLABUS CONTENT AND COURSES

The following analysis of syllabus content for each Examination Board applies to mathematics taken at GCE Advanced level and Scottish Higher as a single subject. Boards update their syllabuses quite frequently so it is very important to use the table only as a guide and to check carefully with the syllabus issued by your Board for the year in which you sit the examination.

For the Advanced Supplementary (AS) Level exam there are currently over 20 syllabuses, covering between half and two thirds of the topics listed in the grid. We suggest you construct your *own* grid for your syllabus, using the dashed column alongside the main table.

Key
AEB   Associated Examining Board
CAM   University of Cambridge Local Examinations Syndicate
LON   University of London Schools Examinations Board
OXF   University of Oxford, Delegacy of Local Examinations
JMB   Joint Matriculation Board
WJEC Welsh Joint Education Committee
SUJB Southern Universities Joint Board for School Examinations
NI   Northern Ireland Schools Examinations Council
O&C   Oxford and Cambridge Schools Examination Board
SMP   School Mathematics Project
MEI   Mathematics in Education and Industry Schools Project
SH   Scottish Higher

PURE MATHS	GCE A-Level Mathematics											SH	AS
	AEB	CAM	LON	OXF	JMB	WJEC	SUJB	NI	O&C	SMP	MEI		
Chapter 3 Algebra													
Algebra of polynomials	√	√	√	√	√	√	√	√	√	√	√	√	
Rational functions	√	√	√	√	√	√	√	√	√	√	√	√	
Remainder, factor theorems	√	√	√	√	√	√	√	√	√	√	√	√	
Quadratic equations	√	√	√	√	√	√	√	√	√	√	√	√	
Inequations	√	√	√	√	√	√	√	√	√	√	√	√	
Partial fractions	√	√	√	√	√	√	√	√	√	√	√	√	
Indices	√	√	√	√	√	√	√	√	√	√	√	√	
Logarithms	√	√	√	√	√	√	√	√	√	√	√	√	
Chapter 4 Series													
Arithmetic progressions	√	√	√	√	√	√	√	√	√	√	√	√	
Geometric progressions	√	√	√	√	√	√	√	√	√	√	√	√	
Binomial theorem	√	√	√	√	√	√	√	√	√	√	√		
Series expansions	√	√	√	√	√	√	√	√	√	√	√		
Summation of series	√	√	√	√	√	√	√	√	√				
Induction		√							√		√		
Chapter 5 Trigonometry													
Angle measure	√	√	√	√	√	√	√	√	√	√	√	√	
Circular functions	√	√	√	√	√	√	√	√	√	√	√	√	
General angles	√	√	√	√	√	√	√	√	√	√	√	√	
Trigonometrical graphs	√	√	√	√	√	√	√	√	√	√	√	√	
Inverse trigonometric functions	√	√	√	√	√	√	√	√	√	√	√	√	
Identities	√	√	√	√	√	√	√	√	√	√	√	√	
Equations	√	√	√	√	√	√	√	√	√	√	√	√	
Sine and cosine rules	√	√	√	√	√	√	√	√	√	√	√		
2- and 3-dimensional problems	√	√	√	√	√	√	√	√	√	√	√	√	
Chapter 6 Basic Differentiation													
Basic rules	√	√	√	√	√	√	√	√	√	√	√	√	
Implicit functions	√	√	√	√	√	√	√	√	√	√	√	√	
Parametric equations	√	√	√	√	√	√	√	√	√	√	√	√	
Logarithmic differentiation	√	√	√	√	√	√	√	√	√	√	√	√	
Differential relationships	√	√	√	√	√	√	√	√	√	√	√	√	
Chapter 7 Applications of Differentiation													
Rates of change	√	√	√	√	√	√	√	√	√	√	√	√	
Tangents and normals	√	√	√	√	√	√	√	√	√	√	√	√	
Maxima and minima	√	√	√	√	√	√	√	√	√	√	√	√	
Newton–Raphson formula	√	√	√	√	√	√			√	√	√	√	
Chapter 8 Elementary Coordinate Geometry													
Straight line and circle	√	√	√	√	√	√	√	√	√	√	√	√	
Curve sketching	√	√	√	√	√	√	√	√	√	√	√	√	
Polar coordinates		√									√		
Transformations	√	√	√		√	√	√	√	√	√	√	√	√
Experimental laws and reduction to linear form	√		√		√	√		√					

PURE MATHS	GCE A-Level Mathematics											SH	AS
	AEB	CAM	LON	OXF	JMB	WJEC	SUJB	NI	O&C	SMP	MEI		
Chapter 9 Basic Integration													
Basic results	✓	✓	✓	✓	✓	✓	✓	✓	✓	✓	✓	✓	
Substitution	✓	✓	✓	✓	✓	✓	✓	✓	✓	✓	✓	✓	
Trigonometric functions	✓	✓	✓	✓	✓	✓	✓	✓	✓	✓	✓	✓	
Integration by parts	✓	✓	✓	✓	✓	✓	✓	✓	✓	✓	✓	✓	
Definite Integrals	✓	✓	✓	✓	✓	✓	✓	✓	✓	✓	✓	✓	
Chapter 10 Applications of Integration													
Areas, volumes of revolution	✓	✓	✓	✓	✓	✓	✓	✓	✓	✓	✓	✓	
Centroids	✓		✓		✓	✓	✓	✓	✓		✓		
Moment of inertia		✓											
Trapezoidal rule	✓	✓	✓	✓	✓	✓	✓	✓	✓	✓	✓		
Simpson's rule	✓				✓	✓	✓	✓	✓	✓	✓		
First-order variables separable	✓	✓	✓	✓	✓	✓	✓	✓	✓	✓	✓		✓
First-order linear					✓						✓		
Chapter 11 Complex Numbers													
Basic rules	✓	✓	✓	✓	✓	✓	✓	✓	✓	✓	✓		
Argand diagram	✓	✓	✓	✓	✓	✓	✓	✓	✓	✓	✓		
Modulus, argument and polar form	✓	✓	✓	✓	✓	✓	✓	✓	✓	✓	✓		
Chapter 12 Vectors													
Basic rules	✓	✓	✓	✓	✓	✓	✓	✓	✓	✓	✓		
Vector geometry	✓	✓	✓	✓	✓	✓	✓	✓	✓	✓	✓		
Scalar product	✓	✓	✓	✓	✓	✓	✓	✓	✓	✓	✓		
Chapter 13 Functions													
Definition	✓	✓	✓	✓	✓	✓	✓	✓	✓	✓	✓		
Inverse function	✓	✓	✓	✓	✓	✓	✓	✓	✓	✓	✓		
Composite functions	✓	✓	✓	✓	✓	✓	✓	✓	✓	✓	✓		
Exponential and logarithmic functions	✓	✓	✓	✓		✓	✓	✓	✓	✓	✓		

MECHANICS	GCE A-Level Mathematics											SH	AS
	AEB	CAM	LON	OXF	JMB	WJEC	SUJB	NI	O&C	SMP	MEI		
	636	9205	371	9850	162	A2							
Chapter 14 Linear Motion													
Uniform acceleration	✓	✓	✓	✓	✓	✓	✓	✓	✓	✓	✓		
Non-uniform acceleration	✓	✓	✓	✓	✓	✓	✓	✓	✓	✓	✓		
Simple harmonic motion	✓			✓	✓	✓	✓	✓	✓	✓	✓		
Chapter 15 Motion in two dimensions													
Projectiles under gravity	✓	✓	✓	✓	✓	✓	✓	✓	✓	✓	✓		
Uniform circular motion	✓	✓		✓	✓	✓	✓	✓	✓	✓	✓		
Motion in a vertical circle under gravity					✓	✓				✓	✓		
Velocity and acceleration as time derivatives of displacement	✓	✓	✓	✓	✓	✓	✓	✓	✓	✓	✓		
Relative motion	✓		✓	✓	✓	✓	✓	✓	✓	✓	✓		

MECHANICS	GCE A-Level Mathematics											SH	AS
	AEB	CAM	LON	OXF	JMB	WJEC	SUJB	NI	O&C	SMP	MEI		
	636	9205	371	9850	162	A2							
Chapter 16 Basic Kinetics													
Newton's laws of motion	✓	✓	✓	✓	✓	✓	✓	✓	✓	✓	✓		
Connected particles	✓	✓	✓	✓	✓	✓	✓	✓	✓	✓	✓		
Work and energy, power	✓	✓	✓	✓	✓	✓	✓	✓	✓	✓	✓		
Work–energy principle	✓	✓	✓	✓	✓	✓	✓	✓	✓	✓	✓		
Forces in circular motion	✓	✓	✓	✓	✓	✓	✓	✓	✓	✓	✓		
Forces in elastic strings	✓	✓	✓	✓	✓	✓	✓	✓	✓	✓	✓		
Chapter 17 Momentum and Impulsive Forces													
Impulse of a force	✓	✓	✓	✓	✓	✓	✓	✓	✓	✓	✓		
Principle of conservation of momentum	✓	✓	✓	✓	✓	✓	✓	✓	✓	✓	✓		
Newton's experiment law for collisions	✓		✓	✓	✓	✓	✓	✓	✓		✓		
Chapter 18 Statics													
Friction	✓	✓	✓	✓	✓	✓	✓	✓	✓	✓	✓		
Forces on particles	✓	✓	✓	✓	✓	✓	✓	✓	✓	✓	✓		
Moments	✓		✓	✓	✓	✓	✓	✓			✓		
Systems of forces in a plane	✓		✓	✓	✓	✓	✓		✓		✓		
Centres of mass of uniform rigid bodies	✓		✓		✓	✓	✓	✓	✓		✓		
Equilibrium of a rigid body	✓		✓	✓	✓	✓	✓	✓	✓		✓		
Chapter 19 Questions and Answers on Mechanics	✓	✓	✓	✓	✓	✓	✓	✓	✓	✓	✓		

STATISTICS	GCE A-Level Mathematics											SH	AS
	AEB	CAM	LON	OXF	JMB	WJEC	SUJB	NI	O&C	SMP	MEI		
	646	9205	374	9850	162	A2							
Chapter 20 Descriptive Statistics													
Frequency distributions, histograms	✓	✓	✓	✓	✓	✓	✓	✓	✓	✓	✓	✓	
Mode, median, IQR	✓	✓	✓	✓	✓	✓	✓	✓	✓	✓	✓	✓	
Mean, variance, standard deviation	✓	✓	✓	✓	✓	✓	✓	✓	✓	✓	✓	✓	
Index numbers and other applications			✓				✓					✓	
Chapter 21 Probability													
Permutations and combinations	✓	✓	✓	✓	✓	✓	✓	✓	✓		✓	✓	
Probability laws	✓	✓	✓	✓		✓	✓	✓	✓	✓	✓		
Conditional probability	✓	✓	✓	✓	✓	✓	✓	✓	✓	✓	✓	✓	
Chapter 22 Discrete Probability Functions													
Binomial distribution	✓	✓	✓	✓	✓	✓	✓	✓	✓	✓	✓	✓	
Poisson distribution	✓	✓	✓	✓	✓	✓	✓						
Geometric distribution			✓										
Uniform distribution	✓	✓	✓	✓	✓	✓	✓	✓	✓	✓	✓		

STATISTICS	GCE A-Level Mathematics											SH	AS
	AEB	CAM	LON	OXF	JMB	WJEC	SUJB	NI	O&C	SMP	MEI		
	646	9205	374	9850	162	A2							
Chapter 23 Continuous Probability Functions													
PDF	√	√	√	√	√	√	√	√	√	√	√	√	
CDF	√	√	√	√	√	√	√	√	√		√		
Normal distribution	√	√	√	√	√	√	√	√	√	√	√	√	√
Exponential distribution			√										
Normal distribution as an approximation of other distributions	√	√	√	√	√	√		√	√	√	√	√	√
Chapter 24 Distribution of Sample Means and Central Limit Theorem													
Central limit theorem	√	√	√	√	√	√	√	√	√	√	√		
Chapter 25 Hypothesis Testing													
Hypothesis tests	√	√	√	√	√				√	√	√		
χ^2 test	√			√			√		√				
Chapter 26 Linear Regression													
Linear regression	√		√	√	√	√	√	√	√		√	√	
Correlation, rank correlation	√		√	√			√		√			√	
Chapter 27 Questions and Answers on Statistics	√	√	√	√	√	√	√	√	√	√	√	√	

Table 1.1 Syllabus coverage chart

GETTING STARTED

You *must* make sure you do the following:

1 Read a number of past papers carefully and be very familiar with their style and requirements.

2 Do work through at least the past five years' papers, especially if there has been no change in the style or format. If there has been a change make sure you have seen the specimen paper issued by the Examination Board.

3 Do a rough calculation to determine how much time you should spend answering a question. If it is a three-hour paper and you have to answer seven questions, then you have approximately 25 minutes per question. If there is a Section A (40 marks) and Section B (60 marks) then you have approximately 70 minutes to answer Section A. Then if you have to answer four questions from Section B you have approximately 25 minutes per question. Papers which have questions of variable length and where you are required to answer all questions usually have a mark assignment at the side of each question. Simply determine the time allocation per mark and multiply that figure by the mark allocation; e.g. full marks $= 100$, mark allocation for a $2\frac{1}{2}$-hour paper is equivalent to 1 mark per $1\cdot5$ minutes; a 6-mark question therefore has a time allocation of 9 minutes.

4 Do get used to working in a neat and orderly manner. Experience shows that those candidates who take care in the presentation of their work make fewer errors. This not only leads to higher marks, but also to the saving of time and less panic should a question not work out nicely.

5 Do get used to showing all of your working. It is your responsibility to show the examiner what you are doing, not the examiner's to guess. Do not cut corners, the examiner cannot give marks for method when little or none appears in an answer to a question. This is particularly relevant to the use of calculators.

6 Do make sure you appear in the examination room in a fresh and composed way; not worn out by last-minute revision. Remember there is no substitute for good steady work throughout your course. If you have done this then you have nothing to fear. One examination paper is very much like another. Indeed, looking at past papers will show that you will almost certainly have seen many of the questions before, apart from a slight twist here and there together with different numerical values.

7 Do make yourself familiar with the formula booklet, but do not learn to depend exclusively upon it. Make sure you learn the formulae and only use the booklet when doubt arises.

ESSENTIAL PRINCIPLES

The day has arrived, you have worked steadily throughout your course so there is no need to panic. Take a few deep breaths to compose yourself and then settle down to enjoy yourself and justify yourself.

Make sure you do the following:

1   Read quickly through the paper to get a feel for the demands of each question; note in particular the questions that you think you will attempt. Do not completely discard, at this stage, any of the questions, for in mathematics you never know how difficult a question is until you have tried it.

> *A good start will give you confidence.*

2   Pick out the question that looks the most familiar and the easiest and attempt this question first. Experience has shown that candidates who get a good start to a paper usually continue in that vein. Never make the mistake of saving your best question for the end. Many papers where all questions must be attempted are already graded with the shorter and easier questions at the beginning. In such cases you may find it best to work through the questions in the order they are given.

3   Read the questions carefully and do make sure you copy out any equations or information given in the question correctly. Pay particular attention to signs, indices and the order of numbers. Each year many marks are lost through the miscopying of information. **But never waste time copying out the question itself.**

4   Do write out your answers in a neat and orderly manner. Paper is not restricted so do not write here, there and everywhere. Make sure your writing is legible, and continue progressively down a page stating exactly what you are doing. You must not leave the examiner to guess.

5   Do not cut corners and do use brackets wherever necessary. More mistakes are made through the omission of brackets than anything else.

> *Make sure you obey all the instructions.*

6   When you have completed a question, read through the question on the question paper again to make sure that you have a) not omitted any part of it, b) given the answer in the form required, e.g. to 3 significant figures or to 3 decimal places or radians rather than degrees, or used $g = 10$ rather than $g = 9 \cdot 8$ as instructed at the beginning of the paper. This is another area where marks are often lost unnecessarily.

7   Remember that unless otherwise instructed you can leave answers in terms of e, π, g etc. In solving a differential equation you need not rearrange your answer in the form $y = f(x)$ unless the wording of the question requires such a rearrangement.

8   Do remember that if an answer is given on the question paper the examiner is particularly careful to make sure your answer has been arrived at in a correct manner. It is no use faking the answer amidst a mound of indecipherable working!

9   Remember that when the question says 'hence or otherwise' the 'hence' is only suggesting a method of solution. It is, however, usually the most suitable method.

10 Do use a diagram (especially in applied mathematics) and transfer the information given in the question to the diagram.

At some stage or other in the examination you are almost certain to experience difficulty with a question. You cannot get the given answer; an equation which you would expect to factorise easily does not or you just get bogged down in a tangled mass of arithmetic and algebraic manipulation. Don't panic and do not continue with your solution, especially if you are running out of the allotted time for that question. Proceed as follows:

1   Go back and read the question again to make sure you have used the correct figures and information. If you have, quickly go through your solution to try to find an error. If you cannot find an error either leave the question until later and come back to it, time permitting, or indeed if you do have time to spare start the question again. Errors in a solution often only come to light when the solution is restarted. *Do not cross out any of your previous working.* Work not crossed out must be marked by the examiner, but crossed out work subsequently replaced by other attempts may not be marked. You will not lose marks by leaving incorrect work which is then replaced by correct work. If you do leave the question in the hope of returning to it later, then leave sufficient space for the later working.

2   If you have difficulty with one of the longer type of questions where intermediate answers are usually given and you cannot get the answer, even after repeated

checking, then do not continue with your answer. Instead, take the answer given on the question paper and try to complete the question. You will not only have easier working, but you will also be able to get all of the remaining method and accuracy marks for the rest of the question. If you continue with your answer you can easily lose most of the ensuing accuracy marks.

WHAT NOT TO DO

1 Do not write your answers in red ink or pencil. Red ink should not be used anywhere in your solutions.
2 Do not waste time:
   a) underlining your answers
   b) writing QED or QEF
   c) writing out the question before starting your answer
   d) drawing diagrams with drawing instruments, but learn instead to draw freehand.
3 Do not cross out any of your working unless you are absolutely certain you do not want it to be looked at.
4 Do not use rough paper, blotting paper or graph paper to work out a solution which is then partly copied into your answer book.
5 Do not write essays telling the examiner how to answer a question. Instead answer the question. Such essays do not gain marks.
6 Do not waste time 'faking' answers. You will not get away with it.

ALGEBRA

GETTING STARTED

An important element of GCE A-level and AS-level is algebra. You will have met many ideas involved in algebra at GCSE-level, but at A-level and AS-level it is necessary to consolidate those ideas and develop them further. In this chapter we inform you of those properties you must certainly commit to memory and be able to use and develop further.

Unless you do develop and enhance these communication skills much of the rest of your course will be more difficult than it need be. Our experience as examiners is that many marks are unnecessarily lost through basic errors.

BASIC ALGEBRAIC OPERATIONS

Arithmetic deals with definitions and processes involving numbers, each of which has a definite value. **Algebra** is the extension of arithmetic to expressions which besides involving ordinary arithmetical numbers, also contain letters (symbols) which often do not have a single definite value; e.g. $7a$, $13a - 2b$, $x^2 - y^2$, $4\alpha + 5\beta^2 - 6\gamma$, (where α, β and γ are Greek letters).

It is essential for you to be able to handle expressions involving letters as confidently as those involving numbers.

ESSENTIAL PRINCIPLES

KEY DEFINITIONS

The following names should be noted:

- **Monomial:**   An algebraic expression consisting of only one term,

$$7x,\ 8x^2,\ 9ab, \frac{-2x}{\sqrt{y}}$$

- **Binomial:**   An algebraic expression consisting of two terms,

$$2x - y,\ x^3 + y^3, \frac{x}{y} - \frac{y}{x^2}$$

- **Power:**
- **Degree:**   When a product is formed by multiplying together letters of the same kind, the number of the letters forming the product is said to be the power (or degree) of that letter;
  $$x.x.x = x^3 \Rightarrow \text{third power, or degree } 3$$
  $$7y^6 \Rightarrow \text{sixth power, or degree } 6$$

- **Index:**   The figure indicating the power is the index,
  $$x^3, \text{ index } 3;\ 7y^6, \text{ index } 6$$

- **Integral term:**   A term consisting of the product of a number of letters and/or numbers, so that only multiplication and neither addition nor subtraction nor division occurs.
  $$6xy^2,\ -3x^{10},\ yz\sqrt{2},\ 7$$

- **Rational term:**   A term in which the indices of the letters are rational.

$$6xy^2, \frac{x}{y^2}, \frac{7}{x}$$

- **Irrational term:**   A term in which one or more of the indices of the letters are irrational.
  $$x^{\sqrt{2}},\ x^2 y^{\sqrt{5}}$$

POLYNOMIALS

A **polynomial** is an algebraic expression consisting of a number of terms, each of which is integral and rational.

$$2x + 3y + 5z,\ x^2 y - 2y^2 x + 3x^2 + 4y - 2$$

A frequently occurring polynomial is one involving a single variable

$$\Rightarrow a_0 + a_1 x + a_2 x^2 + \ldots + a_n x^n = \sum_{r=0}^{n} a_r x^r \text{ where } a_0,\ a_1,\ a_2 \ldots a_n \text{ are constants. It is said}$$

> Polynomials of degree n often occur.

to be a polynomial of degree n, where n is the highest power of x. You must be able to add, subtract, multiply and divide polynomials. Use the same rules as in arithmetic, but leave spaces for missing terms.

Worked example
Add $1 + 2x + x^3$ to $2 - 3x + x^2 - 5x^4$

$$
\begin{array}{l}
1 + 2x + \quad\ x^3 \\
2 - 3x + x^2 \qquad - 5x^4 \\
\hline
3 -\ \ x + x^2 + x^3 - 5x^4
\end{array}
$$

Worked example
Subtract $3x^2 - 2xy + 5y^3$ from $-x^2 + 2xy + 3y^3$

$$
\begin{array}{l}
-x^2 + 2xy + 3y^3 \\
\ 3x^2 - 2xy + 5y^3 \\
\hline
-4x^2 + 4xy - 2y^3
\end{array}
$$

Worked example
Multiply $2 - 3x + x^3$ by $x + 3x^2 - x^3$

$$
\begin{array}{l}
\qquad\qquad\quad 2 - 3x\ +\quad x^3 \\
\qquad\qquad\quad x + 3x^2 -\quad x^3 \\
\hline
\times \quad x \Rightarrow \quad 2x - 3x^2 \qquad\quad + x^4 \\
\times \quad 3x^2 \Rightarrow \qquad\quad 6x^2 - 9x^3 \qquad + 3x^5 \\
\times \ -x^3 \Rightarrow \qquad\qquad\quad - 2x^3 + 3x^4 \qquad - x^6 \\
\hline
\qquad 2x + 3x^2 - 11x^3 + 4x^4 + 3x^5 - x^6
\end{array}
$$

Worked example

Divide $3 + x^2 - 2x^3 + x^5$ by $1 - 2x + x^2$

Write the polynomials in descending powers of x and arrange your work as follows:

$$
\begin{array}{r}
x^3 + 2x^2 + \; x \; + 1 \\
x^2 - 2x + 1 \enclose{longdiv}{x^5 \qquad - 2x^3 + \; x^2 \qquad + 3} \\
\underline{x^5 - 2x^4 + \; x^3} \\
2x^4 - 3x^3 + \; x^2 \\
\underline{2x^4 - 4x^3 + 2x^2} \\
x^3 - \; x^2 \\
\underline{x^3 - 2x^2 + \; x} \\
x^2 - \; x + 3 \\
\underline{x^2 - 2x + 1} \\
x + 2
\end{array}
$$

$$\Rightarrow \quad \text{Answer } x^3 + 2x^2 + x + 1 + \frac{x + 2}{x^2 - 2x + 1}$$

BRACKETS

> *Make sure you know how to use, and remove, brackets.*

Algebraic operations often involve the use of **brackets** and it is absolutely essential for you to become familiar with the use of and the removal of brackets. Remember when removing brackets that if the sign before the bracket is positive the $+$ and $-$ signs inside the bracket are unaltered. If the sign before the bracket is $-$, the $+$ and $-$ signs inside the brackets change to $-$ and $+$ respectively. Remember also that inner brackets should always be removed first.

Worked example

Simplify $\quad 3(2xy + z) - 4(xy - 2z) + 2(z - xy)$
$$\Rightarrow \quad 6xy + 3z - 4xy + 8z + 2z - 2xy = 13z$$

Worked example

Simplify $\quad 3 - x - 2[1 - 3(x - y)]$
$$\Rightarrow \quad 3 - x - 2[1 - 3x + 3y] = 3 - x - 2 + 6x - 6y = 1 + 5x - 6y$$

Worked example

Solve the equation $\quad (x - 2)(x + 3) = x(x - 1)$
$$\Rightarrow \quad x(x + 3) - 2(x + 3) = x(x - 1)$$
$$\Rightarrow \quad x^2 + 3x - 2x - 6 = x^2 - x \Rightarrow x^2 + 3x - 2x - 6 - x^2 + x = 0$$
$$\Rightarrow \quad 2x - 6 = 0 \Rightarrow 2x = 6 \Rightarrow x = 3$$

Worked example

Simplify $\quad \dfrac{2}{x - 3} - \dfrac{3}{x + 2} - \dfrac{(1 - x)}{(x - 3)(x + 2)}$

$$\Rightarrow \quad \frac{2(x + 2) - 3(x - 3) - (1 - x)}{(x - 3)(x + 2)}$$

$$= \frac{2x + 4 - 3x + 9 - 1 + x}{(x - 3)(x + 2)} = \frac{12}{(x - 3)(x + 2)}$$

EXERCISES 3.1

1. Add $(x^3 - 3x^2 - 5x + 2)$, $(2x^3 + x^2 - 2x - 4)$, $(1 - 4x^2 - 7x^3)$
2. Add $(ac - bc - ab)$, $(-6ab - 3bc - 2ac)$, $(4ab + 2bc - 4ac)$
3. Add $(x^4 - 2x^2y^2 + y^4)$, $(-3x^3y + 2x^2y^2)$, $(5x^3y - y^4)$, $(2xy^3 - x^4)$
4. Subtract $5a - 3b + 2c$ from $3a + b - 3c$
5. Subtract $4x^3 - 3x^2 + 2x - 1$ from $5x^3 + x^2 - 3x + 2$
6. Find $(x^2 - 7x + 2) + 2(3 - 2x^2) - 3(x^2 - 5x + 1)$
7. Multiply $2x^3 + 7x^2 + 13x + 26$ by $2x - 3$
8. Multiply $2x + y - z$ by $-3x + 2y + 4z$
9. Divide $1 - x^2 + x^4$ by $1 - x$
10. Divide $7x^3 - 6x^2 + x - 3$ by $x^2 - x + 2$

11 Divide $x^7 - 1$ by $x^2 - 1$

12 Simplify $\dfrac{1+x}{x-1} - \dfrac{3x}{3x-4}$

13 Simplify $\dfrac{1}{x-1} - \dfrac{3}{x} + \dfrac{2}{x+1}$

14 Simplify $\dfrac{5}{(x-z)(y-z)} + \dfrac{1}{(y-x)(z-x)} + \dfrac{1}{(x-y)(z-y)}$

15 Solve the equation $\dfrac{x}{x+1} - \dfrac{1+x}{x} = 0$

16 Solve the equation $\dfrac{1}{2x-1} + \dfrac{1}{2x+1} - \dfrac{1}{x+2} = 0$

EQUATIONS AND IDENTITIES

An equality such as $6(x-2) = 24$, which is true only when $x = 6$, is called an **equation**. An equality such as $x^2 - 4 = (x-2)(x+2)$, which is true for *all* real values of x, is called an **identity** and as such is written

$$x^2 - 4 \equiv (x-2)(x+2)$$

The symbol $\equiv$ meaning 'is identically equal to'.

The following identities frequently occur in mathematics and you should become familiar with them. They can all be proved by multiplying out the brackets.

> *Some familiar identities.*

$$(x+y)^2 \equiv x^2 + 2xy + y^2$$
$$(x-y)^2 \equiv x^2 - 2xy + y^2$$
$$(x+y)^3 \equiv x^3 + 3x^2y + 3xy^2 + y^3$$
$$(x-y)^3 \equiv x^3 - 3x^2y + 3xy^2 - y^3$$
$$(x+y)(x-y) \equiv x^2 - y^2$$
$$(x+y)(y+z)(z+x) \equiv x^2(y+z) + y^2(z+x) + z^2(x+y) + 2xyz$$

Worked example
Find the product $(x+y+2)(x+y-2)$
$(x+y+2)(x+y-2) \equiv [(x+y)+2][(x+y)-2] \equiv (x+y)^2 - 2^2$
$\equiv x^2 + 2xy + y^2 - 4$

Worked example
Find the product $(x-3y+2z)^2$
$(x-3y+2z)^2 \equiv [(x-3y)+2z]^2 \equiv (x-3y)^2 + 2(x-3y).2z + (2z)^2$
$\equiv x^2 - 6xy + 9y^2 + 4xz - 12yz + 4z^2$

Worked example
Find the product $(a-2)^3(a+2)^3$
$(a-2)^3(a+2)^3 \equiv [(a-2)(a+2)]^3 = (a^2-4)^3$
$\equiv (a^2)^3 - 3(a^2)^2 4 + 3(a^2)(4)^2 - 4^3$
$\equiv a^6 - 12a^4 + 48a^2 - 64$

FACTORS

When the above identities are read from right to left, rather than left to right, then the terms in the brackets are said to be the **factors** of the identity.

Thus $x^2 - y^2 \equiv (x-y)(x+y) \Rightarrow x^2 - y^2$ has factors $x-y$ and $x+y$.
$x^3 - 3x^2y + 3xy^2 - y^3$ factorises into $(x-y)^3$.

It is important that you are able to factorise simple polynomials. Unfortunately there is no single direct approach to obtaining factors and the ease with which you carry out the factorisation will depend to a large extent on experience, practice and familiarity with standard identities. However, you should find the following procedures helpful.

1 **Common monomial factor** $\Rightarrow ab + ac \equiv a(b+c)$

Worked example
$2x^2y^2 + x^2y - 3xy^2 \equiv xy(2xy + x - 3y)$

Worked example
$$xy - 3x + 2y^2 - 6y \equiv x(y-3) + 2y(y-3) \equiv (y-3)(x+2y)$$

2   **Difference of two squares** $\Rightarrow x^2 - y^2 \equiv (x-y)(x+y)$

Worked example
$$9x^2 - 4y^2 = (3x)^2 - (2y)^2 \equiv (3x-2y)(3x+2y)$$

Worked example
$$x^4 - y^4 \equiv (x^2)^2 - (y^2)^2 \equiv (x^2-y^2)(x^2+y^2)$$
$$\equiv (x-y)(x+y)(x^2+y^2)$$

3   **Perfect square trinomials** $\Rightarrow x^2 + 2xy + y^2 \equiv (x+y)^2$
$$x^2 - 2xy + y^2 \equiv (x-y)^2$$
i.e. two terms are perfect squares, the third term is twice the product of the square roots of the other two terms.

Worked example
$$4x^2 + 4x + 1 \Rightarrow (2x)^2 + 2.(2x).1 + 1^2 \equiv (2x+1)^2$$

Worked example
$$9x^2 - 48x + 64 \Rightarrow (3x)^2 - 2.(3x).8 + 8^2 \equiv (3x-8)^2$$

4   **Trinomial** $\Rightarrow x^2 + (a+b)x + ab \equiv (x+a)(x+b)$
i.e. coefficient $x^2 = 1$, constant term = product of a and b, coefficient x = sum of a and b, where a and b may be positive or negative.

Worked example
$x^2 + 5x + 6$, $6 = 2\times3,\ 5 = 2+3$,
$\Rightarrow (x+2)(x+3)$

Worked example
$x^2 + x - 6$ $-6 = 3\times(-2),\ 1 = 3+(-2)$,
$\Rightarrow (x+3)(x-2)$

Worked example
$x^2 - 7x + 12$, $12 = (-3)\times(-4),\ -3+(-4) = -7$,
$\Rightarrow (x-3)(x-4)$

5   **Trinomial** $\Rightarrow acx^2 + (ad+bc)x + bd \equiv (ax+b)(cx+d)$
i.e. coefficient $x^2 = ac$, constant term = bd, and the coefficient of $x = ad+bc$, the sum of two numbers whose product is the same as the product of the constant term and the coefficient of the x^2 term, i.e. $abcd$. It is not always easy to find $ad+bc$, but the following examples may help.

Worked example
$6x^2 + 17x + 12$ $\Rightarrow 6\times12 = 72 \Rightarrow$ need two numbers whose sum is 17 and whose product is 72. These are 8 and 9.
$6x^2 + 17x + 12 = 6x^2 + (8x+9x) + 12 = 2x(3x+4) + 3(3x+4)$
$$= (2x+3)(3x+4)$$

Worked example
$8x^2 + 6x - 27$ $\Rightarrow 8\times(-27) = -216 \Rightarrow$ need two numbers whose sum is 6 and whose product is -216. These are 18 and -12.
$8x^2 + 6x - 27 = 8x^2 + 12x + 18x - 27 = 4x(2x-3) + 9(2x-3) = (4x+9)(2x-3)$

Worked example
$12x^2 - 23x + 10$ $\Rightarrow 12\times10 = 120 \Rightarrow$ need two numbers whose sum is -23 and whose product is 120. These are -8 and -15.
$12x^2 - 23x + 10 = 12x^2 - 8x - 15x + 10 = 4x(3x-2) - 5(3x-2) = (4x-5)(3x-2)$

6   **Sum, difference of two cubes**
$$x^3 + y^3 = (x+y)(x^2 - xy + y^2)$$
$$x^3 - y^3 = (x-y)(x^3 + xy + y^2)$$

Note: in each case, the first bracket is the L.H.S. without the indices; and, in the second bracket, the xy term has the opposite sign to that in the first bracket.

Worked example

$$8x^3 - 1 = (2x)^3 - 1^3 = (2x-1)[(2x)^2 + 2x.1 + 1^2]$$
$$= (2x-1)[4x^2 + 2x + 1]$$

Worked example

$$x^{12} + y^{12} = (x^4)^3 + (y^4)^3 = (x^4 + y^4)(x^8 - x^4y^4 + y^8)$$

EXERCISES 3.2

Factorise the following expressions:

1   $6x^2 + 12xy$
2   $6y^3 + 12y^2 - 24y$
3   $x^2 - bx + ax - ab$
4   $a^2 - 9b^2 + a + 3b$
5   $x^2y^2 - 4y^4$
6   $4x^2 - 36$
7   $25x^2 - 9y^2$
8   $1 - 4x^2 + 4x^4$
9   $x^2 - 7x + 12$
10   $2x^2 - 17x + 8$
11   $3x^2 + 13x - 30$
12   $15 + 99x - 42x^2$

13   $3a^2 + 23a + 30$
14   $20a^2 + 9ab - 20b^2$
15   $14y^2 + 45y - 14$
16   $x^2 - xy - 90y^2$
17   $x^4 + x^2 - 2$
18   $(x+2)^2 + 3(x+2) + 2$
19   $a^4 - 10a^2 + 9$
20   $x^6 + 64$
21   $a^9 - b^9$
22   $16x^3 - 54$
23   $x^6 + 9x^3 + 8$
24   $x^6 + 7x^3 - 8$

QUADRATIC EQUATIONS

> *Quadratic equations have two roots, which may be real, equal or complex.*

$ax^2 + bx + c = 0$, where a, b and c are constants is the most general form of a **quadratic equation**. Such equations have two roots which may be real, equal or complex.

They are *real* provided $b^2 - 4ac \geq 0$
They are *equal* provided $b^2 - 4ac = 0$
They are *complex* provided $b^2 - 4ac < 0$

$b^2 - 4ac$, usually denoted by Δ, is called the *discriminant*. The roots of the equation are obtained by factorising or by completing the square.

Worked example

Solve $2x^2 + 5x - 3 = 0$
Factorising $(2x-1)(x+3) = 0 \Rightarrow 2x - 1 = 0$ or $x + 3 = 0$

$$\Rightarrow x = \frac{1}{2} \text{ or } -3$$

Worked example

Solve $2x^2 + 6x - 3 = 0$
Completing the square. Follow the rules:

i) Divide by coefficient of $x^2 \Rightarrow x^2 + 3x - \dfrac{3}{2} = 0$

ii) Take constant term to R.H.S. $\Rightarrow = x^2 + 3x = \dfrac{3}{2}$

iii) Complete square of L.H.S. $\Rightarrow = \left(x + \dfrac{3}{2}\right)^2 = \dfrac{3}{2} + \left(\dfrac{3}{2}\right)^2 = \dfrac{15}{4}$

iv) Take square root of both sides $\Rightarrow x + \dfrac{3}{2} = \pm \dfrac{\sqrt{15}}{2}$

v) Make x the subject $\Rightarrow x = -\dfrac{3}{2} \pm \dfrac{\sqrt{15}}{2}$

GENERAL RESULT

Apply the above method to the equation $ax^2 + bx + c = 0$ to obtain

$$x = \frac{-b \pm \sqrt{b^2 - 4ac}}{2a}$$

This formula may be remembered and quoted in the examination.

Worked example

Solve $5x^2 - 2x - 3 = 0$

$$a = 5,\ b = -2,\ c = -3,\ \Rightarrow x = \frac{-(-2) \pm \sqrt{[(-2)^2 - 4.5.(-3)]}}{2.5}$$

$$\Rightarrow x = \frac{2 \pm \sqrt{64}}{10} = \frac{2 \pm 8}{10} \Rightarrow x = 1 \text{ or } -\frac{3}{5}$$

Worked example

Find the value of K given that $2x^2 + 3x - K = 0$ has equal roots

Discriminant $\Delta \equiv b^2 - 4ac = 3^2 - 4.2(-K)$

For equal roots $\Delta = 0 \Rightarrow 9 + 8K = 0 \Rightarrow K = -\frac{9}{8}$

Worked example

Show that the roots of the equation $x^2 + Kx - (1 - K)$ are real for any real value of K

$\Delta \equiv b^2 - 4ac = K^2 + 4(1 - K) = (K - 2)^2 \geqslant 0$ for all real K

$\Rightarrow$ roots are real

EXERCISES 3.3

1 Solve the equations:
   i) $x^2 - 7x + 12 = 0$
   ii) $2x^2 - 5x + 2 = 0$
   iii) $5x^2 - 4x - 12 = 0$
   iv) $5x^2 - 4x - 2 = 0$
   v) $2x^2 - 13x + 2 = 0$

2 Show that the roots of the equation $9x^2 - 30x + 25 = 0$ are equal

3 Given that a, b and c are real, prove that the roots of the equation $(x + a)(x + b) = c^2$ are real

4 State the condition for the equation $3ax^2 + 3bx + 2c = 0$, $a \neq 0$, to have real roots. Prove that if this condition is satisfied, then the roots of the equation $a^2x^2 + (3b^2 - 4ac)x + 4c^2 = 0$ are real

5 Find the greatest value of $3 + 12x - x^2$

FORMATION OF EQUATIONS

When α and β are the roots of a quadratic equation, the equation may be written as $(x - \alpha)(x - \beta) = 0$

i.e. $x^2 - (\alpha + \beta)x + \alpha\beta = 0$

Note: Coefficient of $x = -(\alpha + \beta) = -$sum of roots

Constant term $= \alpha\beta =$ product of roots

Hence for the equation $ax^2 + bx + c = 0 \Rightarrow x^2 + \frac{b}{a}x + \frac{c}{a} = 0$

Sum of roots $= -\frac{b}{a}$, product of roots $= \frac{c}{a}$

Worked example

Given that α and β are roots of the equation $3x^2 - 4x + 5 = 0$, find an equation whose roots are $\frac{\alpha}{\beta}$ and $\frac{\beta}{\alpha}$

$$3x^2 - 4x + 5 = 0 \Rightarrow \alpha + \beta = \frac{4}{3}, \quad \alpha\beta = \frac{5}{3}$$

$$\text{Sum of new roots} = \frac{\alpha}{\beta} + \frac{\beta}{\alpha} = \frac{\alpha^2 + \beta^2}{\alpha\beta} = \frac{(\alpha+\beta)^2 - 2\alpha\beta}{\alpha\beta} = \frac{\left(\frac{4}{3}\right)^2 - 2.\frac{5}{3}}{\frac{5}{3}} = -\frac{14}{15}$$

$$\text{Product of new roots} = \frac{\alpha}{\beta} \cdot \frac{\beta}{\alpha} = 1$$

Required equation is $x^2 - \left(-\frac{14}{15}\right)x + 1 = 0$ or $15x^2 + 14x + 15 = 0$

❝ Beware of errors. ❞

Do be careful with signs. Use brackets and do not forget the $= 0$. Many students *do* forget to write this in the examination.

Worked example

Find the value of K given that the roots of the equation $2x^2 + 5x - K = 0$ differ by 3

Let the roots be α and $\alpha + 3$

$$\text{Sum of roots} = 2\alpha + 3 = -\frac{5}{2} \Rightarrow \alpha = -\frac{11}{4}$$

$$\text{Product of roots} = \alpha(\alpha+3) = -\frac{K}{2} \Rightarrow K = -2\left(-\frac{11}{4}\right)\left(-\frac{11}{4}+3\right) = \frac{11}{8}$$

Worked example

The roots of the equation $x^2 + 2x - 4 = 0$ are α and β

Find an equation whose roots are $\alpha^2 + \beta$ and $\beta^2 + \alpha$

$x^2 + 2x - 4 = 0 \Rightarrow \alpha + \beta = -2$, $\alpha\beta = -4$

$\text{Sum of new roots} = (\alpha^2 + \beta) + (\beta^2 + \alpha) = (\alpha^2 + \beta^2) + (\alpha + \beta)$

$\qquad\qquad = (\alpha+\beta)^2 - 2\alpha\beta + (\alpha + \beta) = 4 + 8 - 2 = 10$

$\text{Product of new roots} = (\alpha^2 + \beta)(\beta^2 + \alpha) = \alpha^2\beta^2 + \alpha\beta + \alpha^3 + \beta^3$

$\qquad\qquad = (-4)^2 + (-4) + (\alpha+\beta)(\alpha^2 - \alpha\beta + \beta^2)$

$\qquad\qquad = 12 + (\alpha+\beta)[(\alpha+\beta)^2 - 3\alpha\beta] = 12 + (-2)[(-2)^2 - 3(-4)] = -20$

Required equation is $x^2 - 10x - 20 = 0$

Note: Do not attempt to find the roots of the given equation. Instead always find $\alpha + \beta$ and $\alpha\beta$. You must then find the sum and product of the new roots in terms of $\alpha + \beta$ and $\alpha\beta$ and, as in this example, it is frequently necessary to call upon your experience with factors. In particular

$$\alpha^2 + \beta^2 = (\alpha+\beta)^2 - 2\alpha\beta$$

and

$$\alpha^3 + \beta^3 = (\alpha+\beta)(\alpha^2 - \alpha\beta + \beta^2) = (\alpha+\beta)[(\alpha+\beta)^2 - 3\alpha\beta)]$$

EXERCISES 3.4

1 Given that α and β are the roots of the equation $x^2 - 2x + 5 = 0$, find an equation whose roots are i) α^2 and β^2,

$\qquad\qquad$ ii) $\alpha + \dfrac{1}{\alpha}$ and $\beta + \dfrac{1}{\beta}$.

2 Form a quadratic equation with roots which exceed by 3 the roots of the equation $3x^2 - (k-4)x - (2k-1) = 0$

3 Find a quadratic equation whose roots are the squares of the roots of the equation $x^2 - ax + b = 0$. Show that there are four and only four quadratic equations which are unchanged by squaring their roots

4 Given that α and β are the roots of the equation $x^2 + x - 1 = 0$, show that $\alpha^2 = \beta + 2$ and $\beta^2 = \alpha + 2$. Find a quadratic equation whose roots are $\dfrac{\alpha+1}{\beta+1}$, $\dfrac{\beta+1}{\alpha+1}$

BASTARD

THE REMAINDER THEOREM

The **remainder theorem** states that when a polynomial $f(x)$ is divided by $(x - a)$ the remainder is $f(a)$, where $f(a)$ is the value of $f(x)$ when $x = a$.

$f(x) = 3x^3 - x^2 + 2x$ divided by $(x+3)$ gives a remainder
$f(-3) = 3(-3)^3 - (-3)^2 + 2(-3) = -96$.

Check: $\dfrac{3x^3 - x^2 + 2x}{x+3} = 3x^2 - 10x + 32 - \dfrac{96}{x+3} \Rightarrow$ remainder $= -96$

Worked example

Given that $f(x) \equiv x^3 + px^2 + qx + r$, where p, q and r are constants, leaves remainders of 2, 2 and 0 when divided by $x-1$, x and $x+1$ respectively, find the remainder when $f(x)$ is divided by $x-2$

$(x-1) \Rightarrow f(1) = 2 \Rightarrow 1 + p + q + r = 2$

$\quad x \quad \Rightarrow f(0) = 2 \Rightarrow 0 + 0 + 0 + r = 2 \Rightarrow r = 2$ and $p + q = -1$

$(x+1) \Rightarrow f(-1) = 0 \Rightarrow -1 + p - q + r = 0 \Rightarrow p - q = 1 - r = -1$

$\qquad\qquad\qquad\qquad\qquad\qquad\qquad\qquad \Rightarrow p = -1,\ q = 0,\ r = 2$

$\qquad \Rightarrow f(x) = x^3 - x^2 + 2$

$\qquad\quad f(2) = 2^3 - 2^2 + 2 = 6 \Rightarrow$ remainder when $f(x)$ is divided by $(x-2)$ is 6

THE FACTOR THEOREM

For a given function $f(x)$, if $f(a) = 0$, then $(x-a)$ is a *factor* of $f(x)$.

Worked example

Factorise $f(x) \equiv 3x^3 + 8x^2 + 3x - 2$

To factorise an expression of this kind find $f(a)$ where $a = \pm 1, \pm 2$ (or where a is some other factor of the constant term)

$f(+1) = 12 \Rightarrow (x-1)$ is not a factor

$f(-1) = 0 \ \Rightarrow (x+1)$ is a factor

$f(2) \quad \neq 0 \ \Rightarrow (x-2)$ is not a factor

$f(-2) = 0 \ \Rightarrow (x+2)$ is a factor

Having found 2 factors the third can be obtained as follows:

$(x+1)(x+2) = x^2 + 3x + 2$

$\Rightarrow 3x^3 + 8x^2 + 3x - 2 = (x^2 + 3x + 2)(ax + b)$ where a must be 3 to obtain $3x^3$ and b must be -1 to obtain -2

$\Rightarrow 3x^3 + 8x^2 + 3x - 2 = (x+1)(x+2)(3x-1)$

> Ways of obtaining a third factor.

Alternatively, having found one factor, you can divide the polynomial by that factor and factorise the resultant quadratic.

Worked example

$f(x) \equiv 6x^3 + ax^2 + bx + c$, where a, b and c are constants. When $f(x)$ is divided by $x^2 - 4$ the remainder is $23x - 26$. When $f(x)$ is divided by $x+3$ the remainder is -220. Find:

a) $f(x)$,   b) the solution of the equation $f(x) = 0$

$\dfrac{f(x)}{x^2 - 4} = g(x) + \dfrac{23x - 26}{x^2 - 4}$ where $g(x)$ is a linear (first-degree) expression.

$\Rightarrow f(x) = 6x^3 + ax^2 + bx + c = (x^2 - 4)g(x) + 23x - 26$

$\qquad\qquad\qquad\qquad\qquad\qquad = (x-2)(x+2)g(x) + 23x - 26$

Substituting $x = 2\ \Rightarrow 48 + 4a + 2b + c = 20$

Substituting $x = -2 \Rightarrow -48 + 4a - 2b + c = -72$

$f(-3) = -220 \qquad \Rightarrow -162 + 9a - 3b + c = -220$

Solving $\Rightarrow a = -7,\ b = -1,\ c = 2,$

$\qquad \Rightarrow f(x) = 6x^3 - 7x^2 - x + 2$

$f(1) = 0 \Rightarrow (x-1)$ is a factor

$\qquad \Rightarrow 6x^3 - 7x^2 - x + 2 = (x-1)(6x^2 - x - 2) = (x-1)(2x+1)(3x-2)$

$\qquad \Rightarrow f(x) = 0$ when $x = 1,\ -\dfrac{1}{2}$ or $\dfrac{2}{3}$

EXERCISES 3.5

1 Given that $x^3 + px^2 + qx - 24$, where p and q are constants, leaves remainders 56 and 0 on division by $x-5$ and $x+2$ respectively, find p and q and hence solve the equation $x^3 + px^2 + qx - 24 = 0$

2 Factorise $x^3 + 6x^2 + 11x + 6$

3 When the polynomial $f(x)$ is divided by $x-1$ the remainder is 4. When $f(x)$ is divided

by $x+2$ the remainder is 1. Find the remainder when f(x) is divided by $(x-1)(x+2)$

4 Verify that $(2x-1)$ is a factor of $4x^3-3x+1$ and find the other linear factors. Show that when the polynomial f(x) is divided by $(x-a)(x-b)$, $a \neq b$, then the remainder is

$$\frac{x}{a-b}[f(a)-f(b)] + \frac{1}{a-b}[af(b)-bf(a)].$$ Determine the remainder when $4x^3-3x+1$ is

divided by x^2+x-6

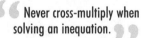

Expression of the form $x>b$, $2x^2-x+3>0$, $\dfrac{x+1}{x-2}>2$ are known as **inequations** or

inequalities. Their solutions consist of a range or ranges of values of the variable concerned. The solution(s) may be represented on a number line as follows.

$x>2$ (number line: 0 1 2, hollow circle at 2)   the hollow circle showing $x=2$ is excluded.

$x \geqslant 2$ (number line: 0 1 2, solid circle at 2)   the solid circle showing $x=2$ is included.

$-1>x \cup x \geqslant 2$ (number line: −1 0 1 2, hollow circle at −1, solid circle at 2)

You should note when dealing with an inequation that if a and b are positive constants then $x>y$ implies:

i) $x \pm a > y \pm a$ i.e. can add or subtract a constant to each side,

ii) $ax > ay$ i.e. can multiply each side by a *positive* constant,

iii) $ax \pm b > ay \pm b$ i.e. can combine i) and ii),

iv) $-x < -y$ i.e. can multiply each side by a minus sign, but the inequality sign must be reversed

> **Never cross-multiply when solving an inequation.**

You are advised *never* to cross-multiply when solving an inequation, but to proceed as follows:

Worked example

Solve $\dfrac{x}{x+1} < \dfrac{x+2}{x+3}$

$\dfrac{x+2}{x+3} - \dfrac{x}{x+1} > 0$ i.e. take all terms to one side

$\dfrac{(x+2)(x+1)-x(x+3)}{(x+3)(x+1)} > 0 \quad \Rightarrow \quad \dfrac{2}{(x+3)(x+1)}$

A *tabular display* can now be used to complete the solution.

Value of x	$x<-3$	$-3<x<-1$	$x>-1$
Sign of $(x+1)$	−ve	−ve	+ve
Sign of $(x+3)$	−ve	+ve	+ve
Sign of $\dfrac{2}{(x+3)(x+1)}$	+ve	−ve	+ve

$\Rightarrow$ Solution is $x<-3 \cup x>-1$

INEQUATIONS INVOLVING MODULUS SIGN

The notation $|x|<2$ means $-2<x<2$, i.e. the numerical value of x is less than 2.

Worked example

Solve $|2x-3|<5$

This means $-5<2x-3<5$ so it is necessary to consider two inequalities $-5<2x-3$ and $2x-3<5$

Now $-5<2x-3 \Rightarrow 2x>-2$, $x>-1$

and $2x-3<5 \Rightarrow 2x<8$, $x<4$

The complete solution is therefore $-1<x<4$

Shown on a number line as   (number line: −1 0 1 2 3 4, hollow circles at −1 and 4)

Worked example

Solve $x^2 - |x| - 6 < 0$

Rearranging $|x| > x^2 - 6$ or $-(x^2 - 6) > x > (x^2 - 6)$

Consider $x > (x^2 - 6) \Rightarrow x^2 - x - 6 < 0 \Leftrightarrow (x-3)(x+2) < 0$

Value of x	$x < -2$	$-2 < x < 3$	$x > 3$
Sign of $(x-3)$	−ve	−ve	+ve
Sign of $(x+2)$	−ve	+ve	+ve
Sign of $(x-2)(x+2)$	+ve	−ve	+ve

$\Rightarrow \quad -2 < x < 3$

Consider $-(x^2 - 6) > x \Rightarrow x^2 + x - 6 < 0 \Leftrightarrow (x+3)(x-2) < 0$

Value of x	$x < -3$	$-3 < x < 2$	$x > 2$
Sign of $(x+3)$	−ve	+ve	+ve
Sign of $(x-2)$	−ve	−ve	+ve
Sign of $(x+3)(x-2)$	+ve	−ve	+ve

$\Rightarrow \quad -3 < x < 2$

Hence the range of x including *both* of the above inequalities is $-3 < x < 3$.

Worked example

For real values of x find the range of values of $\dfrac{4(x-2)}{4x^2 + 9}$

Let $y = \dfrac{4(x-2)}{4x^2 + 9}$

Rearranging $4x^2 y - 4x + (8 + 9y) = 0$

Since x is real, this quadratic must have real roots.

Hence $\Delta \equiv b^2 - 4ac \geqslant 0 \Rightarrow (-4)^2 - 4.4y(8 + 9y) \geqslant 0$

$\Rightarrow 9y^2 + 8y - 1 \leqslant 0 \Leftrightarrow (9y-1)(y+1) \leqslant 0$

Value of y	$y < -1$	$-1 < y < \dfrac{1}{9}$	$y > \dfrac{1}{9}$
Sign of $(9y-1)$	−ve	−ve	+ve
Sign of $(y+1)$	−ve	+ve	+ve
Sign of $(9y-1)(y+1)$	+ve	−ve	+ve

$\Rightarrow \quad -1 \leqslant y \leqslant \dfrac{1}{9} \text{ or } -1 \leqslant \dfrac{4(x-2)}{4x^2 + 9} \leqslant \dfrac{1}{9}$

EXERCISES 3.6

1 Solve i) $x^2 - 5x + 6 \geqslant 2$, ii) $\dfrac{1}{x^2 - 5x + 6} \leqslant \dfrac{1}{2}$

2 Solve $\dfrac{x^2 + 15}{x} > 8$

3 Solve $x^2 - |x| - 12 < 0$

4 Solve $\left| \dfrac{1}{1 + 2x} \right| < 1$

5 Given that $y = \dfrac{1}{(x-1)(x-2)}$ and x is real, show that y cannot lie between 0 and −4

6 Given that $y = \dfrac{x^2 + 2x + K}{2x - 3}$ and x is real, find the greatest value of K for which y can take all real values

RATIONAL ALGEBRAIC FRACTIONS

The division of a polynomial $f(x)$ of degree n, by a polynomial $g(x)$ of degree m gives a **rational algebraic fraction**.

If $n < m$, the fraction is a *proper* fraction, e.g. $\dfrac{x^2+1}{3x^3-2x+4}$

If $n \geqslant m$, the fraction is an *improper* fraction, e.g. $\dfrac{x^3}{x^2-2x-1}$

When the denominator of a proper fraction factorises it is possible to express the fraction as the sum or difference of other proper fractions. This is known as the resolution of the fraction into its *partial fractions*.

PARTIAL FRACTIONS

Three cases need to be considered depending upon the kind of factors in the denominator.

1 **Linear factors:** For each linear factor of the form $(x - \alpha)$ there corresponds a partial fraction of the form $\dfrac{A}{x-\alpha}$

E.g. $\dfrac{ax+b}{(x-\alpha)(x-\beta)} \Rightarrow \dfrac{A}{x-\alpha} + \dfrac{B}{x-\beta}$, where A and B are constants and $\alpha \neq \beta$

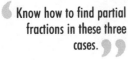
Know how to find partial fractions in these three cases.

2 **Repeated linear factors:** For each repeated linear factor of the form $(x - \alpha)^2$ there correspond partial fractions of the form

$$\dfrac{A}{x-\alpha} + \dfrac{B}{(x-\alpha)^2}$$

E.g. $\dfrac{ax+b}{(x-\alpha)^2(x-\beta)} \Rightarrow \dfrac{A}{x-\alpha} + \dfrac{B}{(x-\alpha)^2} + \dfrac{C}{x-\beta}$

where A, B and C are constants

3 **Quadratic factor:** For each quadratic factor of the form $px^2 + qx + r$ there corresponds a partial fraction of the form $\dfrac{Ax+B}{px^2+qx+r}$

E.g. $\dfrac{ax+b}{(x-\alpha)(px^2+qx+r)} \Rightarrow \dfrac{A}{x-\alpha} + \dfrac{Bx+C}{px^2+qx+r}$, where A, B and C are constants

An improper fraction must be converted into a proper fraction before it can be expressed in partial fraction form. This is done by division.

E.g. $\dfrac{x^2}{(x-1)(x-2)} \equiv 1 + \dfrac{3x-2}{(x-1)(x-2)} \equiv 1 + \dfrac{A}{x-1} + \dfrac{B}{x-2}$

Worked example

Express $\dfrac{x}{x^2+x-2}$ in partial fractions

$$\dfrac{x}{x^2+x-2} = \dfrac{x}{(x+2)(x-1)} \equiv \dfrac{A}{x+2} + \dfrac{B}{x-1} \equiv \dfrac{A(x-1)+B(x+2)}{(x+2)(x-1)}$$

Since the denominators are identical, the numerators must also be identical, i.e. $x \equiv A(x-1) + B(x+2)$

Substituting $x = 1 \Rightarrow 1 = 3B \Rightarrow B = \dfrac{1}{3}$

Substituting $x = -2 \Rightarrow -2 = -3A \Rightarrow A = \dfrac{2}{3}$

$$\Rightarrow \dfrac{x}{x^2+x-2} = \dfrac{\frac{2}{3}}{x+2} + \dfrac{\frac{1}{3}}{x-1}$$

You are advised to check your answer by *reversing* the process.

$$\Rightarrow \frac{\frac{2}{3}}{x+2} + \frac{\frac{1}{3}}{x-1} = \frac{\frac{2}{3}(x-1) + \frac{1}{3}(x+2)}{(x+2)(x-1)} = \frac{x}{x^2+x-2}$$

Worked example

Express in partial fractions $\dfrac{x+8}{x(x^2+4)}$

$$\frac{x+8}{x(x^2+4)} \equiv \frac{A}{x} + \frac{Bx+C}{x^2+4} \equiv \frac{A(x^2+4) + (Bx+C)x}{x(x^2+4)}$$

$\Rightarrow x+8 \equiv A(x^2+4) + (Bx+C)x$

Substituting $x = 0 \Rightarrow 8 = 4A$, $A = 2$

Compare x^2 terms $\Rightarrow 0 = A+B$, $B = -A = -2$

Compare x terms $\Rightarrow 1 = C$

$$\Rightarrow \frac{x+8}{x(x^2+4)} \equiv \frac{2}{x} + \frac{1-2x}{x^2+4}$$

Check R.H.S. $\Rightarrow \dfrac{2(x^2+4) + x(1-2x)}{x(x^2+4)} = \dfrac{8+x}{x(x^2+4)} = \text{L.H.S.}$

Worked example

Express $\dfrac{12x^3-x}{(x-1)(2x-1)^2}$ in partial fractions

$$\frac{12x^3-x}{(x-1)(2x-1)^2} = \frac{12x^3-x}{4x^3-8x^2+5x-1} = 3 + \frac{24x^2-16x+3}{(x-1)(2x-1)^2}$$

$$\frac{24x^2-16x+3}{(x-1)(2x-1)^2} = \frac{A}{x-1} + \frac{B}{2x-1} + \frac{C}{(2x-1)^2} = \frac{A(2x-1)^2 + B(2x-1)(x-1) + C(x-1)}{(x-1)(2x-1)^2}$$

$\Rightarrow 24x^2 - 16x + 3 = A(2x-1)^2 + B(2x-1)(x-1) + C(x-1)$

Substituting $x = 1 \Rightarrow 11 = A$, $\qquad A = 11$

Substituting $x = \dfrac{1}{2} \Rightarrow 1 = -\dfrac{1}{2}C$, $\qquad C = -2$

Compare x^2 terms $\Rightarrow 24 = 4A + 2B$, $B = -10$

Check by comparing x terms L.H.S. $= -16$, R.H.S. $= -4A - 3B + C = -44 + 30 - 2$
$$= -16 = \text{L.H.S.}$$

$$\Rightarrow \frac{12x^3-x}{(x-1)(2x-1)^2} = 3 + \frac{11}{x-1} - \frac{10}{2x-1} - \frac{2}{(2x-1)^2}$$

EXERCISES 3.7

1 Express in partial fractions $\dfrac{1}{(x+1)(x+2)(x+3)}$

2 Express in partial fractions $\dfrac{x+1}{(x-1)^2}$

3 Express in partial fractions $\dfrac{4}{(x+1)(x^2+1)}$

4 Express in partial fractions $\dfrac{4x}{(1-x^2)(1+x)}$

INDICES

The rules for **indices** are:

i) $a.a.a\ldots(m \text{ times}) = a^m$

ii) $a^m.a^n = a^{m+n}$

iii) $\dfrac{a^m}{a^n} = a^{m-n}$

iv) $(a^m)^n = a^{mn}$

v) $a^0 = 1$

vi) $a^{-m} = \dfrac{1}{a^m}$

LOGARITHMS

Definition: If $y = a^x$, $a > 0$, then $x = \log_a y$

LAWS OF LOGARITHMS

> **Some useful laws of logarithms.**

i) $\log_a x + \log_a y = \log_a (xy)$

ii) $\log_a x - \log_a y = \log_a \left(\dfrac{x}{y}\right)$

iii) $\log_a (x)^n = n \log_a x$

iv) $\log_a a = 1$

v) $\log_a 1 = 0$

The laws are proved by using the definition of a logarithm and the rules for indices.

Worked example

Prove $\log_a x - \log_a y = \log_a \left(\dfrac{x}{y}\right)$

Let $\log_a x = A$ and $\log_a y = B$

$\Rightarrow a^A = x$ and $a^B = y$

$\Rightarrow \dfrac{x}{y} = \dfrac{a^A}{a^B} = a^{A-B} \Rightarrow \log_a \left(\dfrac{x}{y}\right) = A - B = \log_a x - \log_a y$

Worked example

Prove the change of base formula $\log_a x = \dfrac{\log_b x}{\log_b a}$

Hence solve $3\log_8 x - \log_x 8 = 2$

Let $\log_a x = y \Rightarrow x = a^y \Rightarrow \log_b x = \log_b (a^y)$

$\Rightarrow \log_b x = y\log_b a$ or $y = \log_a x = \dfrac{\log_b x}{\log_b a}$

$3\log_8 x - \log_x 8 = 2 \Rightarrow 3\log_8 x - \dfrac{\log_8 8}{\log_8 x} = 2$

or $3(\log_8 x)^2 - 2(\log_8 x) - 1 = 0$

$(3\log_8 x + 1)(\log_8 x - 1) = 0 \Rightarrow \log_8 x = -\dfrac{1}{3}$ or 1

$\Rightarrow x = 8^{-\frac{1}{3}} = \dfrac{1}{2}$ or $x = 8^1 = 8$

Worked example

Given that $\log_2(x+1) - 1 = 2\log_2 y$ and $\log_2(x - 5y + 4) = 0$ find x and y

$\log_2(x - 5y + 4) = 0 \Rightarrow x - 5y + 4 = 2^0 = 1 \Rightarrow x = 5y - 3$

$\Rightarrow \log_2(x+1) - 1 = 2\log_2 y \Rightarrow \log_2(5y - 2) - 2\log_2 y = 1$

$\Rightarrow \log_2 \dfrac{5y - 2}{y^2} = 1 \Rightarrow \dfrac{5y - 2}{y^2} = 2$

$\Rightarrow 2y^2 - 5y + 2 = 0 \Longleftrightarrow (2y - 1)(y - 2) = 0 \Rightarrow y = \dfrac{1}{2}$ or 2

When $y = \dfrac{1}{2}$, $x = -\dfrac{1}{2}$, when $y = 2$, $x = 7$

Worked example

Solve for x, to three significant figures
$4^{2x+1} \cdot 5^{x-2} = 6^{1-x}$

Equations such as this where the unknown variable occurs in the index and where there are no plus or minus signs other than those in the index can usually be solved by the use of logarithms. However, you must always be careful when taking logarithms to ensure that you do not fall into the trap of taking logarithms through a plus or minus sign, e.g. $2^x + 3^{2x-1} = 4$. It is not correct to say $\log 2^x + \log 3^{2x-1} = \log 4$. The correct statement is $\log(2^x + 3^{2x-1}) = \log 4$ and this will not lead anywhere.

Returning to our example $4^{2x+1}.5^{x-2} = 6^{1-x} \Rightarrow \log(4^{2x+1}.5^{x-2}) = \log 6^{1-x}$
$\Rightarrow \log 4^{2x+1} + \log 5^{x-2} = \log 6^{1-x} \Rightarrow (2x+1)\log 4 + (x-2)\log 5 = (1-x)\log 6$

Using logarithms to the base $10 \Rightarrow 2.6814x = 1.5741$ or $x = 0.587$

EXERCISES 3.8

1   Given that $\log_2(3x+y) - \log_2(x-3y) = 3$, find the value of $\dfrac{x}{y}$

2   Find the values of x and y such that $y = 2x$ and $\log_2 y + \log_2 x = 3$
3   Solve the equation $\log_2 x + \log_x 2 = 2.5$
4   Solve the equations a) $5(2)^x = 4^{1-x}$, b) $\log_{10} x = \log_5(2x)$

EXAMINATION TYPE QUESTIONS WITH MARKING SCHEME AND GUIDED SOLUTION

1   Use the remainder theorem, or otherwise, to factorise
    $x^3 + 6x^2 + 11x + 6$

Hence express $\dfrac{4x+6}{x^3 + 6x^2 + 11x + 6}$ in partial fractions   *(10 marks)*

$f(x) = x^3 + 6x^2 + 11x + 6$
$f(-1) = -1 + 6 - 11 + 6 = 0 \Rightarrow (x+1)$ is a factor   M1   A1
[There is no point in trying a positive number since all the terms of $f(x)$ are positive]
$f(-2) = -8 + 24 - 22 + 6 = 0 \Rightarrow (x+2)$ is a factor   M1   A1
[Since the constant term in $f(x)$ is 6 it is likely that $(x+3)$ is also a factor]
$f(-3) = -27 + 54 - 33 + 6 = 0 \Rightarrow (x+3)$ is a factor   A1
Hence
$x^3 + 6x^2 + 11x + 6 \equiv (x+1)(x+2)(x+3)$

$\dfrac{4x+6}{x^3 + 6x^2 + 11x + 6} \equiv \dfrac{4x+6}{(x+1)(x+2)(x+3)} = \dfrac{A}{x+1} + \dfrac{B}{x+2} + \dfrac{C}{x+3}$ M1

$\equiv \dfrac{A(x+2)(x+3) + B(x+1)(x+3) + C(x+1)(x+2)}{x^3 + 6x^2 + 11x + 6}$

$\Rightarrow 4x + 6 \equiv A(x+2)(x+3) + B(x+1)(x+3) + C(x+1)(x+2)$
Let $x = -1 \Rightarrow \quad 2 = A(-1+2)(-1+3) = 2A, \ A = 1$ M1   A1
Let $x = -2 \Rightarrow -2 = B(-2+1)(-2+3) = -B, \ B = 2$ A1
Let $x = -3 \Rightarrow -6 = C(-3+1)(-3+2) = 2C, \ C = -3$ A1

Answer $\dfrac{1}{x+1} + \dfrac{2}{x+2} - \dfrac{3}{x+3}$

Check: $1(x+2)(x+3) + 2(x+1)(x+3) - 3(x+1)(x+2)$
        $= 1(x^2 + 5x + 6) + 2(x^2 + 4x + 3) - 3(x^2 + 3x + 2) = 4x + 6$ $\checkmark$
Note: marks are not given for the check, but it only takes a little time and is well worth doing.

2   The roots of the equation $x^2 + px + q = 0$ are α and β
    Form, in terms of p and q, the equation whose roots are $\alpha^3 - p\alpha$ and $\beta^3 - p\beta$ *(8 marks)*

$x^2 + px + q = 0 \Rightarrow \alpha + \beta = -p, \ \alpha\beta = q$ B1
(Do not attempt to calculate α and β. We need the *sum* and *product* of the new roots)

Sum of new roots

$$= (\alpha^3 - p\alpha) + (\beta^3 - p\beta) = \alpha^3 + \beta^3 - p(\alpha + \beta)$$ M1
$$= (\alpha + \beta)(\alpha^2 - \alpha\beta + \beta^2) - p(\alpha + \beta)$$
$$= (\alpha + \beta)[(\alpha + \beta)^2 - 3\alpha\beta] - p(\alpha + \beta)$$ M1
$$= (-p)(p^2 - 3q) - p(-p)$$
$$= -p^3 + 3pq + p^2$$ A1

Product of new roots

$$= (\alpha^3 - p\alpha)(\beta^3 - p\beta)$$ M1
$$= \alpha^3\beta^3 - p\alpha\beta(\alpha^2 + \beta^2) + p^2\alpha\beta$$
$$= \alpha^3\beta^3 - p\alpha\beta[(\alpha + \beta)^2 - 2\alpha\beta] + p^2\alpha\beta$$
$$= q^3 - pq(p^2 - 2q) + p^2q$$
$$= q^3 - p^3q + 2pq^2 + p^2q$$ A1

Required equation is $x^2 - x(p^2 + 3pq - p^3) + q^3 - p^3q + 2pq^2 + p^2q = 0$ M1   A1

---

3   By completing the square or otherwise prove that $x^2 - 2ax + b > 0$ for all values of x if and only if the constants a and b are such that $b > a^2$. Find the range of values of x for which the inequality is broken if $b = a^2 - 4$, $a > 0$

Solve the inequality $\dfrac{x-3}{x+2} > \dfrac{x-2}{3-x}$ *(12 marks)*

$x^2 - 2ax + b = (x - a)^2 + b - a^2$ M1

The minimum value of $(x - a)^2$ is zero when $x = a$. Hence
$(x - a)^2 + b - a^2 > 0$ if and only if $b - a^2 > 0$ i.e. $b > a^2$ A1
When $b = a^2 - 4 \Rightarrow x^2 - 2ax + a^2 - 4 > 0$ provided $(x - a)^2 - 4 > 0$

$$(x - a - 2)(x - a + 2) > 0$$ M1

Value of x	$x < a - 2$	$a - 2 < x < a + 2$	$x > a + 2$
Sign of $(x - a - 2)$	$-$ve	$-$ve	$+$ve
Sign of $(x - a + 2)$	$-$ve	$+$ve	$+$ve
Sign of $(x - a - 2)(x - a + 2)$	$+$ve	$-$ve	$+$ve

   M1

Hence inequality broken when $a - 2 < x < a + 2$ A1   A1
   M1

$$\frac{x-3}{x+2} - \frac{x-2}{3-x} > 0 \Rightarrow \frac{x-3}{x+2} + \frac{x-2}{x-3} > 0$$ M1

$$\frac{(x-3)^2 + (x-2)(x+2)}{(x+2)(x-3)} > 0 \Rightarrow \frac{2x^2 - 6x + 5}{(x+2)(x-3)} > 0$$

or $$\frac{x^2 - 3x + \dfrac{5}{2}}{(x+2)(x-3)} > 0$$

Comparing the numerator with $x^2 - 2ax + b$

$a = \dfrac{3}{2}, b = \dfrac{5}{2} \Rightarrow \dfrac{5}{2} > \left(\dfrac{3}{2}\right)^2$ i.e. numerator > 0 for all x M1   A1

Value of x	$x < -2$	$-2 < x < 3$	$x > 3$
Sign of $(x + 2)$	$-$ve	$+$ve	$+$ve
Sign of $(x - 3)$	$-$ve	$-$ve	$+$ve
Sign of $(x + 2)(x - 3)$	$+$ve	$-$ve	$+$ve

   M1

$\Rightarrow x < -2 \cup x > 3$ A1   A1

EXAMINATION QUESTIONS

1   Find the complete set of values of p for which the roots of the equation
$$2x^2 + px + 3p - 10 = 0$$
are real.

(L)

2   Given that k is a real constant such that $0 < k < 1$, show that the roots of the equation
$$kx^2 + 2x + (1 - k) = 0$$
are

a) always real,

b) always negative.

(L)

> Where $\mathbb{R}$ is the set of real numbers.

3   The function f is defined by
$$f(x) \equiv x^2 + 6x + 20 + k(x^2 - 3x - 12), \quad k \in \mathbb{R}.$$

i) Find the value of k given that the graph of $y = f(x)$ is a straight line.

ii) Find the value of k given that the roots of $f(x) = 0$ are equal in magnitude and opposite in sign.

iii) Find the value of k and the value of p given that the minimum value of $f(x)$ is p and that $f(-2) = p$.

(L)

4   Given that k is real, find the set of values of k for which the roots of the quadratic equation
$$(1 + 2k)x^2 - 10x + (k - 2) = 0$$

a) are real,

b) have a sum which is greater than 5.

(AEB 1988)

5   a) The variables p and q are related by the law
$$q = ap^b,$$ where a and b are constants.
Given that $\ln p = 1 \cdot 32$ when $\ln q = 1 \cdot 73$,
and $\ln p = 0 \cdot 44$ when $\ln q = 1 \cdot 95$ find the values of b and $\ln a$.

b) Given that $y = \log_2 x$ and that
$$\log_2 x - \log_x 8 + \log_2 2^k + k\log_x 4 = 0,$$
prove that
$$y^2 + ky + (2k - 3) = 0.$$

i) Hence deduce the set of values of k for which y is real.

ii) Find the values of x when $k = 1 \cdot 5$.

(AEB 1986)

6   Show that $\log_9 (xy^2) = \dfrac{1}{2}\log_3 x + \log_3 y$.

Hence, or otherwise, solve the simultaneous equations
$$\log_9 (xy^2) = \frac{1}{2}$$
$$(\log_3 x)(\log_3 y) = -3.$$

(AEB 1988)

7   Solve these equations giving your answers to two significant figures:

a) $2\log_x 6 = 7,$

b) $2^{2y+1} = 3^{y-2}.$

(AEB 1989)

8   If α, β are the roots of the quadratic equation
$$x^2 + px + q = 0$$
find the quadratic equation whose roots are $\alpha - \beta$ and $\beta - \alpha$.

(WJEC)

9   Given that $x \neq 2$, find the complete set of values of x for which
$$\frac{3x + 1}{x - 2} > 2.$$

(L)

10 Sketch on the same diagram the graphs of
$$y = 2|x| \text{ and } y = |x-3|.$$
Solve the inequality
$$2|x| \leqslant |x-3|.$$
(JMB)

11 Show that, if x is real, then $x^2 + 6x + 14$ is always positive.
Hence, or otherwise, determine the real values of x for which

i) $\dfrac{(x+3)(x+4)}{(x+1)} > \dfrac{x+8}{3}$;

ii) $\dfrac{(x+3)(x+4)}{(x+1)(x+8)} > \dfrac{1}{3}$.
(NI 1987)

12 Find the values of x which satisfy each of the following inequalities:

i) $|2x+1| < 3$,

ii) $\dfrac{x^2+x-1}{x^2+1} \leqslant \dfrac{1}{2}$.
(WJEC)

13 Express
$$9x^2 - 36x + 52$$
in the form $(Ax-B)^2 + C$, where A, B and C are integers.
Hence, or otherwise, find the set of values taken by $9x^2 - 36x + 52$ for $x \in \mathbb{R}$.
(C)

14 Sketch the graph of $y = |x+2|$ and hence, or otherwise, solve the inequality
$$|x+2| > 2x+1, \; x \in \mathbb{R}.$$
(C)

15 Given that α and β are roots of the equation
$$x^2 - px + 2 = 0,$$
express $\alpha^2 + \beta^2$ in terms of p.
Without solving the given equation, find a quadratic equation whose roots are

$$\alpha^2 + \frac{\alpha}{\beta} \text{ and } \beta^2 + \frac{\beta}{\alpha},$$

giving the coefficients in a simplified form not involving α or β.
(JMB)

16 The roots of the equation
$$x^2 - 2x - 4 = 0$$
are α and β. Without finding α and β, find the quadratic equation with integer coefficients and with roots $\alpha^2\beta$ and $\alpha\beta^2$.
(JMB)

ANSWERS TO EXERCISES

EXERCISES 3.1

1 $-1 - 7x - 6x^2 - 4x^3$

2 $-3ab - 5ac - 2bc$

3 $2x^3y + 2xy^3$

4 $-2a + 4b - 5c$

5 $x^3 + 4x^2 - 5x + 3$

6 $-6x^2 + 8x + 5$

7 $4x^4 + 8x^3 + 5x^2 + 13x - 78$

8 $-6x^2 + 2y^2 - 4z^2 + xy + 11xz + 2yz$

9 $-x^3 - x^2 + \dfrac{1}{1-x}$

10 $7x + 1 - \dfrac{12x+5}{x^2 - x + 2}$

11 $x^5 + x^3 + x + \dfrac{1}{x+1}$

12 $\dfrac{2x-4}{(x-1)(3x-4)}$

13 $\dfrac{3-x}{x(x-1)(x+1)}$

14 $\dfrac{4}{(x-z)(y-z)}$

15 $-\dfrac{1}{2}$

16 $-\dfrac{1}{8}$

EXERCISES 3.2

1 $6x(x+2y)$
2 $6y(y^2+2y-4)$
3 $(x-b)(x+a)$
4 $(a+3b)(a-3b+1)$
5 $y^2(x-2y)(x+2y)$
6 $4(x-3)(x+3)$
7 $(5x-3y)(5x+3y)$
8 $(1-2x^2)^2$
9 $(x-3)(x-4)$
10 $(2x-1)(x-8)$
11 $(3x-5)(x+6)$
12 $(5-2x)(3+21x)$
13 $(3a+5)(a+6)$
14 $(5a-4b)(4a+5b)$
15 $(7y-2)(2y+7)$
16 $(x-10y)(x+9y)$
17 $(x-1)(x+1)(x^2+2)$
18 $(x+3)(x+4)$
19 $(a-1)(a+1)(a-3)(a+3)$
20 $(x^2+4)(x^4-4x^2+16)$
21 $(a-b)(a^2+ab+b^2)(a^6+a^3b^3+b^6)$
22 $2(2x-3)(4x^2+6x+9)$
23 $(x^3+1)(x^3+8)=$
    $(x+1)(x^2-x+1)(x+2)(x^2-2x+4)$
24 $(x^3+8)(x^3-1)=$
    $(x-1)(x^2+x+1)(x+2)(x^2-2x+4)$

EXERCISES 3.3

1 3, 4; ½, 2; −1⅕, 2; 1·15, −0·35;
   6·34, 0·16
4 $3b^2-8ac\geqslant0$
5 39

EXERCISES 3.4

1 $x^2+6x+25=0; 5x^2-12x+20=0$
2 $3x^2-(k+14)x+16+k=0$
3 $x^2-(a^2-2b)x+b^2=0$
4 $x^2+3x+1=0$

EXERCISES 3.5

1 $p=1, q=-14; 4, -2, -3$
2 $(x+1)(x+2)(x+3)$
3 $(x+3)$
4 $(2x-1)(x+1), 25x-23$

EXERCISES 3.6

1 i) $x\geqslant4\cup x\leqslant1$ ii) $x\geqslant4\cup x\leqslant1\cup2\leqslant x\leqslant3$
2 $0<x<3\cup x>5$
3 $-4<x<4$
4 $x>0\cup x<-1$

6 $-\dfrac{21}{4}$

EXERCISES 3.7

1 $\dfrac{1}{2(x+1)}-\dfrac{1}{(x+2)}+\dfrac{1}{2(x+3)}$

2 $\dfrac{2}{(x-1)^2}+\dfrac{1}{x-1}$

3 $\dfrac{2}{x+1}+\dfrac{2-2x}{x^2+1}$

4 $\dfrac{1}{1-x}+\dfrac{1}{1+x}-\dfrac{2}{(1+x)^2}$

EXERCISES 3.8

1 5
2 2, 4
3 $\sqrt{2}$, 4
4 −0·107, 0·1

ANSWERS TO EXAMINATION QUESTIONS

1 $p\leqslant4\cup p\geqslant20$

3 $-1, 2, \dfrac{2}{7}, \dfrac{80}{7}$

4 $-3\leqslant k\leqslant4.5; -\dfrac{1}{2}<k<\dfrac{1}{2}$

5 a) 7·846, −0·25;   b) i) $k\leqslant2, k\geqslant6$;

   ii) $1, \dfrac{1}{4}\sqrt{2}$

6 a) $\left(27, \dfrac{1}{3}\right), \left(\dfrac{1}{9}, \sqrt{27}\right)$

7 1·7, −10·0
8 $x^2-4q-p^2=0$
9 $x>2\cup x<-5$
10 $-3\leqslant x\leqslant1$
11 i) $x>-1$; ii) $x>-1$ or $x<-8$
12 $-2<x<1, -3\leqslant x\leqslant1$
13 $(3x-6)^2+16, \geqslant16$
14 $x<1$
15 $p^2-4, 2x^2-3(p^2-4)x+18=0$
16 $x^2+8x-64=0$

CHAPTER 4

SERIES

ARITHMETIC PROGRESSION OR SERIES

GEOMETRIC SERIES

ARITHMETIC AND GEOMETRIC MEANS

BINOMIAL THEOREM

OTHER EXPANSIONS

SUMMATION OF SERIES

MATHEMATICAL INDUCTION

GETTING STARTED

A set of terms T_1, T_2, T_3... each of which is formed according to a definite pattern is called a **sequence**. When the terms are added then a **series** is formed. If there is a finite number of terms then the series is a **finite series**. If the number of terms is infinite then the series is an **infinite series**.

An **arithmetic progression** or **arithmetic series** is a series in which the difference between any term and the preceding term is always constant; namely

$$a + (a+d) + (a+2d) + (a+3d) + \ldots + [a+(n-1)d] + \ldots$$

where a is the first term and d is the common difference. If the number of terms is n, the nth (or last) term is l and the sum of the first n terms is S_n, then you should remember the following formulae:

$$l = a + (n-1)d, \quad S_n = \frac{n}{2}(a+l) = \frac{n}{2}[2a+(n-1)d]$$

The formula for S_n is easily proved, as is now shown

$$S_n = a + (a+d) + (a+2d) + (a+3d) + \ldots + [a+(n-1)d]$$

Reversing the order of the terms on the R.H.S.

$$S_n = [a+(n-1)d] + [a+(n-2)d] + [a+(n-3)d] + \ldots a$$

Adding the two equations

$$2S_n = [2a+(n-1)d] + [2a+(n-1)d] + \ldots n \text{ times}$$

$$= n[2a+(n-1)d] \Rightarrow S_n = \frac{n}{2}[2a+(n-1)d]$$

or $S_n = \dfrac{n}{2}[a+a+(n-1)d] = \dfrac{n}{2}[a+l]$

ESSENTIAL PRINCIPLES

Worked example

The 9th term of an arithmetic series is -1 and the sum of the first 9 terms is 45. Find i) the common difference, ii) the sum of the first 15 terms.

First write down in algebraic form, the information given

$n = 9$, $l = -1$, $S_9 = 45$

Then look at the formulae and choose the one containing only one unknown.

$$S_n = \frac{n}{2}(a+l) \Rightarrow 45 = \frac{9}{2}[a+(-1)] \Leftrightarrow a = 11$$

$$l = a + (n-1)d \Rightarrow -1 = 11 + 8d \Leftrightarrow d = -\frac{3}{2}$$

$$S_n = \frac{n}{2}[2a+(n-1)d] \Rightarrow S_{15} = \frac{15}{2}[22 + 14\left(-\frac{3}{2}\right)] = \frac{15}{2}$$

Worked example

Find the sum of the first n terms of the arithmetic series $2+5+8+\ldots$ Find the value of n for which the sum of the first $2n$ terms exceeds by 292 the sum of the first n terms.

$2+5+8+\ldots \Rightarrow a = 2$, $d = 3$

$$S_n = \frac{n}{2}[2a+(n-1)d] = \frac{n}{2}[4+3(n-1)] = \frac{n}{2}(3n+1)$$

Replacing n by $2n \Rightarrow S_{2n} = \frac{2n}{2}[3(2n)+1] = n(6n+1)$

But $S_{2n} - S_n = 292 \Rightarrow n(6n+1) - \frac{n}{2}(3n+1) = 292$

$\Rightarrow 9n^2 + n - 584 = 0 \Leftrightarrow (9n+73)(n-8) = 0$

$\Rightarrow n = 8$ since n is positive

Worked example

The sum of n terms of a series is $5n^2$ for $n = 1, 2, 3\ldots$

Show that the series is an arithmetic series

$S_n = 5n^2$, $S_{n-1} = 5(n-1)^2$

$\Rightarrow n$th term T_n is $S_n - S_{n-1} = 5n^2 - 5(n-1)^2 = 10n - 5$

$\Rightarrow T_{n-1} = 10(n-1) - 5$

$\Rightarrow T_n - T_{n-1} = (10n-5) - [10(n-1)-5] = 10$ (a constant)

$\Rightarrow$ series is an arithmetic series

Note: it is not sufficient to say $S_1 = T_1 = 5$

$S_2 = T_1 + T_2 = 5.2^2 = 20 \Rightarrow T_2 = 20 - 5 = 15$

$S_3 = T_1 + T_2 + T_3 = 5.3^2 = 45 \Rightarrow T_3 = 25$

$\Rightarrow T_2 - T_1 = 15 - 5 = 10 = 25 - 15 = T_3 - T_2$

You must show that the difference between *any* term and the preceding term is a constant. Namely, given a formula for S_n, that for S_{n-1} can be deduced. Hence a formula for T_n can be obtained. Then that for T_{n-1} can be deduced and the difference between T_n and T_{n-1} can be shown to be a constant.

GEOMETRIC SERIES

A series in which the ratio of any term to the preceding term is always constant; namely $a + ar + ar^2 + ar^3 + \ldots + ar^n + \ldots$ where a is the first term and r is the common ratio.

If the number of terms is n, the nth (or last) term is T_n and the sum of the first n terms is S_n, then you should remember

$$T_n = ar^{n-1}, \quad S_n = \frac{a(1-r^n)}{(1-r)}$$

Further if $|r| < 1$, i.e. $-1 < r < 1$, then r^n approaches zero as n becomes very large.

Consequently S_n approaches $\dfrac{a}{1-r}$ as n gets very large. We say that the limit of r^n as n

approaches infinity is zero and we write this as

$$\lim_{n \to \infty} r^n = 0$$

Consequently for $|r| < 1$ the geometric series is said to be *convergent* and to converge to a

sum $\dfrac{a}{1-r}$ i.e., for $|r| < 1$, $\lim\limits_{n \to \infty} S_n = S_\infty = \dfrac{a}{1-r}$

The formula for S_n is proved as follows:

> **Proving the formula for summing a geometric series.**

$$S_n = a + ar + ar^2 + \ldots\ldots + ar^{n-2} + ar^{n-1}$$
$$r\,S_n = \qquad ar + ar^2 + ar^3 + \ldots + ar^{n-2} + ar^{n-1} + ar^n$$

$$\Rightarrow S_n(1-r) = a - ar^n \Rightarrow S_n = \frac{a(1-r^n)}{1-r}$$

Worked example

A geometric series with first term 3 converges to sum 2. Find i) the common ratio, ii) the 6th term, iii) the sum of the first 10 terms, giving your answer to 3 decimal places.
Again first interpret the information algebraically. The series converges to 2 is simply another way of saying that the sum of the series to infinity is 2. Hence

$$a = 3, \quad S_\infty = \frac{a}{1-r} = 2$$

$$\Rightarrow \frac{3}{1-r} = 2 \Longleftrightarrow 2 - 2r = 3 \Longleftrightarrow r = -\frac{1}{2}$$

$$T_n = ar^{n-1} \Rightarrow T_6 = 3\left(-\frac{1}{2}\right)^5 = -\frac{3}{32}$$

$$S_n = \frac{a(1-r^n)}{1-r} \Rightarrow S_{10} = \frac{3\left[1-\left(-\frac{1}{2}\right)^9\right]}{1-\left(-\frac{1}{2}\right)} = 2\cdot 004$$

Worked example

The first three terms of an arithmetic series are 1, x, y. Given that 1, x, $-y$ are the first three terms of a geometric series, prove that $x^2 + 2x - 1 = 0$. Hence find the first three terms of each series given that $x > 0$
Arithmetic series $\Rightarrow$ 1, x, $y \Rightarrow 2x = 1 + y$

$$\text{Geometric series } 1, x, (-y) \Rightarrow \frac{x}{1} = \frac{-y}{x} \text{ or } x^2 = -y$$

Solving $\Rightarrow x^2 = 1 - 2x \Longleftrightarrow x^2 + 2x - 1 = 0$
$\Longleftrightarrow (x+1)^2 - 2 = 0 = (x+1-\sqrt{2})(x+1+\sqrt{2})$
$\Longleftrightarrow x = -1 + \sqrt{2}$ or $-1 - \sqrt{2}$ and since $x > 0$, $x = \sqrt{2} - 1 \Rightarrow y = 2\sqrt{2} - 3$
Arithmetic series $\Rightarrow$ 1, $(\sqrt{2} - 1)$, $(2\sqrt{2} - 3)$
Geometric series $\Rightarrow$ 1, $(\sqrt{2} - 1)$, $(3 - 2\sqrt{2})$

Worked example

Find how many terms of the geometric series $10 + 8 + 6\cdot 4 + \ldots$ must be taken to make the sum exceed 49

$$a = 10, \quad r = \frac{4}{5}$$

$$S_n = \frac{a(1-r^n)}{1-r} = \frac{10(1-0\cdot 8^n)}{1-(0\cdot 8)} = 50(1-0\cdot 8^n)$$

Hence $50(1-0\cdot 8^n) > 49 \Rightarrow (0\cdot 8)^n < \dfrac{1}{50}$

Inverting $(1\cdot 25)^n > 50 \Rightarrow n\log(1\cdot 25) > \log 50$
$\Rightarrow 0\cdot 0969 n > 1\cdot 6990 \Rightarrow n > 17\cdot 53$
$\Rightarrow$ least integral value of n is 18.

ARITHMETIC AND GEOMETRIC MEANS

Know the difference between arithmetic and geometric means.

Given that a, b and c are three consecutive terms of an arithmetic series, then $b = \frac{1}{2}(a+c)$ is said to be the **arithmetic mean** of a and c.

Given that x, y, z are three consecutive terms of a geometric series, then $y = \sqrt{(xz)}$ is said to be the **geometric mean** of x and z.

Worked example

The arithmetic mean and the geometric mean of the real numbers x and y, $x>y>0$ are A and G respectively. Prove that $A>G$

$$A = \frac{1}{2}(x+y)$$

$$G = \sqrt{(xy)}$$

$$A - G = \frac{1}{2}(x+y) - \sqrt{(xy)} = \frac{1}{2}[x - 2\sqrt{(xy)} + y]$$

$$= \frac{1}{2}[\sqrt{x} - \sqrt{y}]^2 \text{ and since } x>y>0 \text{ this is real and positive}$$

$$\Rightarrow A - G > 0 \text{ or } A > G$$

Note: whenever you are required to prove one expression, say A, is greater than a second expression, say G, then it is always easier to consider $A - G$ and prove $A - G > 0$.

EXERCISES 4.1

1 Find
   a) the sum of the first 20 terms of the series $1 + 3 + 5 + \ldots + (2n - 1) + \ldots$
   b) the sum of the first 15 terms of the series $1 + 3 + 9 + \ldots + 3^{n-1} + \ldots$

2 The sum of the first 20 terms of an arithmetic series is 45. The sum of the first 40 terms of the series is 290. Find a) the first term, b) the common difference, of the series. Find the number of terms in the series which are less than 100

3 Find the first term and the common difference of an arithmetic series if the sum of the first n terms is $2n^2 + n$

4 Calculate the sum of all the positive integral numbers less than 200 which are divisible by 7

5 The first three consecutive terms of an arithmetic series are $\dfrac{1}{b+c}, \dfrac{1}{c+a}, \dfrac{1}{a+b}$.

   Show that a^2, b^2, c^2 are also three consecutive terms of an arithmetic series

6 The first term of a geometric series is 432. Its fourth term is 128. Calculate a) the common ratio, b) the sum to infinity of the series

7 A piece of string 1 m in length is divided into 10 pieces whose lengths are in geometrical progression. Given that the length of the shortest piece is x cm and the length of the longest piece is $8x$ cm find x correct to 1 decimal place

8 Find the sum of the first n terms of the series whose rth term is $2^r + 2r - 1$

9 Express the recurring decimal $0.3\dot{2}\dot{1}$ in the form $\dfrac{x}{y}$, where x and y are integers with no common factor

BINOMIAL THEOREM

An extremely important series in algebra is the **binomial series** or binomial expansion. It states that

$$(1+x)^n = 1 + \frac{n}{1}x + \frac{n(n-1)}{1.2}x^2 + \frac{n(n-1)(n-2)}{1.2.3}x^3 + \frac{n(n-1)(n-2)(n-3)}{1.2.3.4}x^4 + \ldots$$

The expansion is easy to remember since you should note that in the terms:

i) the factors in n in the *numerator* go down one at a time
   i.e. $n(n-1)(n-2)\ldots$

Ways of remembering the binomial expansion.

ii) the factors in the *denominator* go up one at a time, i.e. $1, 2, 3\ldots$

iii) the powers of x go up one at a time

iv) not counting the x's, there are, in any term, as many terms in the numerator as there are in the denominator and the number of terms in each is the same as the power of x of that term

If n is a positive integer, the expansion terminates after $n+1$ terms and the expansion is valid for all values of x.

If n is not a positive integer, but is a negative integer or a rational number, then the series is an infinite series and the expansion is valid provided $|x|<1$. It is not valid for $|x|\geqslant 1$. The term

$$\frac{n(n-1)(n-2)\ldots .(n-r+1)}{1.2.3.\ldots r}$$

is usually denoted by the symbol $\binom{n}{r}$ so that the binomial expansion can be remembered in the more simplified form of:

$$(1+x)^n = 1+\binom{n}{1}x+\binom{n}{2}x^2+\binom{n}{3}x^3+\ldots+\binom{n}{r}x^r+\ldots$$

$$= 1+\sum_{r=1}^{\infty}\binom{n}{r}x^r \text{ where } \sum \text{ is a summation.}$$

In this instance $\sum_{r=1}^{\infty}\binom{n}{r}x^r$ stands for the summation of all terms of the form $\binom{n}{r}x^r$ when r takes successively the values $1,2,3\ldots$ to infinity.

Certain binomial expansions are used so frequently that you will find it advantageous to commit them to memory. They are:

Some useful expansions to learn.

$$(1+x)^{-1} = 1-x+x^2-x^3+x^4\ldots$$
$$(1-x)^{-1} = 1+x+x^2+x^3+x^4\ldots$$
$$(1+x)^{-2} = 1-2x+3x^2-4x^3\ldots$$
$$(1-x)^{-2} = 1+2x+3x^2+4x^3\ldots$$

Note: when the sign inside the bracket is different from that of the index, the signs of the coefficients in the expansion alternate. When the two signs are both negative, all terms of the expansion are positive. The negative sign inside the bracket makes no difference to the form of the expansion which is obtained by writing $(1-x)^n$ as $[1+(-x)]^n$ and then replacing x by $(-x)$ in the expanded series.

Worked example

Expand a) $(1-3x)^{-2}$, b) $(4+5x)^{\frac{1}{2}}$, in ascending powers of x up to and including the terms in x^3. State for each the range of validity

a) $(1-3x)^{-2} = 1+\dfrac{(-2)}{1}(-3x)+\dfrac{(-2)(-3)}{1.2}(-3x)^2+\dfrac{(-2)(-3)(-4)}{1.2.3}(-3x)^3+\ldots$

$$= 1+6x+27x^2+108x^3+\ldots$$

Valid for $|3x|<1 \Rightarrow -\dfrac{1}{3}<x<\dfrac{1}{3}$

b) $(4+5x)^{\frac{1}{2}} = \left[4\left(1+\dfrac{5}{4}x\right)\right]^{\frac{1}{2}} = 4^{\frac{1}{2}}\left(1+\dfrac{5}{4}x\right)^{\frac{1}{2}} = 2\left(1+\dfrac{5}{4}x\right)^{\frac{1}{2}}$

$$= 2\left[1+\dfrac{\frac{1}{2}}{1}\left(\dfrac{5x}{4}\right)+\dfrac{\left(\frac{1}{2}\right)\left(-\frac{1}{2}\right)}{1.2}\left(\dfrac{5x}{4}\right)^2+\dfrac{\left(\frac{1}{2}\right)\left(-\frac{1}{2}\right)\left(-\frac{3}{2}\right)}{1.2.3}\left(\dfrac{5x}{4}\right)^3+\ldots\right]$$

$$= 2\left[1+\dfrac{5x}{8}-\dfrac{25x^2}{128}+\dfrac{125x^3}{1024}-\ldots\right]$$

$$= 2+\dfrac{5x}{4}-\dfrac{25x^2}{64}+\dfrac{125x^3}{512}-\ldots$$

Valid for $\left|\dfrac{5}{4}x\right|<1 \Rightarrow -\dfrac{4}{5}<x<\dfrac{4}{5}$

Worked example

Given that $f(x) = \dfrac{1+2x}{(x-1)(x-2)^2}$, express $f(x)$ in partial fractions. Hence, or otherwise,

find a quadratic approximation to $f(x)$, given that x is so small that, when $n \geq 3$, x^n can be neglected.

$$\frac{1+2x}{(x-1)(x-2)^2} \equiv \frac{A}{x-1} + \frac{B}{x-2} + \frac{C}{(x-2)^2} \equiv \frac{A(x-2)^2 + B(x-1)(x-2) + C(x-1)}{(x-1)(x-2)^2}$$

$$\Rightarrow 1+2x \equiv A(x-2)^2 + B(x-1)(x-2) + C(x-1)$$

Let $x \to 1 \Rightarrow A = 3$

Let $x \to 2 \Rightarrow C = 5$

Comparing coefficients of x^2, $0 = A+B \Rightarrow B = -A = -3$

$$\Rightarrow \frac{2x+1}{(x-1)(x-2)^2}$$

$$\equiv \frac{3}{x-1} - \frac{3}{x-2} + \frac{5}{(x-2)^2}$$

$$= -3(1-x)^{-1} + \frac{3}{2}\left(1-\frac{x}{2}\right)^{-1} + \frac{5}{4}\left(1-\frac{x}{2}\right)^{-2}$$

$$= -3[1+x+x^2+\ldots] + \frac{3}{2}\left[1+\left(\frac{x}{2}\right)+\left(\frac{x}{2}\right)^2+\ldots\right] + \frac{5}{4}\left[1-2\left(-\frac{x}{2}\right)+3\left(-\frac{x}{2}\right)^2\ldots\right]$$

$$= \left(-3+\frac{3}{2}+\frac{5}{4}\right) + \left(-3+\frac{3}{4}+\frac{5}{4}\right)x + \left(-3+\frac{3}{8}+\frac{15}{16}\right)x^2\ldots$$

$$= -\frac{1}{4} - x - \frac{27}{16}x^2, \text{ neglecting } x^3 \text{ and higher power of } x$$

Worked example

Obtain the first 3 terms in the expansion, in ascending powers of x, of $(1+3x+2x^2)^8$
The expansion can be obtained in one of two ways.

Method I $(1+3x+2x^2)^8 = [(1+2x)(1+x)]^8 = (1+2x)^8 . (1+x)^8$

$$= \left[1+\frac{8}{1}(2x)+\frac{8.7}{1.2}(2x)^2+\frac{8.7.6}{1.2.3}(2x)^3\ldots\right]\left[1+\frac{8}{1}x+\frac{8.7}{1.2}x^2+\frac{8.7.6}{1.2.3}x^3\ldots\right]$$

$$= (1+16x+112x^2+448x^3\ldots)(1+8x+28x^2+56x^3\ldots)$$

$$= 1+24x+(112+28+128)x^2+(448+56+448+896)x^3\ldots$$

$$= 1+24x+268x^2+1848x^3\ldots$$

Method II $(1+3x+2x^2)^8 = [1+(3x+2x^2)]^8$

$$= 1+\frac{8}{1}(3x+2x^2)+\frac{8.7}{1.2}(3x+2x^2)^2+\frac{8.7.6}{1.2.3}(3x+2x^2)^2$$

$$= 1+8(3x+2x^2)+28(9x^2+12x^3\ldots)+56(27x^3\ldots)$$

$$= 1+24x+(16+252x^2)+(336+1512)x^3\ldots$$

$$= 1+24x+268x^2+1848x^3+\ldots$$

EXERCISES 4.2

1   Expand in ascending powers of x, up to and including the term in x^3

a) $(1+8x)^{\frac{1}{4}}$ b) $(4-x)^{-\frac{1}{2}}$ c) $\dfrac{(1+8x)^{\frac{1}{4}}}{(4-x)^{\frac{1}{2}}}$

2   Show that $\dfrac{1}{(1+x+x^2)^2} = \left(\dfrac{1-x}{1-x^3}\right)^2$

Hence, or otherwise, expand $\left(\dfrac{1}{1+x+x^2}\right)^2$ in ascending powers of x up to and including the term in x^7

3   Expand, in ascending powers of x, up to and including the term in x^3, $(1+x)^{\frac{1}{2}}$

The expansion of $\dfrac{(1+x)^{\frac{1}{4}}}{1-px}$, in ascending powers of x, up to and including the term in x^2

is $1+qx^2$, where p and q are constants. Find the values of p and q

4   In the expansion of $(1+ax-3x^2)^6$ in ascending powers of x, the coefficients of x^2 and x^3 are 42 and 20 respectively. Find the value of the constant a

5   Find   i) the coefficient of x^5,   ii) the term independent of x, in the binomial expansion in descending powers of x of $\left(\dfrac{x^2}{2}-\dfrac{3}{x^3}\right)^{10}$

OTHER EXPANSIONS

Other expansions you may find useful.

Certain well-known functions can be expressed as a series of terms in ascending powers of x. Whilst proof of their expansion is not always required in GCE A-level mathematics syllabuses, knowledge of the expansions range of validity and use in approximations is frequently required.

The appropriate series expansions are:

i) $\sin x = x - \dfrac{x^3}{3!} + \dfrac{x^5}{5!} - \dots + (-1)^n \dfrac{x^{2n+1}}{(2n+1)!} + \dots$ valid for all x

ii) $\cos x = 1 - \dfrac{x^2}{2!} + \dfrac{x^4}{4!} - \dots + (-1)^n \dfrac{x^{2n}}{(2n)!} + \dots$ valid for all x

iii) $e^x = 1 + \dfrac{x}{1!} + \dfrac{x^2}{2!} + \dfrac{x^3}{3!} + \dots + \dfrac{x^n}{n!} + \dots$ valid for all x

iv) $\ln(1+x) = x - \dfrac{x^2}{2} + \dfrac{x^3}{3} - \dots + (-1)^{n-1}\dfrac{x^n}{n} + \dots$ valid for $-1 < x \leqslant 1$

From these can be deduced other series. For example

$$\tan x = \dfrac{\sin x}{\cos x} = \dfrac{x - \dfrac{x^3}{3!} + \dfrac{x^5}{5!} - \dots}{1 - \dfrac{x^2}{2!} + \dfrac{x^4}{4!} - \dots} = x + \dfrac{x^3}{3} + \dfrac{2x^5}{15} + \dots$$

Further when x is replaced by $-x$ in the series expansion for $\ln(1+x)$ then

$$\ln(1-x) = (-x) - \dfrac{(-x)^2}{2} + \dfrac{(-x)^3}{3} - \dots + (-1)^{n-1}\dfrac{(-x)^n}{n}$$

$$= -x - \dfrac{x^2}{2} - \dfrac{x^3}{3} - \dots - \dfrac{x^n}{n} - \dots \quad \begin{array}{l}\text{valid for } -1 < (-x) \leqslant 1 \\ \text{i.e. valid for } -1 \leqslant x < 1\end{array}$$

Worked example

Given that x^4 and higher powers may be neglected, show that

$$\dfrac{e^{2x}\ln(1+3x)}{(2+x)} = \dfrac{3x}{2}(1+2x^2)$$

Using the series for e^x with x replaced by $2x$, the series for $\ln(1+x)$ with x replaced by $3x$, together with the binomial expansion of $(2+x)^{-1} = 2^{-1}\left(1+\dfrac{x}{2}\right)^{-1} = 2^{-1}\left(1-\dfrac{x}{2}+\dfrac{x^2}{4}\dots\right)$

$$\Rightarrow \dfrac{e^{2x}\ln(1+3x)}{2+x} = \left(1+\dfrac{2x}{1!}+\dfrac{4x^2}{2!}\dots\right)\left(\dfrac{3x}{1}-\dfrac{9x^2}{2}+\dfrac{27x^3}{3}\dots\right)\dfrac{1}{2}\left(1-\dfrac{x}{2}+\dfrac{x^2}{4}\dots\right)$$

$$= \dfrac{1}{2}\left(3x+\dfrac{3x^2}{2}+6x^3\dots\right)\left(1-\dfrac{x}{2}+\dfrac{x^2}{4}\dots\right)$$

$$= \dfrac{3}{2}x(1+2x^2) \text{ neglecting } x^4 \text{ and higher powers}$$

Note: when answering questions such as this, you should take care not to waste time by expanding series further than necessary. You should note in the above example that, since the logarithmic series commences with an x term, it is not necessary to expand the exponential and binomial series beyond the x^2 terms.

Worked example

Use the series expansion for $\ln(1+x)$ to show that:

i) $\ln\left(\dfrac{1+x}{1-x}\right) = 2\left(x + \dfrac{x^3}{3} + \dfrac{x^5}{5} + \ldots \dfrac{x^{2n+1}}{2n+1} + \ldots\right)$, valid for $-1<x<1$

ii) $\ln m = 2\left[\left(\dfrac{m-1}{m+1}\right) + \dfrac{1}{3}\left(\dfrac{m-1}{m+1}\right)^3 + \dfrac{1}{5}\left(\dfrac{m-1}{m+1}\right)^5 + \ldots + \dfrac{1}{2n+1}\left(\dfrac{m-1}{m+1}\right)^{2n+1} + \ldots\right]$,

valid for $m>0$

Hence, evaluate $\ln 2$ to three decimal places

$$\ln(1+x) = x - \frac{x^2}{2} + \frac{x^3}{3} - \frac{x^4}{4} + \frac{x^5}{5} - \ldots + \frac{(-1)^{2n}x^{2n+1}}{2n+1} + \ldots \tag{1}$$

$$\Rightarrow \ln(1-x) = -x - \frac{x^2}{2} - \frac{x^3}{3} - \frac{x^4}{4} - \frac{x^5}{5} - \ldots - \frac{x^{2n+1}}{2n+1} - \ldots \tag{2}$$

Subtracting (2) from (1)

$$\Rightarrow \ln(1+x) - \ln(1-x) = 2\left[x + \frac{x^3}{3} + \frac{x^5}{5} + \ldots + \frac{x^{2n+1}}{2n+1} + \ldots\right] \tag{3}$$

The series for $\ln(1+x)$ is valid for $-1<x\leqslant 1$
The series for $\ln(1-x)$ is valid for $-1\leqslant x<1$
The series for the combination of these two series is therefore valid for the common part of the two ranges of validity, i.e. $-1<x<1$

Substituting $m = \dfrac{1+x}{1-x}$ in (3), i.e. $m(1-x) = 1+x \Leftrightarrow x = \dfrac{m-1}{m+1}$

$$\Rightarrow \ln m = 2\left[\left(\frac{m-1}{m+1}\right) + \frac{1}{3}\left(\frac{m-1}{m+1}\right)^3 + \frac{1}{5}\left(\frac{m-1}{m+1}\right)^5 + \ldots + \frac{1}{2n+1}\left(\frac{m-1}{m+1}\right)^{2n+1} + \ldots\right]$$

For $-1<x<1$, $m = \dfrac{1+x}{1-x}$ is always positive. Further as $x\to 1$, $m\to\infty$ and as $x\to -1$,

$m\to 0 \Rightarrow$ the series for $\ln m$ is valid for $m>0$
Substitute $m = 2$

$$\Rightarrow \ln 2 = 2\left[\frac{1}{3} + \frac{1}{3^4} + \frac{1}{5}\left(\frac{1}{3^5}\right) + \ldots\right]$$

$$= 2[0\cdot 333333 + 0\cdot 012346 + 0\cdot 000823 + 0\cdot 000065 + \ldots] = 0\cdot 693$$

EXERCISES 4.3

1 Expand $e^{\frac{x}{2}}\ln(1+x)$ in ascending powers of x up to and including the term in x^4. Hence show that for certain values of x to be stated, the series expansion of $e^{\frac{x}{2}}\ln(1+x) + e^{-\frac{x}{2}}\ln(1-x)$ in ascending powers of x commences with a term in x^4. Find this term.

2 Expand $\dfrac{e^x + e^{-x}}{e^{2x}}$ in ascending powers of x, up to and including the term in x^3. State the general term in the expansion

3 Determine m and n so that the coefficients of x^3 and x^4 in the expansion of $(1 + mx + nx^2)\ln(1+x)$ are both zero

Prove that with these values of m and n, the error in taking $\dfrac{x + \frac{1}{2}x^2}{1 + mx + nx^2}$ for $\ln(1+x)$

is $\dfrac{x^5}{180}$, neglecting powers of x higher than the fifth

4 Show that if x is small such that terms in x^n, where $n\geqslant 3$, can be neglected, then
$$\frac{\sin^2 x - x^2\cos x}{x^4} = \frac{1}{6}\left(1 + \frac{x^2}{60}\right)$$

5   Find the expansions, in ascending powers of x, up to and including the terms in x^4 of
a) $e^{-2x} \sin 3x$,   b) $\ln(\cos x)$

SUMMATION OF SERIES

> Here we concentrate on finding the sum of the expansion.

Just as it is often necessary to expand a function in terms of ascending or descending powers of the variable x, say, so it is sometimes necessary to undertake the reverse process; i.e. given the expansion, find its sum. There is no general rule for the summation of a series and much depends upon being able to recognise the series as falling into a particular category. We shall illustrate this by taking examples of the different kinds of summation that you could meet.

Worked example

Find $\displaystyle\sum_{r=1}^{n} (r+1)(r+3)$.

Summation of series of this form can easily be found using one or more of the following known results

$$\sum_{r=1}^{n} r = \frac{1}{2}n(n+1), \quad \sum_{r=1}^{n} r^2 = \frac{1}{6}n(n+1)(2n+1), \quad \sum_{r=1}^{n} r^3 = \left[\frac{1}{2}n(n+1)\right]^2$$

$$\sum_{r=1}^{n} r(r+1)(r+2)\ldots(r+k) = \frac{1}{k+2}n(n+1)(n+2)\ldots(n+k)(n+k+1)$$

Method I $\displaystyle\sum_{r=1}^{n} (r+1)(r+3) = \sum_{r=1}^{n} (r^2+4r+3) = \sum_{r=1}^{n} r^2 + 4\sum_{r=1}^{n} r + 3n$

$$\Rightarrow \sum_{r=1}^{n} (r+1)(r+3) = \frac{1}{6}n(n+1)(2n+1) + 4.\frac{1}{2}n(n+1) + 3n$$

$$= \frac{1}{6}n(2n^2 + 15n + 31)$$

Method II $\displaystyle\sum_{r=1}^{n} (r+1)(r+3) = \sum_{r=1}^{n} r(r+1) + \sum_{r=1}^{n} 3r + 3n$

$$= \frac{1}{3}n(n+1)(n+2) + 3.\frac{1}{2}n(n+1) + 3n$$

$$= \frac{1}{6}n[2n^2 + 15n + 31]$$

Worked example

Find   i) $\displaystyle\sum_{r=1}^{n} \frac{1}{r(r+2)}$ ii) $\displaystyle\sum_{r=1}^{\infty} \frac{1}{r(r+2)}$

Series of this form can be found by using partial fractions

Let $T_n = \dfrac{1}{r(r+2)} = \dfrac{\frac{1}{2}}{r} - \dfrac{\frac{1}{2}}{r+2}$

$$\Rightarrow 2T_n = \frac{1}{r} - \frac{1}{r+2}$$

Taking this result and summing the series for $r = 1, 2, 3 \ldots n \Rightarrow$

$$2T_1 = \frac{1}{1} - \frac{1}{\cancel{3}}$$

$$2T_2 = \frac{1}{2} - \frac{1}{\cancel{4}}$$

$$2T_3 = \frac{1}{\cancel{3}} - \frac{1}{\cancel{5}}$$

$$2T_4 = \frac{1}{\cancel{4}} - \frac{1}{\cancel{6}}$$

$$2T_{n-2} = \frac{1}{\cancel{n-2}} - \frac{1}{\cancel{n}}$$

$$2T_{n-1} = \frac{1}{\cancel{n-1}} - \frac{1}{n+1}$$

$$2T_n = \frac{1}{\cancel{n}} - \frac{1}{n+2}$$

Summing, it can be seen that certain terms in the first column cancel with terms in the second column to give

$$2\sum_{r=1}^{n} T_r = 2\sum_{r=1}^{n} \frac{1}{r(r+2)} = 1 + \frac{1}{2} - \frac{1}{n+1} - \frac{1}{n+2} = \frac{3n^2+5n}{2(n+1)(n+2)}$$

or $\displaystyle\sum_{r=1}^{n} \frac{1}{r(r+2)} = \frac{3n^2+5n}{4(n+1)(n+2)}$

Do make sure that you write down a sufficient number of first and last terms to avoid missing out part of the answer.

For ii) we let $n \to \infty$

Rewriting

$$\frac{3n^2+5n}{4(n+1)(n+2)} = \frac{3+\dfrac{5}{n}}{4\left(1+\dfrac{1}{n}\right)\left(1+\dfrac{2}{n}\right)}$$

it can be seen that

$$\sum_{r=1}^{\infty} \frac{1}{r(r+2)} = \frac{3}{4.1.1} = \frac{3}{4}$$

For series which do not fall into either of these categories it is usually necessary to recognise a connection between the given series and a known standard series for, say, e^{ax}, $\ln(1+ax)$ or $(1+ax)^n$

Worked example

Sum the infinite series

$$S = 1 - \frac{2}{3} + \frac{4}{3.6} - \frac{8}{3.6.9} + \ldots + (-1)^n \frac{2^n}{3.6.9\ldots(3n)}$$

You should note that there are factors of 2 in each term of the numerator and factors of 3 in each term of the denominator. Take out these factors and it should help in recognising the form of the series.

$$S = 1 - \left(\frac{2}{3}\right) + \frac{1}{1.2}\left(\frac{2}{3}\right)^2 - \frac{1}{1.2.3}\left(\frac{2}{3}\right)^3 + \ldots$$

Further, the alternate signs can also be incorporated with the factors and the factorial notation used in the other terms.

$$\Rightarrow S = 1 + \frac{1}{1!}\left(-\frac{2}{3}\right) + \frac{1}{2!}\left(-\frac{2}{3}\right)^2 + \frac{1}{3!}\left(-\frac{2}{3}\right)^3 + \ldots$$

You should now recognise this as being of the form

$$1 + \frac{x}{1!} + \frac{x^2}{2!} + \frac{x^3}{3!} + \ldots \qquad \text{where } x = -\frac{2}{3}$$

$$\Rightarrow S = e^{-\frac{2}{3}}$$

MATHEMATICAL INDUCTION

When the answer to the sum of the first n terms of a series is given then the proof by the method of **mathematical induction** can be used. This method depends upon assuming that the result to be proved is true for n and then using this assumption to prove that the same result is true when n is replaced by $n+1$. The truth of the result is then established independently for the lowest possible value of n, say for instance when $n = 1$. Hence the result is shown to be true for $n = 1+1 = 2$, $n = 2+1 = 3$, etc., i.e. for integral n

Worked example

Where $\mathbb{Z}$ is the set of integers.

Prove by induction that for $n \in \mathbb{Z}^+$, $\displaystyle\sum_{r=1}^{n} r^3 = \left[\frac{1}{2}n(n+1)\right]^2$

Assume that $\displaystyle\sum_{r=1}^{n} r^3 = \left[\frac{1}{2}n\,(n+1)\right]^2$ (1)

Then $\displaystyle\sum_{r=1}^{n+1} r^3 = \sum_{r=1}^{n} r^3 + (n+1)^3 = \left[\frac{1}{2}n(n+1)\right]^2 + (n+1)^3$

$$= \left[\frac{1}{2}(n+1)\right]^2 \left[n^2 + 4(n+1)\right] = \left[\frac{1}{2}(n+1)\right]^2 (n+2)^2$$

$$= \left[\frac{1}{2}(n+1)(n+2)\right]^2 = \left[\frac{1}{2}(\overline{n+1})(\overline{n+1}+1)\right]^2$$

which is the same as (1) except that n is replaced by $n+1$
$\Rightarrow$ if the result is true for n it is also true for $n+1$

But when $n = 1$,   L.H.S. $= \displaystyle\sum_{r=1}^{1} r^3 = 1$

$$\text{R.H.S.} = \left[\frac{1}{2}(1)(1+1)\right]^2 = 1 = \text{L.H.S.}$$

Therefore the result is true for $n = 1$
Therefore the result is true for $1+1 = 2$, $2+1 = 3$, ... i.e. all $n \in \mathbb{Z}^+$

The method of induction is a very powerful tool and can be used to prove results other than those concerned with the summation of series.

Worked example

Prove that for any positive integer n, $n^3 + 6n^2 + 8n$ is divisible by 3
Assume $n^3 + 6n^2 + 8n$ is divisible by 3. Then for $n \in \mathbb{Z}^+$ and $k \in \mathbb{Z}^+$
$n^3 + 6n^2 + 8n = 3k$
Now $(n+1)^3 + 6(n+1)^2 + 8(n+1) - (n^3 + 6n^2 + 8n) = 3n^2 + 12n + 15 = 3(n^2 + 4n + 5)$
$\Rightarrow (n+1)^3 + 6(n+1)^2 + 8(n+1) = n^3 + 6n^2 + 8n + 3(n^2 + 4n + 5)$
$$= 3k + 3(n^2 + 4n + 5) = 3(k + n^2 + 4n + 5)$$
$\Rightarrow$ if $n^3 + 6n^2 + 8n$ is divisible by 3, $(n+1)^3 + 6(n+1)^2 + 8(n+1)$ is also divisible by 3
But when $n = 1$, $n^3 + 6n^2 + 8n = 15$ which is divisible by 3
The result is true for $n = 1 \Rightarrow$ the result is true for $n = 1+1 = 2$, $n = 2+1 = 3$, ..., i.e. for all $n \in \mathbb{Z}^+$

EXERCISES 4.4

1   Find   a) $\displaystyle\sum_{1}^{n} (r+1)^3$ b) $\displaystyle\sum_{r=1}^{n} r(r+1)(2r+1)$

2   Find   a) $\displaystyle\sum_{r=1}^{n} \frac{1}{4r^2-1}$ b) $\displaystyle\sum_{r=1}^{n} \frac{1}{(3r-2)(3r+1)}$

3   Given that

$$f(r) = \frac{1}{r(r+1)}$$

prove that

$$f(r) - f(r+1) = \frac{2}{r(r+1)(r+2)}$$

Hence, or otherwise show that

$$\frac{1}{1.2.3.} + \frac{1}{2.3.4.} + \ldots + \frac{1}{n(n+1)(n+2)} = \frac{n(n+3)}{4(n+1)(n+2)}$$

4   Find the sum to infinity of the following series

a) $1 + \dfrac{4}{2!} + \dfrac{4^2}{3!} + \dfrac{4^3}{4!} + \ldots + \dfrac{4^n}{(n+1)!} + \ldots$

b) $\dfrac{1}{1.3^2} - \dfrac{1}{2.3^3} + \dfrac{1}{3.3^4} - \ldots + (-1)^{n+1}\dfrac{1}{n.3^{n+1}} + \ldots$

c) $1 + 2\left(\dfrac{1}{2}\right) + 3\left(\dfrac{1}{2}\right)^2 + 4\left(\dfrac{1}{2}\right)^3 + \ldots + n\left(\dfrac{1}{2}\right)^{n-1} + \ldots$

5 Prove by induction

a) $\displaystyle\sum_{r=1}^{n} \dfrac{1}{r(r+1)} = \dfrac{n}{n+1}$

b) $\displaystyle\sum_{r=1}^{n} r(r!) = (n+1)! - 1$

c) $\displaystyle\sum_{r=1}^{n} (r+1)2^{r-1} = 2^n n$

6 Show that for $n \epsilon \mathbb{Z}^+$, $n > 1$ that $5^n - 4n - 1$ is divisible by 16

EXAMINATION TYPE QUESTIONS WITH MARKING SCHEME AND GUIDED SOLUTION

1 An arithmetic series is such that the sum of the second and fourth terms is equal to the seventh term. Further, the ninth term is 3 less than three times the third term. Find the sum of the first 20 terms of the series. *(7 marks)*

Let the series be $a + (a+d) + (a+2d) + (a+3d) + \ldots$

$\Rightarrow T_2 = (a+d)$, $T_4 = a+3d$, $T_7 = a+6d \Rightarrow a+6d = 2a+4d$ M1

$\Rightarrow a = 2d$ A1

$T_3 = a+2d$, $T_9 = a+8d \Rightarrow a+8d = 3(a+2d) - 3$ M1

$\Rightarrow 2a = 2d+3$ A1

Solving $\Rightarrow a = 3$, $d = 1\frac{1}{2}$ A1 (both)

$S_{20} = \dfrac{n}{2}[2a+(n-1)d] = 10(6+28\frac{1}{2}) = 345$ M1   A1

2 Expand, in a series of ascending powers of x, up to and including the term in x^3, the expression $\dfrac{x+1}{\sqrt{(4-x)}}$. State the set of values of x for which your expansion is valid *(6 marks)*

$\dfrac{x+1}{\sqrt{(4-x)}} = (x+1)(4-x)^{-\frac{1}{2}} = (x+1)\left(1-\dfrac{x}{4}\right)^{-\frac{1}{2}} \cdot \dfrac{1}{2}$ M1

$= \dfrac{1}{2}(x+1)\left[1 + \dfrac{\left(-\frac{1}{2}\right)}{1}\left(-\dfrac{x}{4}\right) + \dfrac{\left(-\frac{1}{2}\right)\left(-\frac{3}{2}\right)}{1.2}\left(-\dfrac{x}{4}\right)^2 \right.$

$\left. + \dfrac{\left(-\frac{1}{2}\right)\left(-\frac{3}{2}\right)\left(-\frac{5}{2}\right)}{1.2.3}\left(-\dfrac{x}{4}\right)^3 + \ldots\right]$ M1

$= \dfrac{1}{2}(1+x)\left[1 + \dfrac{x}{8} + \dfrac{3x^2}{128} + \dfrac{5x^3}{1024} + \ldots\right]$ M1

$= \dfrac{1}{2}\left[1 + \dfrac{9}{8}x + \dfrac{19x^2}{128} + \dfrac{29x^3}{1024} + \ldots\right]$ M1

$= \dfrac{1}{2} + \dfrac{9}{16}x + \dfrac{19x^2}{256} + \dfrac{29x^3}{2048} + \ldots$ A2 (1, 0)

Valid for $|x| < 4$ B1

EXAMINATION QUESTIONS

1   The first four terms of an arithmetic progression are 2, $a-b$, $2a+b+7$ and $a-3b$ respectively, where a and b are constants. Find a and b and hence determine the sum of the first 30 terms of the progression.　　　　　　　　　　　　　(L)

2   i) Given that a, b and c are the first three terms respectively in an arithmetic series, and that $b = 3c$, find b and c in terms of a.
　　ii) Given that e, f and g are the first three terms respectively in a geometric series, and that $f = g^3$, find f and g in terms of e.　　　　　　　　　(L)

3   A geometric progression has first term 1 and common ratio $\frac{1}{2}\sin 2\theta$.
　　a) Find the sum of the first 10 terms in the case when $\theta = \pi/4$, giving your answer to 3 decimal places.
　　b) Given that the sum to infinity is 4/3, find the general solution for θ in radians.
　　　　　　　　　　　　　　　　　　　　　　　　　　　(AEB 1988)

4   The sum of the first and second terms of a geometric progression is 108 and the sum of the third and fourth terms is 12. Find the two possible values of the common ratio and the corresponding values of the first term.　　　　　　　　(AEB 1989)

5   All the terms of a certain geometric series are positive.
The first term is a and the second term is $a^2 - a$. Find the set of values of a for which the series converges.
Given that $a = 5/3$,

　　a) find the sum of the first 10 terms of the series, giving your answer to 2 decimal places,
　　b) show that the sum to infinity of the series is 5,
　　c) find the least number of terms of the series required to make their sum exceed 4·999.　　　　　　　　　　　　　　　　　　　　(AEB 1987)

6   Show that
$$6\sum_{r=1}^{n} r(r+2) = n(n+1)(2n+7).$$
　　　　　　　　　　　　　　　　　　　　　　　　　　　(L)

7
Write down and simplify the first three terms in the series expansion of $\left(1 + \dfrac{x}{3}\right)^{-\frac{1}{2}}$

in ascending powers of x.
State the set of values of x for which the series is valid.
Given that x is so small that terms in x^3 and higher powers of x may be neglected, show that
$$e^{-x}\left(1 + \frac{x}{3}\right)^{-\frac{1}{2}} = 1 - \frac{7}{6}x + \frac{17}{24}x^2.$$
　　　　　　　　　　　　　　　　　　　　　　　　　　　(AEB 1987)

8   When $(1 + px)^q$ is expanded in ascending powers of x, the coefficients of x and x^2 are -6 and 27 respectively. Find
　　a) the value of p and the value of q,
　　b) the coefficient of x^3 in the expansion,
　　c) the set of values of x for which the expansion is valid.　　　　　(L)

9   Expand $(1 - 2\sin\theta)^{-4}$ in ascending powers of $\sin\theta$ up to and including the term in $\sin^3\theta$. Find the complete set of values of θ, in the interval $0° < \theta < 360°$, for which the expansion is valid.　　　　　　　　　　　　(AEB 1986)

10   a) A certain geometric progression has 30 as its first term. Find the least value of the common ratio if the sum to infinity is not less than 36.
　　b) The tenth term of an arithmetic progression is 32. The sum to 20 terms is 670. If the progression is summed to n terms, find the least value of n for which this sum exceeds 1550.　　　　　　　　　　　　　　　(NI 1987)

11   A geometric series has first term a and common ratio r, where $|r| < 1$. The sum to infinity of the series is 8. The sum to infinity of the series obtained by adding all the odd-numbered terms (i.e. 1st term + 3rd term + 5th term + ...) is 6. Find the value of r.   (C)

12   Find $\sum_{r=0}^{n} (2n + 1 - 2r)$ in terms of n.   (JMB)

13   Evaluate in terms of ln 2
   i) $\ln 2 + \ln(2^2) + \ldots \ln(2^n) + \ldots + \ln(2^{100})$,
   ii) $\sum_{n=1}^{\infty} (\ln 2)^n$.   (JMB)

14   a) Obtain the expansion of $(16 + y)^{\frac{1}{2}}$ in ascending powers of y up to and including the term in y^2.
      State the set of values of y for which the expansion is valid.
      Hence show that if k^3 and higher powers of k are neglected
$$\sqrt{(16 + 4k + k^2)} = 4 + \frac{k}{2} + \frac{3k^2}{32}$$

   b) Prove that, for all real values of k, the roots of the quadratic equation $x^2 - (2 + k)x - 3 = 0$ are real.
      Show further that the roots are of opposite signs.

   c) Show that the positive root of the quadratic equation $x^2 - (2 + k)x - 3 = 0$ is $\frac{1}{2}(2 + k + \sqrt{(16 + 4k + k^2)})$.
      Hence obtain an expression for the positive root, when k is small, in ascending powers of k up to and including the term in k^2.   (AEB 1987)

15   a) Evaluate the term which is independent of x in the expansion of
$$\left(x - \frac{1}{x^2} \right)^{12}.$$

   b) Expand
$$(1 - x)^2 \sqrt{1 + 2x}$$
      in ascending powers of x as far as the term in x^3. For which values of x is this expansion valid?   (NI 1987)

16   In the expansion of
$$\frac{1}{\sqrt{(1 + ax)}} - \frac{1}{1 + 2x}$$
   in ascending powers of x, the first non-zero term is the term in x^2. Find the value of the constant a and hence find the terms in x^2 and x^3.   (JMB)

ANSWERS TO EXERCISES

EXERCISES 4.1

1   a) 400   b) 7174453

2   $-\dfrac{5}{2}, \dfrac{1}{2}, 205$

3   3, 4
4   2842

6   $\dfrac{2}{3}$, 1296

7   2·9
8   $2^{n+1} + n^2 - 2$

9   $\dfrac{53}{165}$

EXERCISES 4.2

1   a) $1 + 2x - 6x^2 + 28x^3$;

   b) $\dfrac{1}{2} + \dfrac{1}{16}x + \dfrac{3}{256}x^2 + \dfrac{5}{2048}x^3$

   c) $\dfrac{1}{2} + \dfrac{17}{16}x - 2\dfrac{221}{256}x^2 + 13\dfrac{1333}{2048}x^3$

2   $1 - 2x + x^2 + 2x^3 - 4x^4 + 2x^5 + 3x^6 - 6x^7$

3   $1 + \dfrac{x}{2} - \dfrac{x^2}{8}; -\dfrac{1}{2}, -\dfrac{1}{8}$

4   -2

5   $-25\dfrac{5}{16}, 265\dfrac{25}{32}$

EXERCISES 4.3

1   $x + \dfrac{5}{24}x^3 - \dfrac{1}{8}x^4$, $-1 < x < 1$, $-\dfrac{x^4}{4}$

2   $2 - 4x + 5x^2 - \dfrac{14}{3}x^3 + \ldots + \dfrac{(-x)^n}{n!}(3^n + 1)$

3   $m = 1$, $n = \dfrac{1}{6}$

5   a) $3x - 6x^2 + \dfrac{3}{2}x^3 + 5x^4$;   b) $-\dfrac{x^2}{2} - \dfrac{x^4}{12}$

EXERCISES 4.4

1   a) $\dfrac{1}{4}(n^2 + 3n)(n^2 + 3n + 4)$;

   b) $\dfrac{1}{2}n(n+1)^2(n+2)$

2   a) $\dfrac{n}{2n+1}$;   b) $\dfrac{n}{3n+1}$

4   a) $\dfrac{1}{4}(e^4 - 1)$;   b) $\dfrac{1}{3}\ln\dfrac{4}{3}$;   c) 4

ANSWERS TO EXAMINATION QUESTIONS

1   2, -3, 1365

2   $\dfrac{3a}{5}$, $\dfrac{a}{5}$, , $e^{\frac{3}{5}}$, $e^{\frac{1}{5}}$

3   a) $1\cdot998$;   b) $\dfrac{1}{2}n\pi + (-1)^n\,\dfrac{\pi}{12}$

4   $\pm\dfrac{1}{3}$, 81, 162

5   $1 < a < 2$;   a) $4\cdot91$ c) 22

7   $1 - \dfrac{x}{6} + \dfrac{x^2}{24}$, $|x| < 3$

8   a) 3, -2;   b) -108;   c) $|x| < \dfrac{1}{3}$

9   $1 + 8\sin\theta + 40\sin^2\theta + 160\sin^3\theta$,
    $0° < \theta < 30°$, $150° < \theta < 210°$,
    $330° < \theta < 360°$

10   $\dfrac{1}{6}$, $\dfrac{1}{32}$

11   $\dfrac{1}{3}$

12   $(n+1)^2$

13   $5050\ln 2$, $\dfrac{\ln 2}{1 - \ln 2}$

14   a) $4 + \dfrac{1}{8}y - \dfrac{1}{512}y^2$, $|y| < 16$;

   c) $3 + \dfrac{3}{4}k + \dfrac{3}{64}k^2$

15   a) 495;   b) $1 - x - \dfrac{3}{2}x^2 + \dfrac{5}{2}x^3 + \ldots$, $|x| \leqslant \dfrac{1}{2}$

16   -4, $10x^2 + 12x^3$

CHAPTER 5

TRIGONOMETRIC FUNCTIONS

GETTING STARTED

ANGLE MEASURES

Angles are measured in either degrees or radians.

- **Degrees:** One complete revolution = 360°.
  One quarter of a complete revolution = 90° = 1 right angle.
  1° = 60 minutes, i.e. 60′, 1 minute = 60 seconds = 60″

- **Radians:** One complete revolution = 2π radians = $2\pi^c$
  One radian is the angle subtended at the centre of a circle by an arc of the circle equal in length to the radius of the circle.

- **Note:**

Degrees	0°	30°	45°	60°	90°	180°	360°
Radians	0^c	$\dfrac{\pi^c}{6}$	$\dfrac{\pi^c}{4}$	$\dfrac{\pi^c}{3}$	$\dfrac{\pi^c}{2}$	π^c	$2\pi^c$

MENSURATION OF A CIRCLE

Note: for a circle, radius r.

Area of circle = πr^2, circumference of circle = $2\pi r$.

Arc length of $AB = r\theta^c$, θ measured in radians.

Area of sector $ACB = \dfrac{1}{2}r^2\theta^c$

Fig. 5.1

Area of segment (minor) $ABD = \dfrac{1}{2}r^2(\theta^c - \sin\theta^c)$

CIRCULAR FUNCTIONS

For any acute angle $\theta = AOB$, say, of a right-angled triangle OAB

$$\sin\theta = \frac{\text{Opposite}}{\text{Hypotenuse}} = \frac{AB}{OB} = \frac{y}{r}$$

$$\cos\theta = \frac{\text{Adjacent}}{\text{Hypotenuse}} = \frac{OA}{OB} = \frac{x}{r}$$

$$\tan\theta = \frac{\text{Opposite}}{\text{Adjacent}} = \frac{AB}{OA} = \frac{y}{x}$$

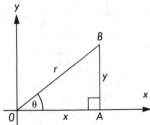

Fig. 5.2 Circular functions

Further, $\operatorname{cosec}\theta = \dfrac{1}{\sin\theta}$, $\sec\theta = \dfrac{1}{\cos\theta}$, $\cot\theta = \dfrac{1}{\tan\theta}$

ESSENTIAL PRINCIPLES

TRIGONOMETRIC RATIOS FOR A GENERAL ANGLE

Angles measured in an anticlockwise sense from the positive x-axis are positive.
Angles measured in a clockwise sense from the positive x-axis are negative.

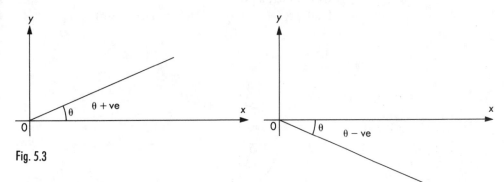

Fig. 5.3

Worked example

Show in relation to a cartesian diagram, angles of

a) $125°$,   b) $-60°$,   c) $\dfrac{5\pi^{c}}{4}$ d) $-210°$

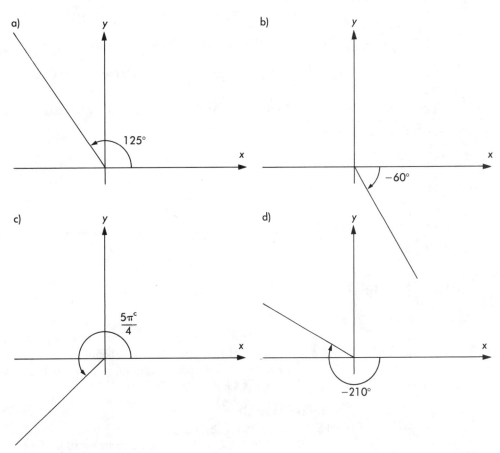

Fig. 5.4

QUADRANTS

The cartesian axes divide a plane into four quadrants.

$$0° \rightarrow 90° \quad \text{1st quadrant,} \quad 90° \rightarrow 180° \quad \text{2nd quadrant.}$$
$$180° \rightarrow 270° \quad \text{3rd quadrant,} \quad 270° \rightarrow 360° \quad \text{4th quadrant.}$$

2nd quadrant	1st quadrant
3rd quadrant	4th quadrant

x negative y positive	x positive y positive
x negative y negative	x positive y negative

The trigonometric ratio of any angle is then obtained by determining the quadrant connected with the angle, the sign of x or y within that quadrant and the associated acute angle made with the positive (or negative) x-axis. Irrespective of in which quadrant the angle lies, r is always taken as positive.

Worked example

Find the value of a) sin 240°, b) cos (−30°)

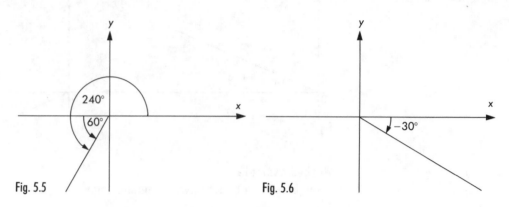

Fig. 5.5   Fig. 5.6

a) 240° ⇒ 3rd quadrant in which y is negative

$$\Rightarrow \sin \theta = \frac{y}{r} \Rightarrow \sin 240° = -\sin 60° = -\frac{\sqrt{3}}{2} \text{ or } -0{\cdot}8660$$

b) −30° ⇒ 4th quadrant in which x is positive

$$\Rightarrow \cos \theta = \frac{x}{r} \Rightarrow \cos(-30°) = \cos 30° = \frac{\sqrt{3}}{2} \text{ or } 0{\cdot}8660$$

> **The CAST diagram is a useful aid.** "

A useful aid is the CAST diagram showing which trigonometric ratios are *positive* in each quadrant.

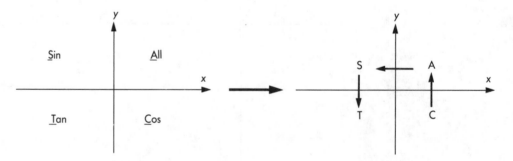

Fig. 5.7

Thus   a) sin 240° ⇒ third quadrant ⇒ tangent only positive
⇒ sin 240° = −sin 60° = −0·8660
b) cos (−30°) ⇒ fourth quadrant ⇒ cosine only positive
⇒ cos (−30°) = cos 30° = 0·8660

Worked example

Find   a) tan 740°,   b) sin 320°,   c) sec 600°
a) 740° = 360° + 360° + 20° ⇒ 1st quadrant ⇒ *all* positive
⇒ tan 740° = tan 20° = 0·3640
b) 320° ⇒ 4th quadrant ⇒ cosine only positive
⇒ sin 320° = −sin 40° = −0·6428
c) 600° = 360° + 240° ⇒ 240° in 3rd quadrant ⇒ tan only positive

$$\Rightarrow \sec 600° = \frac{1}{\cos 600°} = -\frac{1}{\cos 60°} = -2$$

COMMONLY USED RATIOS

The **trigonometrical ratios** associated with the angles 30°, 45° and 60° are used very frequently in problems involving trigonometry. Their exact values can be easily obtained using either an equilateral triangle (of side two units) or an isosceles right-angled triangle.

> **Some widely used ratios.**

$$AD^2 = 2^2 - 1^2 = 3, \qquad AD = \sqrt{3}$$

$$\sin 60° = \frac{\sqrt{3}}{2} \qquad\qquad \cos 60° = \frac{1}{2} \qquad\qquad \tan 60° = \sqrt{3}$$

$$\sin 30° = \frac{1}{2} \qquad\qquad \cos 30° = \frac{\sqrt{3}}{2} \qquad\qquad \tan 30° = \frac{1}{\sqrt{3}}$$

$$AB^2 = 1^2 + 1^2 = 2 \qquad AB = \sqrt{2}$$

$$\sin 45° = \frac{1}{\sqrt{2}} \text{ or } \frac{\sqrt{2}}{2} \quad \cos 45° = \frac{1}{\sqrt{2}} \text{ or } \frac{\sqrt{2}}{2} \qquad \tan 45° = 1$$

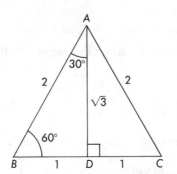

Fig. 5.8

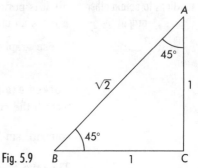

Fig. 5.9

GRAPHS OF THE TRIGONOMETRICAL FUNCTIONS

Using the above procedure for calculating the trigonometric ratios of any angle it is possible to sketch the graphs of the trigonometric functions. They are shown in Fig. 5.10.

—————— $y = \sin x$

— — — — — $y = \cos x$

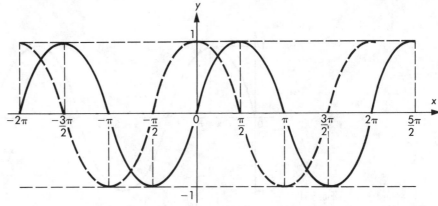

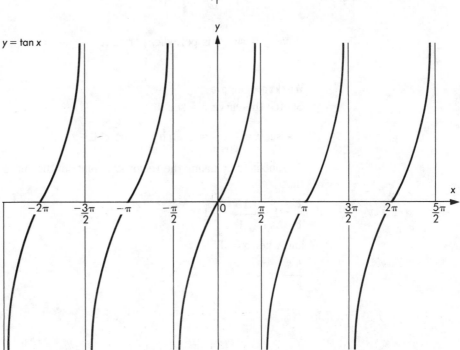

Fig. 5.10

You should note:

For the graphs of the *sine* and *cosine* functions
  i) they are continuous, i.e. no breaks
  ii) they are periodic, period 2π
  iii) they lie entirely within the range -1 to $+1$
       i.e. $-1 \leqslant \sin x \leqslant 1$, $-1 \leqslant \cos x \leqslant 1$

For the graph of the *tangent* function

> You can use graphs of the sine, cosine and tangent functions to obtain other graphs.

  i) it is not continuous, i.e. undefined for $x = n\pi + \dfrac{\pi}{2}$

  ii) it is periodic, period π
  iii) has an unlimited range, i.e. $-\infty \leqslant \tan x \leqslant \infty$

These graphs can be used to obtain the graphs of other trigonometric functions.

Worked example

Sketch the curve $y = \sin 2x$, $0 \leqslant x \leqslant 2\pi$

compare $\sin 2x$ with $\sin \theta$ i.e. $\sin 2x \equiv \sin \theta$ when $x \equiv \dfrac{\theta}{2}$

Thus the graph of $y = \sin 2x$ is the same as that of $y = \sin \theta$ where θ is replaced by $2x$ or alternatively the x values correspond to $\dfrac{\theta}{2}$ values.

Thus $y = \sin 2x$ is as shown in Fig. 5.11.

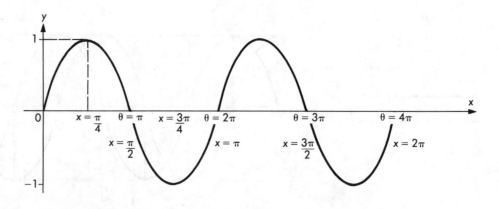

Fig. 5.11

Note: $y = \sin 2x$ is periodic, period π

Worked example

Sketch the curve $y = \sec x$

$y = \sec x = \dfrac{1}{\cos x} \Rightarrow$ sketch $y = \cos x$, $0 \leqslant x \leqslant 2\pi$

i.e. obtained by taking the reciprocal points of the curve $y = \cos x$ and noting in particular that

$$\frac{1}{1} = 1, \quad \frac{1}{-1} = -1$$

and when $x = 0$

$$\frac{1}{x} \to \pm \infty$$

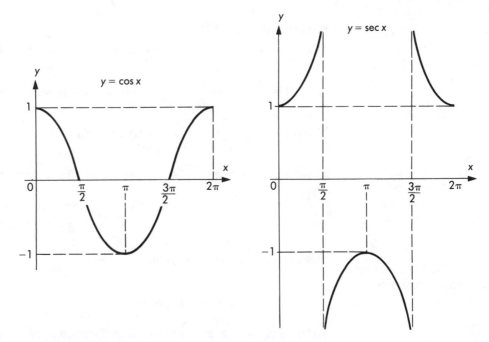

Fig. 5.12

INVERSE TRIGONOMETRIC FUNCTIONS

$y = \arcsin x$ is another way of writing $x = \sin y$ i.e. $y = \sin x$ with the x and the y interchanged. Thus $y = \arcsin x$ is, (subject to limitations on x), the inverse of the function $y = \sin x$. Similarly $y = \arccos x \Rightarrow x = \cos y$ is, (subject to limitations on x), the inverse of the function $y = \cos x$, and $y = \arctan x \Rightarrow x = \tan y$ is, (subject to limitations on x), the inverse of the function $y = \tan x$. Again their graphs can be obtained from the graphs of the corresponding trigonometric functions $y = \sin x$, $y = \cos x$ and $y = \tan x$. Thus $y = \arctan x \Rightarrow \tan y = x$, whose graph can be deduced from that of $y = \tan x$.

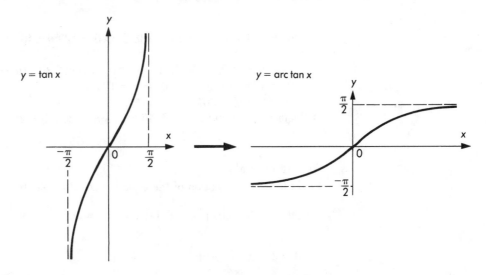

Fig. 5.13

Note: an alternative notation is sometimes used for the inverse trigonometric functions, namely $\arcsin x = \sin^{-1} x$, $\arccos x = \cos^{-1} x$, $\arctan x = \tan^{-1} x$. You must not, however, confuse $\sin^{-1} x$ etc, with $\dfrac{1}{\sin x}$ etc, since $\sin^{-1} x \neq \dfrac{1}{\sin x}$, $\dfrac{1}{\sin x}$ would be written as $(\sin x)^{-1}$.

GENERAL VALUES OF TRIGONOMETRIC EQUATIONS

It will be seen that as the trigonometric functions are periodic there are many solutions to an equation of the form $\sin x = \dfrac{1}{2}$ i.e. there are many angles whose sine is $\dfrac{1}{2}$. Reference to the graphs will show that the general solution of such equations can be summarised in the following formulae.

> ❝ Useful formulae in solving trigonometric equations. ❞

$$\sin x = a \Rightarrow x = n\pi + (-1)^n \alpha, \quad \text{where } n \in \mathbb{Z} \text{ and } \sin \alpha = a$$
$$\cos x = a \Rightarrow x = 2n\pi \pm \alpha, \quad \text{where } n \in \mathbb{Z} \text{ and } \cos \alpha = a$$
$$\tan x = a \Rightarrow x = n\pi + \alpha, \quad \text{where } n \in \mathbb{Z} \text{ and } \tan \alpha = a$$

Worked example

Find the general solution of the equations a) $\sin \theta = \dfrac{1}{2}$ b) $\cos \theta = -\dfrac{1}{2}$, c) $\tan 3\theta = 1$

a) $\sin \theta = \dfrac{1}{2} \Rightarrow \theta = \dfrac{\pi}{6} \Rightarrow$ general solution is $\theta = n\pi + (-1)^n \dfrac{\pi}{6}$

b) $\cos \theta = -\dfrac{1}{2} \Rightarrow \theta = \dfrac{2\pi}{3} \Rightarrow$ general solution is $\theta = 2n\pi \pm \dfrac{2\pi}{3}$

c) $\tan 3\theta = 1 \Rightarrow 3\theta = \dfrac{\pi}{4} \Rightarrow$ general solution is $3\theta = n\pi + \dfrac{\pi}{4}$

$$\text{or } \theta = \frac{n\pi}{3} + \frac{\pi}{12}$$

> Make sure you avoid a very common error.

Note: it would be wrong to say $3\theta = \dfrac{\pi}{4} \Rightarrow \theta = \dfrac{\pi}{12}$

$$\Rightarrow \text{general solution is } \theta = n\pi + \frac{\pi}{12}$$

You must always leave the division by the 3 until the general value has been taken. Do not make this common mistake.

Worked example

Solve the equation $4 \cos^2 3\theta - \cos 3\theta - 3 = 0$ for $0° \leqslant \theta \leqslant 180°$

$4\cos^2 3\theta - \cos 3\theta - 3 = (4\cos 3\theta + 3)(\cos 3\theta - 1) = 0$

$\Rightarrow \cos 3\theta = -\dfrac{3}{4}$ or $\cos 3\theta = 1$

$\cos 3\theta = -\dfrac{3}{4} \Rightarrow 3\theta = 360°n \pm 138 \cdot 6° \Rightarrow \theta = 120°n \pm 46 \cdot 2°$

$\qquad \Rightarrow \theta = 46 \cdot 2°, \ 166 \cdot 2°$ (from +ve sign) or $73 \cdot 8°$ (from −ve sign).

$\cos 3\theta = 1 \Rightarrow 3\theta = 360°n \pm 0° \Rightarrow \theta = 120°n$

$\qquad \Rightarrow \theta = 0°, \ 120°$

$\Rightarrow$ Required solution is $\theta = 0°, \ 46 \cdot 2°, \ 73 \cdot 8°, \ 120°, \ 166 \cdot 2°$

Worked example

Find the general solution of the equation $\cos 2\theta = \sin \theta$

$\cos 2\theta = \sin \theta = \cos \left(\dfrac{\pi}{2} - \theta \right)$ since $\sin \theta = \cos \left(\dfrac{\pi}{2} - \theta \right)$

$\Rightarrow 2\theta = 2n\pi \pm \left(\dfrac{\pi}{2} - \theta \right)$

$2\theta = 2n\pi + \left(\dfrac{\pi}{2} - \theta \right) \Rightarrow 3\theta = 2n\pi + \dfrac{\pi}{2}$ or $\theta = \dfrac{2}{3}n\pi + \dfrac{\pi}{6}$

$2\theta = 2n\pi - \left(\dfrac{\pi}{2} - \theta \right) \Rightarrow \theta = 2n\pi - \dfrac{\pi}{2}$

General solution is $\theta = \dfrac{2}{3}n\pi + \dfrac{\pi}{6}$ or $2n\pi - \dfrac{\pi}{2}$

EXERCISES 5.1

1 Find the values of   i) $\sin 150°$,   ii) $\cos 570°$,   iii) $\tan (-240°)$

2 Evaluate   i) $\arcsin \left(\dfrac{\sqrt{3}}{2} \right)$,   ii) $\arccos \left(-\dfrac{1}{2} \right)$,   iii) $\arctan (-1)$

3 Sketch the curve $y = \cos 2x$ for $0 \leqslant x \leqslant \pi$. Hence find an approximation to the smallest positive root of the equation $\cos 2x = x$

4   A chord of a circle divides the circle into two segments whose areas are in the ratio $5:1$. Show that the angle 2θ radians subtended by the chord at the centre of the circle satisfies the equation

$\sin 2\theta = 2\theta - \dfrac{\pi}{3}$. Solve the equation graphically for values of θ within the range

$0 \leqslant \theta \leqslant \dfrac{\pi}{2}$

5   Solve, for $0° < x < 360°$, $\cos\left(\dfrac{3x}{4}\right) = \tan 163°$

6   Find all the angles between $0°$ and $360°$ which satisfy the equations
    i) $\cos(2y + 30°) = 0\cdot342$,   ii) $3\sin^2\theta - 4\sin\theta - 4 = 0$

7   Find the general solution of the equation   i) $\cos 2x = \sin\left(\dfrac{7\pi}{6}\right)$,   ii) $\sin 2\theta = \cos 3\theta$

8   Solve, for $-180° < \theta < 180°$, the equation $\tan^2\theta - \tan\theta - 2 = 0$.

TRIGONOMETRIC IDENTITIES

Make sure you know, and can use, the formulae.

Most examining boards issue books of formulae in which are listed the trigonometric formulae that you are expected to know and use in the examination. Do not fall into the trap of depending upon the booklet for the formulae, but learn them by heart. Experience shows that those candidates who do not know the formulae invariably fail to use them at the appropriate moment in a question and consequently fail to obtain many marks for that question. By all means use the booklet to check when in doubt about a term or its sign, but do not become too dependent upon the booklet.

The fundamental trigonometric formulae upon which many of the identities and relations are based are the addition formulae for sine and cosine.

$\sin(A \pm B) = \sin A \cos B \pm \cos A \sin B$
$\cos(A \pm B) = \cos A \cos B \mp \sin A \sin B$

(Look carefully at the signs. Note the $\mp$ on the R.H.S. of the second formula.) Learn these along with the following results which can be deduced from them.

Some important results to learn.

$\tan(A \pm B) = \dfrac{\tan A \pm \tan B}{1 \mp \tan A \tan B}$

$\sin 2A = 2\sin A \cos A$
$\cos 2A = \cos^2 A - \sin^2 A = 2\cos^2 A - 1 = 1 - 2\sin^2 A$

$\sin A + \sin B = 2\sin\left(\dfrac{A+B}{2}\right)\cos\left(\dfrac{A-B}{2}\right)$

$\sin A - \sin B = 2\cos\left(\dfrac{A+B}{2}\right)\sin\left(\dfrac{A-B}{2}\right)$

$\cos A + \cos B = 2\cos\left(\dfrac{A+B}{2}\right)\cos\left(\dfrac{A-B}{2}\right)$

$\cos A - \cos B = 2\sin\left(\dfrac{A+B}{2}\right)\sin\left(\dfrac{B-A}{2}\right)$ $\left(\text{Note } \dfrac{B-A}{2}\right)$

The one fundamental identity that you must know is

$\cos^2 x + \sin^2 x \equiv 1$

From it you should be able to deduce that

$1 + \tan^2 x \equiv \sec^2 x$ and $1 + \cot^2 x \equiv \csc^2 x$

Using the identity in conjunction with the listed formulae you should be able to prove other useful formulae.

Namely $\sin 2A \equiv \dfrac{2\tan A}{1 + \tan^2 A}$

$$\cos 2A \equiv \frac{1-\tan^2 A}{1+\tan^2 A}$$

$$\tan 2A \equiv \frac{2\tan A}{1-\tan^2 A}$$

$$\sin 3A \equiv 3\sin A - 4\sin^3 A$$
$$\cos 3A \equiv 4\cos^3 A - 3\cos A$$

Worked example

Prove that $\cos 3x = 4\cos^3 x - 3\cos x$

Hence solve, for $0 \leqslant x \leqslant 2\pi$, the equation $\cos 3x + 2\cos x = 0$

$\cos 3x = \cos(2x+x) = \cos 2x \cos x - \sin 2x \sin x$
$$= (2\cos^2 x - 1)\cos x - (2\sin x \cos x)\sin x$$
$$= 2\cos^3 x - \cos x - 2\cos x(1 - \cos^2 x) = 4\cos^3 x - 3\cos x$$

$\cos 3x + 2\cos x = 0 \Rightarrow 4\cos^3 x - 3\cos x + 2\cos x = 0$

$$\Rightarrow \cos x (4\cos^2 x - 1) = 0 \Longleftrightarrow \cos x = 0 \text{ or } \pm\frac{1}{2}$$

$$\cos x = 0 \Rightarrow x = \frac{\pi}{2} \text{ or } \frac{3\pi}{2}$$

$$\cos x = \frac{1}{2} \Rightarrow x = \frac{\pi}{3} \text{ or } \frac{5\pi}{3}$$

$$\cos x = -\frac{1}{2} \Rightarrow x = \frac{2\pi}{3} \text{ or } \frac{4\pi}{3}$$

$$\Rightarrow x = \frac{\pi}{3}, \frac{\pi}{2}, \frac{2\pi}{3}, \frac{4\pi}{3}, \frac{3\pi}{2} \text{ or } \frac{5\pi}{3}$$

Note: in proving the bookwork realise that $\cos 3x$ is required entirely in terms of $\cos x$. Hence break $3x$ down to $2x + x$ and always be prepared to use the three very important and most frequently used identities

$$\cos 2x \equiv 2\cos^2 x - 1; \quad \sin 2x \equiv 2\sin x \cos x; \quad \cos^2 x + \sin^2 x \equiv 1$$

Worked example

Express $3\cos\theta - \sin\theta$ in the form $r\sin(\theta + \alpha)$ where $r > 0$ and $0° < \alpha < 360°$

Hence, or otherwise, solve for $0° \leqslant \theta \leqslant 360°$, the equation $3\cos\theta - \sin\theta = 2$

To express $3\cos\theta - \sin\theta$ in the form $r\sin(\theta + \alpha)$ equate the two expressions; thus,

$3\cos\theta - \sin\theta = r\sin(\theta + \alpha) = r\sin\theta\cos\alpha + r\cos\theta\sin\alpha$

For the two sides to be equal for all values of θ the coefficient of $\cos\theta$ on the L.H.S. must equal the coefficient of $\cos\theta$ on the R.H.S. Similarly for the coefficients of $\sin\theta$

Equating coefficients of $\cos\theta \Rightarrow 3 = r\sin\alpha$

Equating coefficients of $\sin\theta \Rightarrow -1 = r\cos\alpha$

The two equations in two unknowns, r and α, can now be solved.

Squaring and adding them

$\Rightarrow 3^2 + (-1)^2 = r^2(\sin^2\alpha + \cos^2\alpha) = r^2 \Rightarrow r^2 = 10, \ r = \sqrt{10}$

Dividing $\Rightarrow \dfrac{3}{-1} = \dfrac{r\sin\alpha}{r\cos\alpha} \Longleftrightarrow \tan\alpha = -3$

There are two angles between $0°$ and $360°$ with tangent equal to (-3) and it is essential that the correct one is chosen. Since r is positive we know that the cosine is negative and the sine is positive. Hence α must be an angle in the second quadrant $\Rightarrow \alpha = 180° - 71·57° = 108·43°$

$\Rightarrow 3\cos\theta - \sin\theta = \sqrt{10}\sin(\theta + 108·43°)$.

To solve the given equation we now proceed as follows.

$3\cos\theta - \sin\theta = \sqrt{10}\sin(\theta + 108·43°) = 2$

$$\Rightarrow \sin(\theta + 108·43°) = \frac{2}{\sqrt{10}} = 0·6325$$

$\Rightarrow \theta + 108·43° = n \times 180° + (-1)^n \, 39·25°$ (remember you must go immediately to the general value)

$n = 1 \Rightarrow \theta = 180° - 39·25° - 108·43° = 32·32°$

$n = 2 \Rightarrow \theta = 360° + 39\cdot25° - 108\cdot43° = 290\cdot82°$
The required solutions are $\theta = 32\cdot3°$ or $290\cdot8°$ to 1 decimal place.

Note: an alternative method of solving an equation such as $3\cos\theta - \sin\theta = 2$ is to use the formula known as the *tan half angle formula*, i.e.

$$\sin 2A = \frac{2t}{1+t^2}, \quad \cos 2A = \frac{1-t^2}{1+t}$$

where $t = \tan A$, A being half of the original angle $2A$

$$3\cos\theta - \sin\theta = 2 \Rightarrow \frac{3(1-t^2)}{1+t^2} - \frac{2t}{1+t^2} = 2$$

$$\Rightarrow 3 - 3t^2 - 2t = 2 + 2t^2 \Longleftrightarrow 5t^2 + 2t - 1 = 0$$

$$\Longleftrightarrow t = \frac{-2 \pm \sqrt{(4+20)}}{10} = 0\cdot2899 \text{ or } -0\cdot6899$$

$$\Rightarrow \frac{\theta}{2} = n \times 180° + 16\cdot17° \text{ or } n \times 180° - 34\cdot60°$$

$$\Rightarrow \theta = n \times 360° + 32\cdot34° \text{ or } n \times 360° - 69\cdot20°$$

$$\Rightarrow \theta = 32\cdot3° \text{ or } 290\cdot8° \text{ in the given range, to 1 decimal place}$$

Worked example

Find in radians, the general solution of the equation $\sin x + \sin 3x + \sin 5x = 0$
When solving equations such as this, in which you are asked to consider the sum or difference of two or more sine or cosine terms, always expect to use the formulae for $\sin A \pm \sin B$ or $\cos A \pm \cos B$. Take care when pairing to make sure that you can bring in any remaining terms as factors at a later stage.

$\sin x + \sin 3x + \sin 5x = 0 \Rightarrow (\sin x + \sin 5x) + \sin 3x = 0$
$\Rightarrow 2\sin 3x \cos 2x + \sin 3x = 0 \Rightarrow \sin 3x (2\cos 2x + 1) = 0$
$\Rightarrow \sin 3x = 0$ or $2\cos 2x + 1 = 0$

$$\sin 3x = 0 \Rightarrow 3x = n\pi, \quad x = \frac{n\pi}{3}$$

$$2\cos 2x + 1 = 0 \Rightarrow \cos 2x = -\frac{1}{2} \Rightarrow 2x = 2n\pi \pm \frac{2\pi}{3} \Rightarrow x = n\pi \pm \frac{\pi}{3}.$$

Notes:
i) make sure you do not forget the factor $\sin 3x = 0$. Many candidates cancel the $\sin 3x$ terms and consequently omit the solutions from $\sin 3x = 0$
ii) the pairings $\sin x + \sin 3x$ or $\sin 3x + \sin 5x$ would not have produced any common factors since
   $\sin x + \sin 3x = 2\sin 2x \cos x$ and $\sin 3x + \sin 5x = 2\sin 4x \cos x$

> **Make sure you avoid very common errors.**

Worked example

Prove that $\arctan A + \arctan B = \arctan\left(\dfrac{A+B}{1-AB}\right)$

Given that $\arctan(x^2) + \arctan(2) = \dfrac{3\pi}{4}$, find, without using tables, the value of x

$$\tan[\arctan A + \arctan B] = \frac{\tan(\arctan A) + \tan(\arctan B)}{1 - \tan(\arctan A).\tan(\arctan B)}$$

$$= \frac{A+B}{1-AB}$$

$$\Rightarrow \arctan A + \arctan B = \arctan\left(\frac{A+B}{1-AB}\right)$$

$$\arctan(x^2) + \arctan(2) = \arctan\left(\frac{x^2+2}{1-2x^2}\right)$$

$$\Rightarrow \frac{x^2+2}{1-2x^2} = \tan\frac{3\pi}{4} = -1$$

$$\Rightarrow x^2 + 2 = -1 + 2x^2 \Longleftrightarrow x^2 = 3, \ x = \pm\sqrt{3}$$

EXERCISES 5.2

1 Solve, for $0° \leqslant \theta \leqslant 360°$, $\cos 2\theta - 3\cos \theta + 2 = 0$
2 Find the general solution of the equation $2\sin \theta = (\sqrt{3})\tan \theta$
3 Solve for $0° \leqslant \theta \leqslant 360°$, $5\sin^2 \theta + \sin \theta \cos \theta = 3$
4 Solve for $0° \leqslant t \leqslant 360°$, $\cos t + \cos 2t + \cos 3t = 0$
5 Prove that $\cos 3x = 4\cos^3 x - 3\cos x$ and hence or otherwise solve the equation
   $4\cos 3x - 4\cos 2x + 9\cos x = 4$ for $0° \leqslant x \leqslant 360°$
6 Find the general solution of the equation $\sin 5\theta - \sin 4\theta - \sin 3\theta + \sin 2\theta = 0$
7 Express $4\sin x - 3\cos x$ in the form $R\sin (x - \alpha)$, where $R > 0$ and $0° < \alpha < 90°$
   Hence a) solve the equation $4\sin x - 3\cos x = 3$

         b) find the greatest and least values of $\dfrac{1}{4\sin x - 3\cos x + 6}$

8 Express $\cos \theta - (\sqrt{3})\sin \theta$ in the form $R\cos (\theta + \alpha)$, where $R > 0$ and $0 < \alpha < \dfrac{\pi}{2}$

   Hence or otherwise solve the equation $\cos \theta - (\sqrt{3})\sin \theta = \sqrt{2}$ for $0 < \theta < 2\pi$

9 Prove that $\tan 2\theta = \dfrac{2 \tan \theta}{1 - \tan^2 \theta}$, $\cos 2\theta = \dfrac{1 - \tan^2 \theta}{1 + \tan^2 \theta}$

   Show that $\sec 2\theta + \tan 2\theta = \tan \left(\dfrac{\pi}{4} + \theta \right)$

   Solve for $0 < 2\theta < \dfrac{\pi}{2}$, the equation $\sec 2\theta + \tan 2\theta = \cot 4\theta$

TRIGONOMETRY OF THE TRIANGLE

> Know when to use the sine and cosine rules.

Many practical problems involve work with a plane triangle. If the triangle is a right-angled triangle then there is usually little difficulty since the trigonometric ratios for sine, cosine or tangent, together with Pythagoras's theorem will normally provide a solution to the problem. If the triangles are *not* right-angled then it is necessary to use the sine rule or cosine rule, depending upon the information given.

- **Use the sine rule when either:**
  i) two angles and one side are given
  or
  ii) two sides and a non-included angle are given
- **Use the cosine rule when either:**
  i) three sides are given,
  or
  ii) two sides and an included angle are given.

SINE AND COSINE RULES

Consider any triangle ABC. A circle can be drawn passing through the points A, B and C. Then if R is the radius of this circle

Sine rule $\Rightarrow \dfrac{a}{\sin A} = \dfrac{b}{\sin B} = \dfrac{c}{\sin C} = 2R$

Cosine rule $\Rightarrow$

$a^2 = b^2 + c^2 - 2bc \cos A$ or $\cos A = \dfrac{b^2 + c^2 - a^2}{2bc}$

$b^2 = a^2 + c^2 - 2ac \cos B$ or $\cos B = \dfrac{a^2 + c^2 - b^2}{2ac}$

$c^2 = a^2 + b^2 - 2ab \cos C$ or $\cos C = \dfrac{a^2 + b^2 - c^2}{2ab}$

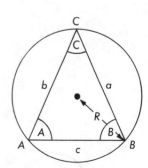

Fig. 5.14

Care must be taken when using the sine rule to find an angle to ensure that the smallest unknown angle is found first since it must be remembered that $\sin(\pi - x) = \sin x$. When using the cosine rule remember that the cosine of an obtuse angle is negative.

Worked example

A and B are two points on one bank of a straight river such that AB is 649 m. C is on the other bank and angles CAB, CBA are respectively 48°31′ and 75°25′. Find the width of the river.

$$\angle ACB = 180° - (48°31' + 75°25') = 180° - 123°56' = 56°4'$$

$$\text{Sine rule} \Rightarrow \frac{AC}{\sin 75°25'} = \frac{649}{\sin 56°4'} \Rightarrow AC = 757 \cdot 0 \, \text{m}$$

Width of river $= AC \sin 48°31' = 757 \sin 48°31' = 567 \cdot 1 \, \text{m} \approx 567 \, \text{m}$

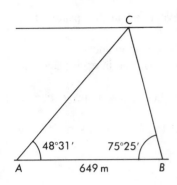

48°31′   75°25′

A   649 m   B

Fig. 5.15

Worked example

Use the sine rule to show that for any plane triangle ABC

$$\tan \frac{A}{2} \tan \frac{B-C}{2} = \frac{b-c}{b+c}$$

Three ships at sea are such that the bearings of ships B and C from A are 36° and 247° respectively. Given that B is 120 km from A and C is 234 km from A calculate the bearing and distance of ship B from ship C.

$$\frac{b-c}{b+c} = \frac{\sin B - \sin C}{\sin B + \sin C} = \frac{2 \cos\left(\frac{B+C}{2}\right) \sin\left(\frac{B-C}{2}\right)}{2 \sin\left(\frac{B+C}{2}\right) \cos\left(\frac{B-C}{2}\right)}$$

$$= \cot\left(\frac{B+C}{2}\right) \tan\left(\frac{B-C}{2}\right) = \cot\left(\frac{\pi}{2} - \frac{A}{2}\right) \tan\left(\frac{B-C}{2}\right)$$

$$= \tan \frac{A}{2} \tan\left(\frac{B-C}{2}\right)$$

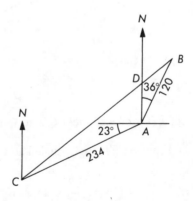

Fig. 5.16

In solving problems such as this the first step is to draw an accurate diagram of the situation marking all known angles and distances.

$b = 234$, $c = 120$, $\angle BAC = 149°$

$$\Rightarrow \tan\left(\frac{149°}{2}\right) \tan\left(\frac{B-C}{2}\right) = \frac{234 - 120}{234 + 120} = \frac{114}{354}$$

$$\Rightarrow \tan\left(\frac{B-C}{2}\right) = 0 \cdot 0893 \Rightarrow \frac{B-C}{2} = 5 \cdot 1°$$

$\Rightarrow \angle B - \angle C = 10 \cdot 2°$

But $\angle B + \angle C = 180° - 149° = 31° \Rightarrow \angle B = 20 \cdot 6°$, $\angle C = 10 \cdot 4°$

Hence $\angle CDA = 36° + 20 \cdot 6° = 56 \cdot 6°$

i.e. bearing of ship B from ship C is 56·6°

$$\text{From } \triangle ABC \Rightarrow \frac{BC}{\sin 149°} = \frac{234}{\sin 20 \cdot 6°} \Rightarrow BC = 342 \cdot 5$$

$\Rightarrow$ Distance of ship B from ship C is approximately 342·5 km

THREE-DIMENSIONAL PROBLEMS

" Learn how to draw three-dimensional diagrams. "

In solving such problems always try to draw a good *three-dimensional diagram* and mark clearly on it the values of all known angles, particularly the right angles. Do not try to depend upon a series of two-dimensional diagrams as this often leads to incorrect figures. Remember:

i)   a line L perpendicular to a plane is perpendicular to every line in that plane which intersects L

ii) two non-parallel planes meet in a line, normally called the common line

iii) the angle between two planes is the angle between two lines, one in each plane and both perpendicular to the common line

iv) the angle between a line and a plane is the angle between the line and its projection on the plane

v)   the line of greatest slope in a plane is a line perpendicular to the line of intersection of the plane and a horizontal plane

Worked example

A pole AB of length $10\,$m is held in a vertical position with B on horizontal ground by three equal stays AC, AD and AE fixed so that CDE forms an equilateral triangle of side $12\,$m. Show that

a) the length of each stay is $2\sqrt{37}\,$m

b) the inclination of each stay to the horizontal is θ where $\tan\theta = (5\sqrt{3})/6$

c) the angle contained between the planes ACD and CDE is ϕ where $\tan\phi = (5\sqrt{3})/3$.

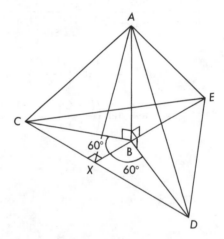

Fig. 5.17

Let EB produced meet CD at X, then $CX = XD = 6\,$m

a) From $\triangle CXB$, right-angled at X, $\dfrac{CX}{CB} = \sin 60° = \dfrac{\sqrt{3}}{2}$

$$\Rightarrow CB = \frac{6.2}{\sqrt{3}} = 4\sqrt{3}\,\text{m}$$

From $\triangle ABC$, right-angled at B, $AC^2 = AB^2 + BC^2$
$$\Rightarrow AC^2 = 10^2 + (4\sqrt{3})^2 = 148 \Rightarrow AC = \sqrt{148}\,\text{m} = 2\sqrt{37}\,\text{m}$$
$\Rightarrow$ each stay is of length $2\sqrt{37}\,$m

b) From $\triangle ABC$, $\tan\angle ACB = \dfrac{AB}{BC} = \dfrac{10}{4\sqrt{3}} = \dfrac{5\sqrt{3}}{6}$

$\Rightarrow$ inclination of each stay to horizontal is θ

where $\tan\theta = \dfrac{5\sqrt{3}}{6}$.

c) The angle contained between the planes ACD and CDE is $\angle AXB$

From $\triangle CXB$ right-angled at X, $\dfrac{CX}{BX} = \tan 60° = \sqrt{3}$

$$\Rightarrow BX = \frac{6}{\sqrt{3}} = 2\sqrt{3}\,\text{m}$$

From $\triangle AXB$ right-angled at B, $\tan \angle AXB = \dfrac{AB}{BX} = \dfrac{10}{2\sqrt{3}}$

$$\Rightarrow \tan \phi = \dfrac{5}{\sqrt{3}} = \dfrac{5\sqrt{3}}{3}.$$

Worked example

Towns A and B are 350 and 690 m respectively above town C, which is at sea level. Town B is due north of C and A lies to the east of B and C. Given that the elevation of B from C is $10 \cdot 3°$, that of A from C $6 \cdot 7°$ and that of B from A is $4 \cdot 6°$, calculate the bearing of A from C.

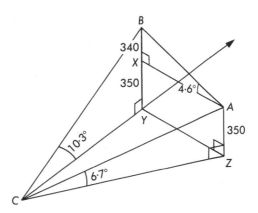

Fig. 5.18

From $\triangle ACZ$ right-angled at Z, $CZ = (350 \cot 6 \cdot 7°)$m $= 2979 \cdot 4$ m
From $\triangle ABX$ right-angled at X, $AX = (340 \cot 4 \cdot 6°)$m $= 4225 \cdot 8$ m $= YZ$
From $\triangle CBY$ right-angled at Y, $CY = (690 \cot 10 \cdot 3°)$m $= 3796 \cdot 8$ m
From $\triangle CYZ$, using the cosine rule

$$\cos \angle YCZ = \dfrac{CY^2 + CZ^2 - YZ^2}{2CY.CZ} = 0 \cdot 24023$$

$$\Rightarrow \angle YCZ = 76 \cdot 1°$$
$$\Rightarrow \text{bearing of } A \text{ from } C \text{ is } 76 \cdot 1°$$

EXERCISES 5.3

1   The angle of elevation of the top of a church tower from two points A and B on a horizontal road in direct line with the tower are α and β respectively, where $\beta > \alpha$. Given that $AB = d$ m, show that the height of the tower is $\dfrac{d \sin \alpha \sin \beta}{\sin (\beta - \alpha)}$.

2   An aeroplane is observed simultaneously from two points A and B on horizontal ground, A being at a distance c due north of B. From A the bearing of the aeroplane is θ at an elevation α. From B the bearing of the aeroplane is ϕ. Show that i) the height of the aeroplane is $\dfrac{c \tan \alpha \sin \phi}{\sin (\theta - \phi)}$ ii) the angle of elevation of the aeroplane from B is $\arctan \left(\dfrac{\tan \alpha \sin \phi}{\sin \theta} \right)$.

3   A right pyramid, vertex V stands on a square base $ABCD$. Each side of the base is 12 cm and the edges VA, VB, VC and VD are each 15 cm. Find the angle between the two faces VAB and VBC

4   A right pyramid $VABC$ has vertex V and an equilateral triangular base ABC of side 6 cm. Given that VA is 10 cm calculate the angle between i) the face VAB and the face VBC; ii) the edge VA and the face VBC

5   A gate $2\frac{1}{3}$ m wide and $1\frac{1}{3}$ m high is supported on a vertical post. The gate swings through an angle of 50°. Find the angle between the new position of a diagonal of the gate and the old position of a horizontal bar of the gate

EXAMINATION TYPE QUESTIONS WITH MARKING SCHEME AND GUIDED SOLUTION

1 Solve the equation $\cos x + \cos 3x = \cos 2x + \cos 4x$ for $0° \leqslant x \leqslant 180°$ *(8 marks)*

$2\cos 2x \cos x = 2\cos 3x \cos x$	M1	A1
$\Rightarrow 2\cos x (\cos 2x - \cos 3x) = 0$		
$\Rightarrow 2\cos x . 2\sin\dfrac{5x}{2}\sin\dfrac{x}{2} = 0$	M1	A1
$\Leftrightarrow \cos x = 0$ or $\sin\dfrac{5x}{2} = 0$ or $\sin\dfrac{x}{2} = 0$		A1

$$\cos x = 0 \Leftrightarrow x = 90°$$
$$\sin\frac{5x}{2} = 0 \Leftrightarrow \frac{5x}{2} = 0,\ 180°,\ 360°,\ \Leftrightarrow x = 0°,\ 72°,\ 144° \qquad \text{M1}$$
$$\sin\frac{x}{2} = 0 \Leftrightarrow \frac{x}{2} = 0,\ \Leftrightarrow x = 0°$$
$$\Leftrightarrow x = 0°,\ 90°,\ 72°,\ 144°,\ 180° \qquad\qquad \text{A2} \quad (1,0)$$

2 Express $y \equiv 3\cos x + \sin x$ in the form $R\cos(x - \alpha)$, where $R > 0$ and $0° < \alpha < 90°$
Deduce the maximum and minimum values of y and find the values of x in $-180° \leqslant x \leqslant 180°$ for which $y = 2$
Sketch the curve $y = 3\cos x + \sin x$ for $-270° \leqslant x \leqslant 270°$ *(12 marks)*

$$3\cos x + \sin x \equiv R\cos x \cos\alpha + R\sin x \sin\alpha$$
$$\Rightarrow \begin{array}{l} 3 = R\cos\alpha \\ 1 = R\sin\alpha \end{array} \Leftrightarrow R^2 = 3^2 + 1^2 = 10,\ R = \sqrt{10} \qquad \text{M1} \quad \text{A1}$$

Dividing $\tan\alpha = \frac{1}{3} \Rightarrow \alpha \approx 18\cdot43°$ A1
$\Rightarrow y = 3\cos x + \sin x = \sqrt{10}\cos(x - 18\cdot43°)$
The maximum value of y occurs when $\cos(x - 18\cdot43°) = 1$
i.e. $y_{\max} = \sqrt{10}$ M1   A1
Similarly $y_{\min} = -\sqrt{10}$

$y = 2 \Rightarrow 2 = \sqrt{10}\cos(x - 18\cdot43°)$,
$\Rightarrow \cos(x - 18\cdot43°) = 0\cdot6325 \Leftrightarrow x - 18\cdot43° = 360°n \pm 50\cdot77°$ M1   M1
$x = 360°n + 69\cdot2°$ or $x = 360°n - 32\cdot34°$
$\Rightarrow x = 69\cdot2°$ or $-32\cdot3°$ (to 1 decimal place)   A1   A1
$y = 3\cos x + \sin x = \sqrt{10}\cos(x - 18\cdot43°)$, a cosine wave curve of amplitude $\sqrt{10}$ displaced in the negative direction of the x-axis through a distance equivalent to $18\cdot43°$

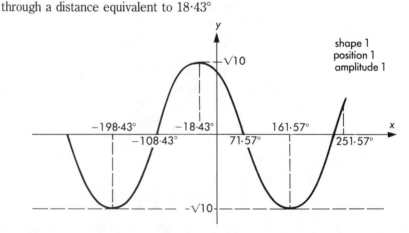

Fig. 5.19

3 A triangle ABC is drawn on a plane which makes an angle θ with the horizontal. A and B are on the same level and C is below them. CA, CB makes angles α, β with the horizontal plane. Show that $AC\sin\alpha = BC\sin\beta$ and by applying the sine rule to $\triangle ABC$ prove that $\sin\theta = \dfrac{AB\sin\beta}{AC\sin C}$
Deduce that $\sin^2\theta \sin^2 C = \sin^2\alpha + \sin^2\beta - 2\sin\alpha \sin\beta \cos C$

    (20 marks)

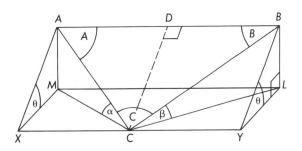

Fig. 5.20

Diagrams and angles B2 (1,0)

XYLM is the horizontal plane, *AM* and *BL* are perpendicular to the horizontal plane.
The perpendicular from *C* to *AB* meets *AB* at *D*

From $\triangle ACM$ right-angled at M $AM = AC\sin\alpha$ M1   A1

From $\triangle BCL$ right-angled at L $BL = BC\sin\beta$ A1

Since AB is horizontal $AM = BL \Rightarrow AC\sin\alpha = BC\sin\beta$ M1   A1

From $\triangle BYL$ right-angled at L $\sin\theta = \dfrac{BL}{BY} \Rightarrow BY = \dfrac{BL}{\sin\theta}$ M1   A1

From $\triangle CBD$ right-angled at D $\sin B = \dfrac{CD}{BC} = \dfrac{BY}{BC} = \dfrac{BL}{BC\sin\theta}$ M1   A1

From $\triangle ABC$ sine rule $\Rightarrow \dfrac{\text{Sin } C}{AB} = \dfrac{\text{Sin } B}{AC}$ M1

$\Rightarrow \dfrac{\text{Sin } C}{AB} = \dfrac{BL}{BC\sin\theta} \cdot \dfrac{1}{AC} = \dfrac{BC\sin\beta}{BC.AC\sin\theta}$ since $BL = BC\sin\beta$ M1

$\Rightarrow \dfrac{\text{Sin } C}{AB} = \dfrac{\text{Sin }\beta}{AC\sin\theta}$ or $\sin\theta = \dfrac{AB\sin\beta}{AC\sin C}$ A2 (1,0)

Squaring $\sin^2\theta = \dfrac{AB^2\sin^2\beta}{AC^2\sin^2 C}$ M1

$\Rightarrow \sin^2\theta\sin^2 C = \dfrac{AB^2\sin^2\beta}{AC^2} = \dfrac{\sin^2\beta}{AC^2}\left[AC^2 + BC^2 - 2.AC.BC\cos C\right]$ M1

$\Rightarrow \sin^2\theta\sin^2 C = \sin^2\beta\left[1 + \left(\dfrac{BC}{AC}\right)^2 - 2\left(\dfrac{BC}{AC}\right)\cos C\right]$ M1

$= \sin^2\beta\left[1 + \left(\dfrac{\sin\alpha}{\sin\beta}\right)^2 - 2\left(\dfrac{\sin\alpha}{\sin\beta}\right)\cos C\right]$ A1

$\Rightarrow \sin^2\theta\sin^2 C = \sin^2\beta + \sin^2\alpha - 2\sin\alpha\sin\beta\cos C$ A1

EXAMINATION QUESTIONS

1 Prove that
$$\cos(A + B)\cos(A - B) \equiv \cos^2 A + \cos^2 B - 1.$$ (L)

2 Express $y = \cos\theta° - (\sqrt{3})\sin\theta°$ in the form
$y = R\cos(\theta° + \alpha°)$, where $R > 0$ and $0 < \alpha < 90$.
Hence find

a) the general value of θ for which $y = 1$,

b) the least value of $\dfrac{1}{y^2}$ (L)

3 The outputs of two signal generators A and B are combined and the resultant waveform displayed on an oscilloscope. Given that the signal from A is given by $y_1 = 2\sin x$ and that from B by $y_2 = 4\cos x$, express the resultant, $y = y_1 + y_2$, in the form $R\sin(x + \alpha)$, where $R > 0$ and $0 < \alpha < \dfrac{\pi}{2}$, giving R and α to 3 significant figures. (L)

4 Points A, B and C are the corners of a triangular field which has $\angle ABC = 90°$ and $AC = 60$ m. The point D lies on AC and the point E lies on AB. A straight path joins D to E so that $\angle ADE = 90°$ and $AD = BE = 10$ m. Given that $\angle BAC = x°$,

a) find, in terms of x, the lengths of AE and AB.

b) Hence prove that
$$6\cos^2 x° - \cos x° - 1 = 0.$$

c) Determine the values of $\cos x°$ which satisfy this equation and hence find the values of x, $0 \le x \le 360$, which also satisfy this equation.

Calculate the length, in m to 1 decimal place,

d) of the path DE,

e) of a straight fence which joins B to D.   (L)

5   Find all the roots of the equation
$$\cos 2\theta = 2 \cos \theta$$
lying between $0°$ and $360°$, giving your answers to the nearest degree.   (WJEC)

6   Prove the identity
$$\tan \theta + \cot \theta \equiv 2 \operatorname{cosec} 2\theta.$$
Find, in radians, all the solutions of the equation
$$\tan x + \cot x = 8 \cos 2x$$
in the interval $0 < x < \pi$.   (AEB 1989)

7   a) Starting from the identity $\cos(A+B) \equiv \cos A \cos B - \sin A \sin B$, prove the identify $\cos 2\theta \equiv 2\cos^2 \theta - 1$.

b) Find the general solution of the equation $\sin \theta + \tan \theta \cos 2\theta = 0$, giving your answer in radians in terms of π.

c) Prove the identity $2\cos^2 \theta - 2\cos^2 2\theta \equiv \cos 2\theta - \cos 4\theta$.

d) By substituting $\theta = \pi/5$ in the identity in c), prove that
$$\cos\left(\frac{\pi}{5}\right) - \cos\left(\frac{2\pi}{5}\right) = \frac{1}{2}.$$

e) Hence find the value of $\cos \pi/5$ in the form $a + b\sqrt{5}$, stating the values of a and b.   (AEB 87)

8   Determine all of the angles between $0°$ and $180°$ which satisfy the equation
$$\cos 5\theta + \cos 3\theta = \sin 3\theta + \sin \theta.$$ (NI 87)

9   Given that $3\cos \theta + 4 \sin \theta \equiv R \cos(\theta - \alpha)$, where $R > 0$ and $0 \le \alpha \le \pi/2$, state the value of R and the value of $\tan \alpha$.

For each of the following equations, solve for θ in the interval $0 \le \theta \le 2\pi$ and give your answers in radians correct to one decimal place.

a) $3 \cos \theta + 4 \sin \theta = 2$,

b) $3 \cos 2\theta + 4 \sin 2\theta = 5 \cos \theta$.

The curve with equation $y = \dfrac{10}{3 \cos x + 4 \sin x + 7}$ between $x = -\pi$ and $x = \pi$, cuts the y-axis at A, has a maximum point at B and a minimum point at C. Find the coordinates of A, B and C.   (AEB 99)

10   Express $\dfrac{\sin 3\theta}{\sin \theta}$ in terms of $\cos \theta$.

Hence show that if $\sin 3\theta = \lambda \sin 2\theta$, where λ is a constant, then either $\sin \theta = 0$ or $4 \cos^2 \theta - 2\lambda - 1 = 0$.

Determine the general solution, in degrees, of the equation
$$\sin 3\theta = 3 \sin 2\theta.$$ (AEB 89)

11   A pyramid $VABCD$ has a horizontal square base $ABCD$, of side $6a$. The vertex V of the pyramid is at a height $4a$, vertically above the centre of the base. Calculate, in degrees, to one decimal place, the acute angle between

a) the edge VA and the horizontal,

b) the plane VAB and the horizontal,

c) the planes VBA and VBC.   (AEB 88)

12   The base of a pyramid is a square of side a. Each face makes an angle of $\dfrac{\pi}{3}$ radians with the base. Show that

i) the height of the pyramid is $\dfrac{a\sqrt{3}}{2}$, and

ii) the angle between two adjacent sloping faces is $\cos^{-1}\left(-\tfrac{1}{4}\right)$ radians.   (NI 87)

13   If θ is real, show that

i) $|\sin\theta + \cos\theta| \leqslant \sqrt{2}$, and

ii) $|\tan\theta + \cot\theta| \geqslant 2$.   (NI 87)

14   a) Show that
$$\tan 4\theta = \frac{4\tan\theta(1-\tan^2\theta)}{\tan^4\theta - 6\tan^2\theta + 1}$$
Hence show that $\tan 22\tfrac{1}{2}°$ is a root of the equation $t^4 - 6t^2 + 1 = 0$.
Deduce that
$$\tan 22\tfrac{1}{2}° = \sqrt{3 - \sqrt{8}}$$
and obtain a similar expression for $\tan 67\tfrac{1}{2}°$.
Use your results to show that
$$\sec^2 22\tfrac{1}{2}° + \sec^2 67\tfrac{1}{2}° = 8.$$

b) AB is a chord of a circle centre O which subtends an angle 2θ radians at O ($\theta < \tfrac{1}{2}\pi$). If AB divides the circle into two regions, one having twice the area of the other, show that θ satisfies the equation.
$$6\theta - 3\sin 2\theta - 2\pi = 0.$$
Hence show that θ lies between $1\cdot30$ and $1\cdot31$.   (WJEC)

15   a) The length a of the side BC of a triangle ABC is related to the lengths b, c of the sides CA, AB and the included angle $\widehat{BAC}$ (measured in radians) by the cosine formula
$$a^2 = b^2 + c^2 - 2bc\cos A.$$
If A is increased by 1% and b, c held constant, show that a increases by $x\%$ where
$$x \approx \frac{Abc\sin A}{a^2}.$$

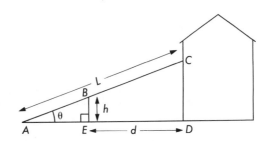

b) The above figure shows a ladder AC of length L leaning against the side of a house. The ladder, which is inclined at an angle θ to the horizontal AD, also rests on the top of a wall BE of height h situated at a distance d from the house. Obtain an expression for L in terms of h, d and θ and show that its minimum value as θ varies is
$$(h^{2/3} + d^{2/3})^{3/2}$$
(WJEC)

16   In the triangle ABC, $AB = 8$, $BC = 7$, $\widehat{BAC} = 60°$. Given that $AC < 4$, find AC.   (C)

17   Express $\cos x + \sin x$ in the form $R\cos(x - \alpha)$, where $R \geqslant 0$ and $0 \leqslant \alpha < 2\pi$, giving the value of R and the value of α.

i) Sketch the graph of $y = \cos x + \sin x$ for $-2\pi \leqslant x \leqslant 2\pi$.

ii) By drawing appropriate lines on your graph, determine the number of roots, lying in the interval $-2\pi \leqslant x \leqslant 2\pi$, of each of the following equations:
   a) $\cos x + \sin x = -\tfrac{1}{2}$;
   b) $\cos x + \sin x = 2$.

iii) Find the set of values of x, lying in the interval $-\pi \leqslant x \leqslant \pi$, for which $|\cos x + \sin x| > 1$.

18 Express $3\cos x + 4\sin x$ in the form $R\cos(x-\alpha)$, where $R>0$ and $-\pi<\alpha\leqslant\pi$. Give α correct to four decimal places.

The function f is defined for real x by
$$f(x) = 5 + 3\cos x + 4\sin x.$$

i) Find the values of x in the interval $-\pi<x\leqslant\pi$ for which $f(x)=7$, giving the solutions correct to three decimal places.

ii) State the greatest and least values of $f(x)$. Find, to three decimal places, the corresponding values of x in the interval $-\pi<x\leqslant\pi$. Sketch the graph of $y=f(x)$ for $-\pi<x\leqslant\pi$.

iii) The graph of $y=f(x)$ may be obtained from the graph of $y=\cos x$ by means of transformations. Describe suitable transformations in detail and in the order in which they are used. (JMB)

ANSWERS TO EXERCISES

EXERCISES 5.1

1 $\dfrac{1}{2}, -\dfrac{\sqrt{3}}{2}, -\sqrt{3}$

2 $\dfrac{\pi}{3}, \dfrac{2\pi}{3}, -\dfrac{\pi}{4}$

3 0·51

4 0·98

5 143·7°

6 i) 20°, 130°, 200°, 310°; ii) 221·8°, 318·2°

7 i) $n\pi \pm \dfrac{\pi}{3}$; ii) $\theta = 2n\pi - \dfrac{\pi}{2}$ or $\dfrac{1}{5}\left(2n\pi + \dfrac{\pi}{2}\right)$

8 −45°, 63·4°, 135°, −116·6°

EXERCISES 5.3

3 100·9°

4 63·2°, 30·7°

5 56·1°

EXERCISES 5.2

1 0°, 60°, 300°, 360°

2 $n\pi, 2n\pi, \pm\dfrac{\pi}{6}$

3 45°, 123·7°, 225°, 303·7°

4 45°, 120°, 135°, 225°, 240°, 315°

5 41·4°, 90°, 104·5°, 255·5°, 270°, 318·6°

6 $n\pi, \dfrac{2n\pi}{7}$

7 $5\sin(x-36·9°)$; a) 73·7°, 180°; b) $1, \dfrac{1}{11}$

8 $\dfrac{17\pi}{12}, \dfrac{23\pi}{12}$

9 $\dfrac{\pi}{10}$

ANSWERS TO EXAMINATION QUESTIONS

2 $2\cos(\theta + 60°)$; a) $360°n$ or $360°n - 120°$; b) $\frac{1}{4}$

3 $4·7\sin(x + 1·11^c)$

4 a) $\dfrac{10}{\cos x}$, $60\cos x$; c) $\cos x = -\frac{1}{3}$ or $\frac{1}{2}$, 60°, 109·5°, 250·5°, 300°
   d) 17·3; e) 26·5

5 111°, 249°

6 $\dfrac{\pi}{24}, \dfrac{5\pi}{24}, \dfrac{13\pi}{24}, \dfrac{17\pi}{24}$

7 b) $2n\pi \pm \dfrac{\pi}{3}$, $n\pi$; e) $a = b = \frac{1}{4}$

8 15°, 75°, 90°, 135°

9 $5, \frac{4}{3}$; a) 2·1, 6·1; b) 0·3, 0·9, 2·4, 4·5;
   $A(0,1)$; $B(-2·2, 5)$; $C(0·9, \frac{5}{6})$

10 $4\cos^2\theta - 1$, $360°n \pm 98·7°$

11 43·3°, 53·1°, 68·9°

14 $\sqrt{(3 + \sqrt{8})}$

15 $h\cosec\theta + d\sec\theta$

16 3

17 $\sqrt{2}, \dfrac{\pi}{4}$ (ii) a) 4, b) 0 (iii) $0 < x < \dfrac{\pi}{2} \cup -\pi < x < -\dfrac{\pi}{2}$

18 $5\cos(x - 0·9273^c)$; i) $2·087^c$, $-0·232^c$;
   ii) 10,0; $0·927^c, -2·214^c$

DIFFERENTIATION

GETTING STARTED

Calculus forms a very important part of all GCE A-level syllabuses. You must make sure you know the following basic results and are able to apply the basic rules of differentiation. You will find various notations are used for the derivative of a function. The most common one being that if, say, y is given as a function of x, i.e. $y = f(x)$, then the derivative of y with respect to x is $\frac{dy}{dx}$. Be prepared, however, to see and use the alternatives of $f'(x)$, y', $\dot{y}$ or $\frac{d}{dx} f(x)$.

BASIC RESULTS

$y = f(x)$	$\frac{dy}{dx} = f'(x)$
C, a constant	0
x^n, $n \neq 0$	nx^{n-1}
$\sin x$	$\cos x$
$\cos x$	$-\sin x$
$\tan x$	$\sec^2 x$
$\ln x$	$\frac{1}{x}$
e^x	e^x
$\arcsin x$	$\frac{1}{\sqrt{(1-x^2)}}$
$\arccos x$	$\frac{-1}{\sqrt{(1-x^2)}}$
$\arctan x$	$\frac{1}{1+x^2}$

The GCE boards normally issue a list of standard results, but you should not become dependent upon such a list. You must know the basic results and be able to use them with ease, otherwise you will inevitably lose valuable time and marks by failing to realise what is required of a question at an appropriate moment.

ESSENTIAL PRINCIPLES

BASIC RULES OF DIFFERENTIATION

The basic rules of differentiation include

1 $y = cf(x)$, c a constant, $\Rightarrow \dfrac{dy}{dx} = cf'(x)$;

e.g. $y = 7x^3 \Rightarrow \dfrac{dy}{dx} = 21x^2$

2 $y = u \pm v$, u and v functions of x, $\Rightarrow \dfrac{dy}{dx} = \dfrac{du}{dx} \pm \dfrac{dv}{dx}$

e.g. $y = 3\sin x + 2x^5 \Rightarrow \dfrac{dy}{dx} = 3\cos x + 10x^4$

3 $y = u\,v \Rightarrow \dfrac{dy}{dx} = u\dfrac{dv}{dx} + v\dfrac{du}{dx}$

e.g. $y = x^2 \tan x \Rightarrow \dfrac{dy}{dx} = x^2(\sec^2 x) + 2x\tan x$

4 $y = \dfrac{u}{v} \Rightarrow \dfrac{dy}{dx} = \dfrac{v\dfrac{du}{dx} - u\dfrac{dv}{dx}}{v^2}$

e.g. $y = \dfrac{2x^3}{1-x^2} \Rightarrow \dfrac{dy}{dx} = \dfrac{(1-x^2)\,6x^2 - 2x^3\,(-2x)}{(1-x^2)^2} = \dfrac{6x^2 - 2x^4}{(1-x^2)^2}$

5 $y = f(u)$ where $u = g(x)$, i.e. a function of a function of $x \Rightarrow \dfrac{dy}{dx} = \dfrac{dy}{du} \cdot \dfrac{du}{dx}$ often referred to as the chain rule

e.g. $y = \sin 3x$ i.e. $y = \sin u$ where $u = 3x$

$\Rightarrow \dfrac{dy}{dx} = \cos u \cdot 3 = 3\cos 3x$

Note:

a) a special case of this rule is $\dfrac{dy}{dx} = \left(\dfrac{1}{\dfrac{dx}{dy}} \right)$

e.g. $x = \sin 3y \Rightarrow \dfrac{dx}{dy} = 3\cos 3y \Rightarrow \dfrac{dy}{dx} = \left(\dfrac{1}{\dfrac{dx}{dy}}\right) = \dfrac{1}{3\cos 3y}$

b) the rule can be extended $\Rightarrow \dfrac{dy}{dx} = \dfrac{dy}{du} \cdot \dfrac{du}{dv} \cdot \dfrac{dv}{dx}$

e.g. $y = \cos^2(3x)$ i.e. $y = u^2$ where $u = \cos v$ and $v = 3x$

$\Rightarrow \dfrac{dy}{du} = 2u = 2\cos v = 2\cos 3x$, $\dfrac{du}{dv} = -\sin v = -\sin 3x$, $\dfrac{dv}{dx} = 3$

$\Rightarrow \dfrac{dy}{dx} = \dfrac{dy}{du} \cdot \dfrac{du}{dv} \cdot \dfrac{dv}{dx} = 2\cos 3x\,(-\sin 3x)\cdot 3 = -6\cos 3x \sin 3x$

This rule enables the table of derivatives to be generalised.

y	$\dfrac{dy}{dx}$
$u^n = [f(x)]^n$	$n[f(x)]^{n-1}\,f'(x)$
$\sin f(x)$	$f'(x)\cos f(x)$
$\cos f(x)$	$-f'(x)\sin f(x)$
$\tan f(x)$	$f'(x)\sec^2 f(x)$
$\ln f(x)$	$\dfrac{f'(x)}{f(x)}$
$e^{f(x)}$	$f'(x)e^{f(x)}$
$\arcsin f(x)$	$\dfrac{f'(x)}{\sqrt{[1-(f(x))^2]}}$
$\arccos f(x)$	$\dfrac{-f'(x)}{\sqrt{[1-(f(x))^2]}}$
$\arctan f(x)$	$\dfrac{f'(x)}{1+(f(x))^2}$

Worked examples

You should examine the following worked examples and then try as many similar examples as you can.

1 $y = 7(x+2)^4 \Rightarrow \dfrac{dy}{dx} = 7.4 \,.\, (x+2)^3 \,.\, 1 = 28\,(x+2)^3$

2 $y = \sqrt{(2+x^3)} = (2+x^3)^{\frac{1}{2}} \Rightarrow \dfrac{dy}{dx} = \frac{1}{2}(2+x^3)^{-\frac{1}{2}} \,.\, 3x^2 = \dfrac{3x^2}{2\sqrt{(2+x^3)}}$

3 $y = e^{4x} - e^{-3x} \Rightarrow \dfrac{dy}{dx} = 4e^{4x} - (-3)\,e^{-3x} = 4e^{4x} + 3e^{-3x}$

4 $y = \tan 2x \Rightarrow \dfrac{dy}{dx} = (\sec^2 2x)\,.\, 2 = 2\sec^2 2x$

5 $y = \sin 3x \cos 2x \Rightarrow \dfrac{dy}{dx} = 3\cos 3x \cos 2x + \sin 3x\,(-2\sin 2x)$

$$= 3\cos 3x \cos 2x - 2\sin 3x \sin 2x$$

6 $y = 2\sin^3(4x) \Rightarrow \dfrac{dy}{dx} = 2.3\sin^2(4x)\,.\,(\cos 4x)(4) = 24\sin^2 4x \cos 4x$

7 $y = \arcsin(3x) \Rightarrow \dfrac{dy}{dx} = \dfrac{3}{\sqrt{(1-(3x)^2)}} = \dfrac{3}{\sqrt{(1-9x^2)}}$

8 $y = \dfrac{e^{2x}}{\sin 3x} \Rightarrow \dfrac{dy}{dx} = \dfrac{\sin 3x \,.\, 2e^{2x} - e^{2x} \,.\, 3\cos 3x}{\sin^2 3x}$

With practice students usually become efficient at differentiation. Usually marks are lost because:

i) the standard results have not been learnt thoroughly
ii) sufficient care has not been taken in 'talking oneself' through the necessary operations.

First you should recognise the form of the function and then 'talk yourself' through the basic rules to be used.

In example 2, $y = \sqrt{(2+x^3)}$ you should first recognise y as a function of x raised to the power of $\frac{1}{2}$. Differentiation therefore involves multiplying by the power $(\frac{1}{2})$, reducing the power by 1, $(2+x^3)^{-\frac{1}{2}}$, and finally multiplying by the derivative of the expression inside the bracket, $(3x^2)$.

Thus $\dfrac{dy}{dx} = \frac{1}{2}\,.\,(2+x^3)^{-\frac{1}{2}}\,.\,(3x^2) = \dfrac{3x^2}{2\sqrt{(2+x^3)}}$

Derivatives involving products or quotients simply add extra lines to the working.

In example 8; $y = \dfrac{e^{2x}}{\sin 3x}$ you should recognise the quotient rule is required.

$$= \dfrac{(\text{denominator})\,(\text{derivative of numerator}) - (\text{numerator})\,(\text{derivative of denominator})}{(\text{denominator})^2}$$

For $\dfrac{d}{dx}(e^{2x})$, the derivative of an exponential, is the exponential, (e^{2x}), multiplied by the derivative of the index, (2)

$$\Rightarrow \dfrac{d}{dx}(e^{2x}) = 2e^{2x}$$

For $\dfrac{d}{dx}(\sin 3x)$, the derivative of a sine is a cosine, $(\cos 3x)$, multiplied by the derivative of $3x$, (3)

$$\Rightarrow \dfrac{d}{dx}\sin 3x = 3\cos 3x$$

$$\Rightarrow \dfrac{dy}{dx} = \dfrac{\sin 3x \,.\, 2e^{2x} - e^{2x} \,.\, 3\cos 3x}{\sin^2 3x}$$

EXERCISES 6.1

Differentiate with respect to the appropriate variable

1 $(1-4x)^3$ $-12\cancel{}(1-4x)^2$

2 $\dfrac{x^3(3x-2)^{-1}}{\cancel{8x-2}}$

3 $x^2 - \sqrt{(1+x)}$

4 $\sqrt{\left(\dfrac{1-x}{1+x}\right)}$

5 $\sin 7t$

6 $\operatorname{cosec} 3x$

7    $\sin 4x \cos 3x$

8    $\tan^3 2x$

9    $\dfrac{t}{\sin t}$

10    $\dfrac{1 + \cos x}{1 - \cos x}$

11    $\sin 2x \cos^2 x$

12    $\ln \sin 2x$

13    $x \ln x$

14    $x^2 \ln\left(\dfrac{1}{x}\right)$

15    $x^2 e^{3x}$

16    $\dfrac{e^{2x}}{1 + 2e^{3x}}$

17    $e^{\sqrt{x}}$

18    $(1 + x^2) e^{-2x}$

19    $\sin^3\left(\dfrac{\pi}{3} - 2x\right)$

20    $\arcsin 2x$

21    $\arctan 3t$

22    $x^2 \arcsin x$

23    $\ln \dfrac{1 + \cos x}{1 - \cos x}$

24    $\dfrac{1}{\sin x + \cos x}$

HIGHER DERIVATIVES

When $y = f(x)$ and $\dfrac{dy}{dx} = g(x)$, then $g(x)$ can be differentiated with respect to x to give $\dfrac{d^2y}{dx^2}$, the second derivative of y with respect to x. You should note the different positions of the 2's in the numerator and denominator.

$\dfrac{d^2y}{dx^2} = \dfrac{d}{dx}\left(\dfrac{dy}{dx}\right)$, read as 'differentiate $\dfrac{dy}{dx}$ with respect to x'.

Similarly, for higher derivatives $\dfrac{d^3y}{dx^3}, \dfrac{d^4y}{dx^4}, \ldots \dfrac{d^ny}{dx^n}$ the third, fourth, $\ldots$ nth derivatives respectively.

Alternative notations include:

$f'(x), f''(x), f'''(x), \ldots f^{(n)}(x),$
or $y'(x), y''(x), y'''(x), \ldots y^{(n)}(x),$
or $y_1(x), y_2(x), y_3(x), \ldots y_n(x)$

Worked example

Find $\dfrac{d^3y}{dx^3}$ when $y = \ln x$

$\dfrac{dy}{dx} = \dfrac{1}{x}, \quad \dfrac{d^2y}{dx^2} = -\dfrac{1}{x^2}, \quad \dfrac{d^3y}{dx^3} = \dfrac{2}{x^3}$

Worked example

Given that $y = Ax^3 + B\ln x + C$ where A, B and C are constants, show that $\dfrac{d^3y}{dx^3} = \dfrac{2}{x^2}\dfrac{dy}{dx}$

$\dfrac{dy}{dx} = 3Ax^2 + \dfrac{B}{x}, \quad \dfrac{d^2y}{dx^2} = 6Ax - \dfrac{B}{x^2}$

$\dfrac{d^3y}{dx^3} = 6A + \dfrac{2B}{x^3} = \dfrac{2}{x^2}\left[3Ax^2 + \dfrac{B}{x}\right] = \dfrac{2}{x^2}\dfrac{dy}{dx}$

Worked example

Given that $y = \sin 3x$ show that $\dfrac{dy}{dx} = 3\sin\left(3x + \dfrac{\pi}{2}\right)$, and deduce that $\dfrac{d^3y}{dx^3} = 3^3 \sin 3\left(x + \dfrac{\pi}{2}\right)$

$y = \sin 3x \Rightarrow \dfrac{dy}{dx} = 3\cos 3x = 3\sin\left(3x + \dfrac{\pi}{2}\right)$

Repeating the process $\dfrac{d^2y}{dx^2} = 3.3\cos\left(3x + \dfrac{\pi}{2}\right) = 3^2\sin\left(3x + \dfrac{\pi}{2} + \dfrac{\pi}{2}\right)$

$= 3^2\sin(3x + \pi)$

$\Rightarrow \dfrac{d^3y}{dx^3} = 3^2.3\cos(3x + \pi)$

$= 3^3\sin\left(3x + \pi + \dfrac{\pi}{2}\right) = 3^3 \sin 3\left(x + \dfrac{\pi}{2}\right)$

IMPLICIT FUNCTIONS

A relationship which cannot be arranged in the explicit form $y = f(x)$, e.g. $x^3 - 3y^4 - y^2 - 2x = 0$, is said to define an **implicit function** y of x. The derivative of such a function with respect to x is found by applying the normal rules for derivatives, but it should be noted that the chain rule is required to differentiate a function of y with respect to x.

Thus $\dfrac{d}{dx} f(y) = \dfrac{dy}{dx} \dfrac{d}{dy} f(y) = f'(y) \dfrac{dy}{dx}$

Worked example

$x^3 + 3y^4 - y^2 - 2x = 0 \Rightarrow \dfrac{d}{dx}(x^3) + \dfrac{d}{dx}(3y^4) - \dfrac{d}{dx}(y^2) - \dfrac{d}{dx}(2x) = 0$

$\Rightarrow 3x^2 + 12y^3 \dfrac{dy}{dx} - 2y \dfrac{dy}{dx} - 2 = 0 \Leftrightarrow \dfrac{dy}{dx} = \dfrac{2 - 3x^2}{12y^3 - 2y}$

Worked example

Find $\dfrac{dy}{dx}$ in terms of x and y when $\ln y = y \ln x$, $(x > 0, \ y > 0)$

$\dfrac{d}{dx}(\ln y) = \dfrac{d}{dx}(y \ln x) \Rightarrow \dfrac{1}{y}\dfrac{dy}{dx} = \dfrac{dy}{dx}\ln x + \dfrac{y}{x}$

$\Rightarrow \left(\dfrac{1}{y} - \ln x\right)\dfrac{dy}{dx} = \dfrac{y}{x} \Rightarrow \dfrac{dy}{dx} = \dfrac{y^2}{x(1 - y\ln x)}$

PARAMETRIC EQUATIONS

It is often more convenient to express a relation between two variables, x and y say, by introducing a third independent variable, t, say. For example, the equation $y^2 = 4x$ can be replaced by the equations $x = t^2$, $y = 2t$, since whatever the value of t, the equations $x = t^2$, $y = 2t$ satisfy the relationship $y^2 = 4x$. The equations $x = t^2$, $y = 2t$ are referred to as **parametric equations**, the independent variable t is the **parameter**.

The derivatives of functions given parametrically are obtained by use of the chain rule, thus:

$$\dfrac{dy}{dx} = \dfrac{dy}{dt} \cdot \dfrac{dt}{dx} \quad \text{and} \quad \dfrac{dt}{dx} = \dfrac{1}{\left(\dfrac{dx}{dt}\right)}$$

Worked example

Find $\dfrac{dy}{dx}$ when $x = t^2$, $y = 2t$

$\Rightarrow \dfrac{dx}{dt} = 2t, \ \dfrac{dy}{dt} = 2 \Rightarrow \dfrac{dy}{dx} = \dfrac{dy}{dt} \cdot \dfrac{dt}{dx} = \dfrac{2}{2t} = \dfrac{1}{t}$

Worked example

Given that $x = 3\cos\theta - \cos 3\theta$, $y = 3\sin\theta - \sin 3\theta$, show that $\dfrac{dy}{dx} = \tan 2\theta$

$\dfrac{dx}{d\theta} = -3\sin\theta + 3\sin 3\theta, \qquad \dfrac{dy}{d\theta} = 3\cos\theta - 3\cos 3\theta$

$\dfrac{dy}{dx} = \dfrac{dy}{d\theta} \cdot \dfrac{d\theta}{dx} = \dfrac{3(\cos\theta - \cos 3\theta)}{3(\sin 3\theta - \sin\theta)} = \dfrac{2\sin 2\theta \sin\theta}{2\cos 2\theta \sin\theta} = \tan 2\theta$

Second and higher derivatives of functions given parametrically can be obtained once the first derivative has been found. Care must be taken however to ensure that the chain rule is applied correctly.

Worked example

Given that $x = 3\cos\theta - \cos 3\theta$, $y = 3\sin\theta - \sin 3\theta$ show that $\dfrac{d^2y}{dx^2} = \dfrac{1}{3\cos^3 2\theta \sin\theta}$

Using the result of the previous worked example we found

$\dfrac{dx}{d\theta} = 3(\sin 3\theta - \sin\theta)$ and $\dfrac{dy}{dx} = \tan 2\theta$

Hence $\dfrac{d}{dx}\left(\dfrac{dy}{dx}\right) = \dfrac{d}{dx}(\tan 2\theta) \Rightarrow \dfrac{d^2y}{dx^2} = \dfrac{d\theta}{dx} \cdot \dfrac{d}{d\theta}(\tan 2\theta)$

$$\Rightarrow \frac{d^2y}{dx^2} = \frac{1}{3\,(\sin 3\theta - \sin \theta)} \cdot 2\sec^2 2\theta = \frac{1}{6\cos 2\theta \sin \theta} \cdot \frac{2}{\cos^2 2\theta} = \frac{1}{3\cos^3 2\theta \sin \theta}$$

❝ A common error. ❞

Warning: do not fall into the trap that many students fall into. Namely $\dfrac{\text{‘}d^2y}{dx^2} = \dfrac{d^2y}{d\theta^2} \Big/ \dfrac{d^2x\text{’}}{d\theta^2}$.

This is *not* a correct application of the chain rule and in any case the 2 in $\dfrac{d^2x}{d\theta^2}$ would be in an incorrect position. $\dfrac{d^2y}{dx^2}$ is not $\dfrac{d^2y}{d^2x}$; $\dfrac{d^2y}{d^2x}$ has no meaning.

LOGARITHMIC DIFFERENTIATION

When the independent variable, say x, occurs in the index of any function other than the exponential function, (e^x), then it is necessary to use logarithmic differentiation, i.e. to take logarithms of the function before differentiating.

Worked example

Differentiate $y = a^x$, where a is a constant
$$y = a^x \Rightarrow \ln y = \ln a^x = x \ln a$$
$$\Rightarrow \frac{d}{dx}(\ln y) = \frac{d}{dx}(x \ln a) \Longleftrightarrow \frac{1}{y}\frac{dy}{dx} = 1 \cdot \ln a$$
$$\Rightarrow \frac{dy}{dx} = y \ln a = a^x \ln a$$

Worked example

Differentiate $y = (x + 2)^{x+1}$
$$\Rightarrow \ln y = \ln (x + 2)^{x+1} = (x + 1) \ln (x + 2)$$
$$\Rightarrow \frac{1}{y}\frac{dy}{dx} = 1 \ln (x + 2) + \frac{x+1}{x+2}$$
$$\Rightarrow \frac{dy}{dx} = y\left[\ln (x + 2) + \frac{x+1}{x+2}\right] = \left[\ln(x + 2) + \frac{x+1}{x+2}\right](x + 2)^{x+1}$$

Logarithmic differentiation is also useful when required to find the derivatives of functions which involve products or quotients

Worked example

Find $\dfrac{dy}{dx}$ given that $y = \dfrac{e^{2x}\sin 3x}{\sqrt{(1-x^2)}}$

$$y = \frac{e^{2x}\sin 3x}{\sqrt{(1-x^2)}} \Rightarrow \ln y = \ln\left[\frac{e^{2x}\sin 3x}{(1-x^2)^{\frac{1}{2}}}\right] = \ln e^{2x} + \ln \sin 3x - \ln (1-x^2)^{\frac{1}{2}}$$

$$\ln y = 2x + \ln \sin 3x - \tfrac{1}{2}\ln (1-x^2)$$
$$\Longleftrightarrow \frac{1}{y}\frac{dy}{d\theta} = 2 + \frac{3\cos 3x}{\sin 3x} - \frac{1}{2}\cdot\frac{(-2x)}{1-x^2}$$
$$\Longleftrightarrow \frac{dy}{dx} = \left[2 + \frac{3\cos 3x}{\sin 3x} + \frac{x}{1-x^2}\right]\frac{e^{2x}\sin 3x}{\sqrt{(1-x^2)}}$$

FORMATION OF DIFFERENTIAL RELATIONSHIPS

A-level questions on differentiation often involve the formation or verification of differential relationships. Verification can always be shown by determining the necessary derivatives and substituting them into the given relationship directly. However, it is often easier to determine the relationship.

Worked example

Given that $y = e^{3x} \sin 4x$ show that $\dfrac{d^2y}{dx^2} - 6\dfrac{dy}{dx} + 25y = 0$

$$\frac{dy}{dx} = 3e^{3x}\sin 4x + e^{3x}4\cos 4x$$

$$\frac{d^2y}{dx^2} = 9e^{3x}\sin 4x + 12e^{3x}\cos 4x + 12e^{3x}\cos 4x - 16e^{3x}\sin 4x$$

$$= 24e^{3x}\cos 4x - 7e^{3x}\sin 4x$$

$$\Leftrightarrow \frac{d^2y}{dx^2} - 6\frac{dy}{dx} + 25y = e^{3x}[(24\cos 4x - 7\sin 4x) - 6(3\sin 4x + 4\cos 4x) + 25\sin 4x]$$

$$= e^{3x}[(24-24)\cos 4x + (-7-18+25)\sin 4x]$$
$$= 0$$

Alternatively $y = e^{3x}\sin 4x \Rightarrow ye^{-3x} = \sin 4x$

$$\Rightarrow \frac{dy}{dx}e^{-3x} - 3ye^{-3x} = 4\cos 4x$$

$$\Rightarrow \left(\frac{d^2y}{dx^2}e^{-3x} - 3\frac{dy}{dx}e^{-3x}\right) - 3\left(\frac{dy}{dx}e^{-3x} - 3ye^{-3x}\right) = -16\sin 4x = -16ye^{-3x}$$

$$\Rightarrow \frac{d^2y}{dx^2} - 3\frac{dy}{dx} - 3\frac{dy}{dx} + 9y = -16y$$

or $\dfrac{d^2y}{dx^2} - 6\dfrac{dy}{dx} + 25y = 0$

EXERCISES 6.2

1   Express $\dfrac{1}{(x-1)(2-x)}$ in partial fractions. Hence or otherwise find $\dfrac{d^2y}{dx^2}$ given that
$$y = \frac{1}{(x-1)(2-x)}$$

2   Find $\dfrac{d^3y}{dx^3}$ when $y = \ln x$

3   Find $\dfrac{dy}{dx}$ in terms of x and y when
   a) $x^2 + 3xy - y^2 = 2$
   b) $xy + \sin y = 2$
   c) $(x-1)^2 + (y-3)^2 = 4$

4   Given that $y^3 e^{-2x^2} = xe^{4x}$ prove that $\dfrac{dy}{dx} = \dfrac{y(2x+1)^2}{3x}$

5   Find $\dfrac{dy}{dx}$ when a) $y = 3^x$, b) $y = x^x$, c) $y = \tan^x x$

6   Given that $e^{2y} = \dfrac{(2-x)^{\frac{1}{2}}\cos 3x}{(1+x)^x}$, find the value, when $x = 0$ of $\dfrac{dy}{dx}$

7   The parametric equations of a curve are
   $x = a(\theta + \sin\theta)$, $y = a(1 - \cos\theta)$. Show that $\dfrac{dy}{dx} = \tan\dfrac{\theta}{2}$

8   Given that $x = \cot\theta$, $y = \sin^2\theta$ show that
   a) $\dfrac{dy}{dx} = -2\sin^3\theta\cos\theta$,   b) $\dfrac{d^2y}{dx^2} = 2\sin^3\theta\sin 3\theta$

9   Given that $y = \cos\ln(1+x)$, prove that
   a) $(1+x)\dfrac{dy}{dx} = -\sin\ln(1+x)$,   b) $(1+x)^2\dfrac{d^2y}{dx^2} + (1+x)\dfrac{dy}{dx} + y = 0$

10   Given that $y = \dfrac{\sin x}{x^2}$, $x > 0$, prove that $x^2\dfrac{d^2y}{dx^2} + 4\dfrac{dy}{dx} + (x^2+2)y = 0$

EXAMINATION QUESTIONS

1   For the curve with equation $x^3 + 2y^3 + 3xy = 0$, find the gradient at the point $(2, -1)$.

(C)

2   Differentiate with respect to x:
   i) $\dfrac{x}{2x+1}$;
   ii) $\sin^{-1}(x^2)$.

(C)

3 A curve has parametric equations
$$x = 2t - \ln(2t), \quad y = t^2 - \ln(t^2),$$
where $t > 0$. Find the value of t at the point on the curve at which the gradient is 2. (C)

4 Given that $\quad x = e^{3t} \cos 3t,$
$$y = e^{3t} \sin 3t,$$

show that $\quad \dfrac{dy}{dx} = \tan(3t + \tfrac{1}{4}\pi)$ (C)

5 Differentiate with respect to x:

i) $(4x - 1)^{20}$
ii) $\tan^{-1}(\sqrt{x})$. (C)

6 Differentiate with respect to x:

i) $\dfrac{1}{(3 - 4x)^2}$

ii) $\cos(\sqrt{x})$. (C)

ANSWERS TO EXERCISES

EXERCISES 6.1

1 $-12(1 - 4x)^2$

2 $\dfrac{6x^2(x - 1)}{(3x - 2)^2}$

3 $2x - \dfrac{1}{2(1 + x)^{\frac{1}{2}}}$

4 $\dfrac{-1}{(1 - x)^{\frac{1}{2}}(1 + x)^{\frac{3}{2}}}$

5 $7\cos 7t$

6 $-3\cot 3x \cosec 3x$

7 $4\cos 4x \cos 3x - 3\sin 4x \sin 3x$

8 $6\tan^2 2x \sec^2 2x$

9 $\dfrac{\sin t - t \cos t}{\sin^2 t}$

10 $\dfrac{-2\sin x}{(1 - \cos x)^2}$

11 $\cos 4x + \cos 2x$

12 $2\cot 2x$

13 $1 + \ln x$

14 $-2x\ln x - x$

15 $xe^{3x}(2 + 3x)$

16 $2e^{2x}(1 - e^{3x})(1 + 2e^{3x})^{-2}$

17 $\dfrac{e^{\sqrt{x}}}{2\sqrt{x}}$

18 $-2x(1 + x)e^{-2x}$

19 $-3\sin\left(\dfrac{\pi}{3} - 2x\right)\sin\left(\dfrac{2\pi}{3} - 4x\right)$

20 $2(1 - 4x^2)^{-\frac{1}{2}}$

21 $3(1 + 9t^2)^{-1}$

22 $2x\arcsin x + x^2(1 - x^2)^{-\frac{1}{2}}$

23 $-2\cosec x$

24 $(\sin x - \cos x)(1 + \sin 2x)^{-1}$

EXERCISES 6.2

1 $\dfrac{1}{x - 1} - \dfrac{1}{x - 2}; \dfrac{2}{(x - 1)^3} + \dfrac{2}{(2 - x)^3}$

2 $\dfrac{2}{x^3}$

3 a) $\dfrac{2x + 3y}{2y - 3x};$ b) $\dfrac{-y}{x + \cos y};$ c) $\dfrac{x - 1}{3 - y}$

5 a) $3^x \ln 3;$ b) $x^x(1 + \ln x);$
   c) $(\ln\tan x + 2x\cosec 2x)\tan^x x$

6 $-\tfrac{1}{8}$

ANSWERS TO EXAMINATION QUESTIONS

1 $-\dfrac{3}{4}$

2 $\dfrac{1}{(2x + 1)^2}, \dfrac{2x}{\sqrt{(1 - x^4)}}$

3 $t = 2$

5 $80(4x - 1)^{19}, \dfrac{1}{2\sqrt{x}(1 + x)}$

6 $\dfrac{8}{(3 - 4x)^3}, \dfrac{-1}{2\sqrt{x}(1 - x)^{\frac{1}{2}}}$

APPLICATIONS OF DIFFERENTIATION

GETTING STARTED

Given $y = f(x)$, then $\dfrac{dy}{dx}$ measures the rate of change of y with respect to x. Hence, for a relationship $s = f(t)$, where s is distance travelled in time t along a straight line from some fixed point on the line, $\dfrac{ds}{dt}$ is a measure of the rate of change of distance s with respect to time $t \Rightarrow speed$, say $v = \dfrac{ds}{dt}$.

Similarly $\dfrac{dv}{dt}$ is a measure of the rate of change of speed with respect to time $t \Rightarrow acceleration$, say $a, = \dfrac{dv}{dt}$

Note: $a = \dfrac{dv}{dt}$ may be written in other forms:

$$a = \frac{d}{dt}(v) = \frac{d}{dt}\left(\frac{ds}{dt}\right) = \frac{d^2s}{dt^2}$$

or $\quad a = \dfrac{dv}{dt} = \dfrac{dv}{ds} \cdot \dfrac{ds}{dt} = v\dfrac{dv}{ds}$ since $\dfrac{ds}{dt} = v$

Worked example

A particle moves in a straight line so that its distance x metres from a fixed point O on the line Ox at time t seconds, $t \geq 0$, is given by $x = 2t^3 - 15t^2 + 36t$. Find

a) the speed with which the particle leaves O
b) the times when the particle is at rest
c) the distances of the particle from O when at rest
d) the time when the acceleration of the particle is zero

$$x = 2t^3 - 15t^2 + 36t \Rightarrow \frac{dx}{dt} = 6t^2 - 30t + 36$$

When $t = 0$, $\dfrac{dx}{dt} = 36 \Rightarrow$ particle leaves O with speed $36\,\text{m s}^{-1}$

Particle is at rest when $\dfrac{dx}{dt} = 0 \Rightarrow 6t^2 - 30t + 36 = 0$

$\Rightarrow 6(t^2 - 5t + 6) = 0 = 6(t-2)(t-3) \Longleftrightarrow t = 2$ or 3
Particle at rest when $t = 2$ and when $t = 3$.
When $t = 2 \Rightarrow x = (2.2^3 - 15.2^2 + 36.2) = 28$
When $t = 3 \Rightarrow x = (2.3^3 - 15.3^2 + 36.3) = 27$

Particle moves at constant speed when acceleration is zero

$\Rightarrow a = \dfrac{d^2x}{dt^2} = 12t - 30$ and this is zero when $t = 2\tfrac{1}{2}$

ESSENTIAL PRINCIPLES

RATES OF CHANGE

Worked example

Given that $y = 2\cos 2\theta$ and θ increases steadily at $2{\cdot}5$ radians per second find the rate at which y is increasing when $\theta = \dfrac{7\pi}{12}$

In examples such as this it is often advisable to write down in mathematical terms: i) the information *given* in the question, ii) the information *required*. Once this is done it is usually quite easy to see a *connection* between the two.

Given: $y = 2\cos 2\theta, \dfrac{d\theta}{dt} = 2{\cdot}5$

> ❝ Note the information given, that required, and any connections. ❞

Required: $\dfrac{dy}{dt}$ when $\theta = \dfrac{7\pi}{12}$

Connection: chain rule $\Rightarrow \dfrac{dy}{dt} = \dfrac{dy}{d\theta} \cdot \dfrac{d\theta}{dt}$

$y = 2\cos 2\theta \Rightarrow \dfrac{dy}{d\theta} = -4\sin 2\theta$

$\Rightarrow \dfrac{dy}{dt} = -4\sin 2\theta . 2{\cdot}5 = -10\sin 2\theta$

When $\theta = \dfrac{7\pi}{12}, \quad \dfrac{dy}{dt} = -10\sin\left(\dfrac{7\pi}{6}\right) = -10\left(-\dfrac{1}{2}\right) = 5{\cdot}0$

$\Rightarrow y$ is increasing at the rate of $5{\cdot}0$ units per second.

This example required only a simple application of the chain rule, but you must be prepared to meet slightly more complicated problems.

Worked example

A spherical raindrop is formed by condensation. In an interval of 40 seconds its volume increases at a constant rate from $0{\cdot}032\,\text{mm}^3$ to $0{\cdot}256\,\text{mm}^3$. Find the rate at which the surface area of a raindrop is increasing when its radius is $0{\cdot}5\,\text{mm}$

Given: raindrop spherical $\Rightarrow$ radius $r\,\text{mm}$, volume $V = \dfrac{4}{3}\pi r^3\,\text{mm}^3$

Surface area $A = 4\pi r^2\,\text{mm}^2$

Volume increases (at constant rate) by $(0{\cdot}256 - 0{\cdot}032)\,\text{mm}^3 = 0{\cdot}224\,\text{mm}^3$ in 40 s

$\Rightarrow \dfrac{dV}{dt} = \dfrac{0{\cdot}224}{40} = 0{\cdot}0056\,\text{mm}^3\,\text{s}^{-1}$.

Required: $\dfrac{dA}{dt}$ when $r = 0{\cdot}5\,\text{mm}$

Connection: $\dfrac{dA}{dt} = \dfrac{dA}{dV} \cdot \dfrac{dV}{dt} = \dfrac{dA}{dr} \cdot \dfrac{dr}{dV}\dfrac{dV}{dt}$ (extended chain rule)

$V = \dfrac{4\pi r^3}{3} \Rightarrow \dfrac{dV}{dr} = \dfrac{4\pi}{3} \cdot 3r^2 = 4\pi r^2$

$\dfrac{dA}{dr} = 4\pi . 2r = 8\pi r$

$\Rightarrow \dfrac{dA}{dt} = 8\pi r . \dfrac{1}{4\pi r^2} . 0{\cdot}0056 = \dfrac{0{\cdot}0112}{r}$

$\Rightarrow$ when $r = 0{\cdot}5\,\text{mm} \Rightarrow \dfrac{dA}{dt} = \dfrac{0{\cdot}0112}{0{\cdot}5}\,\text{mm}^2\,\text{s}^{-1} = 0{\cdot}0224\,\text{mm}^2\,\text{s}^{-1}$

$\Rightarrow$ surface area is increasing at the rate of $0.0224\,\text{mm}^2\,\text{s}^{-1}$

SMALL INCREASES, APPROXIMATIONS, ERRORS

Given $y = \text{f}(x)$ and δx, δy are small increases ($\delta x \equiv$ small amount of x) in x and y respectively, then $\dfrac{dy}{dx} = \lim\limits_{\delta x \to 0}\dfrac{\delta y}{\delta x}$

In other words when x is very small $\dfrac{dy}{dx} \approx \dfrac{\delta y}{\delta x} \Rightarrow \delta y \approx \dfrac{dy}{dx}\delta x$, which can be used as an approximation to estimate the change in the value of a function due to a small change δx in the independent variable x.

Worked example

The radius of a sphere is decreased by $0 \cdot 2\%$. Calculate the appropriate percentage change in the volume of the sphere.

When the radius of the sphere is r, the volume is $V = \dfrac{4}{3}\pi r^3 \, \text{cm}^3$

If r decreases by $0 \cdot 2\%$ then $\delta r \, \text{cm} = \dfrac{0 \cdot 2 \, r}{100} \, \text{cm} \Rightarrow \delta r = -\dfrac{0 \cdot 2 r}{100}$, the negative sign indicating the decrease in r

$$\frac{dV}{dr} = \frac{4\pi}{3} \cdot 3r^2 \Rightarrow \delta V \approx \frac{dV}{dr}\delta r = 4\pi r^2 \cdot \left(-\frac{0 \cdot 2 \, r}{100}\right)$$

$$\Rightarrow \delta V = -\frac{8}{1000} \cdot \pi r^3$$

$$\Rightarrow \text{percentage change in } V = \frac{\delta V}{V} \times 100 = \left(-\frac{8\pi r^3}{1000}\right)\left(\frac{3}{4\pi r^3}\right) \times 100 = -0 \cdot 6$$

$$\Rightarrow \text{the volume decreases by } 0 \cdot 6\%$$

Worked example

A container full of milk takes the form of an inverted right circular cone of height $10 \, \text{m}$ and base radius $4 \, \text{m}$. Find, in cm s^{-1}, the rate at which the milk level in the container is falling when the height of milk in the container is $5 \, \text{m}$, given that milk is flowing from the container at the rate of $100 \, \text{cm}^3 \, \text{s}^{-1}$.

Let height of milk be $h \, \text{m}$ when surface radius is $r \, \text{m}$.

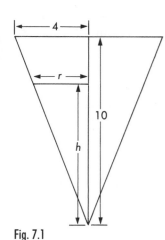

Fig. 7.1

Similar triangles $\Rightarrow \dfrac{r}{h} = \dfrac{4}{10} \Longleftrightarrow r = \dfrac{2h}{5}$

Given: $\dfrac{dV}{dt} = 100 \, \text{cm}^3 \, \text{s}^{-1}$, $V = \dfrac{\pi}{3} r^2 h = \dfrac{\pi}{3}\left(\dfrac{2h}{5}\right)^2 h = \dfrac{4\pi h^3}{75}$

$$\Rightarrow \frac{dV}{dh} = \frac{4\pi}{75} \cdot 3h^2 = \frac{4\pi h^2}{25}$$

Required: $\dfrac{dh}{dt} = \dfrac{dh}{dV} \cdot \dfrac{dV}{dt} = \dfrac{25}{4\pi h^2} \cdot \dfrac{dV}{dt}$

$$\Rightarrow \text{when } h = 5 \, \text{m} = 500 \, \text{cm and } \frac{dV}{dt} = 100 \, \text{cm}^3 \, \text{s}^{-1} \Rightarrow \frac{dh}{dt} = \frac{25}{4\pi} \cdot \frac{1}{500^2} \cdot 100 \, \text{cm s}^{-1}$$

$$\Rightarrow \frac{dh}{dt} = \frac{1}{400\pi} \, \text{cm s}^{-1}$$

$\Rightarrow$ milk level is dropping at the rate of $\dfrac{1}{400\pi} \, \text{cm s}^{-1}$

Note: you must be careful with the units and change from metres to centimetres towards the end of your calculation.

EXERCISES 7.1

1 A particle moves along the x-axis starting from the origin O at time $t = 0$. Its displacement after t seconds is S metres where $S = 15te^{-3t}$. Find the initial speed of the particle and determine, correct to 3 significant figures, the distance from O of the particle when it comes to rest and its acceleration at that point

2 The surface area of a sphere is decreasing at a rate of $0 \cdot 4 \, \text{m}^2 \text{s}^{-1}$ when the radius is $0 \cdot 5 \, \text{m}$. Calculate the rate of decrease of the volume of the sphere at this instant

3 The acceleration g due to gravity is calculated by measuring the time of oscillation t of a simple pendulum of length l. Given that $g = \dfrac{4\pi^2 l}{t^2}$ and that l can be measured accurately, but that it is likely that there may be errors of $\pm 3\%$ in the measurement of t, show that the probable error range in g is $\mp 6\%$. Explain the significance of the signs of the error ranges

4 Using differential calculus show that $\sqrt[5]{244} \approx 3\dfrac{1}{405}$

TANGENTS AND NORMALS

You should remember that:
i) the gradient of the tangent to the curve $y = f(x)$ at the point (x_1, y_1) on the curve is given by the value of $\dfrac{dy}{dx}$ when $x = x_1$, $y = y_1$

ii) two lines having gradients m_1 and m_2 are perpendicular when $m_1 m_2 = -1$
⇒ the gradient of the normal to the curve $y = f(x)$ at the point (x_1, y_1) on the curve is given by the value of $\left(\dfrac{-1}{\frac{dy}{dx}}\right)$ when $x = x_1$, $y = y_1$

iii) an equation of a straight line having gradient m and passing through the point (x_1, y_1) is $y - y_1 = m(x - x_1)$.

These results are used in most questions involving tangents and normals.

Worked example

Find equations of the tangent and normal to the curve $y = 3x^2 - 9x + 4$ at the point $(1, -2)$

$y = 3x^2 - 9x + 4 \Rightarrow \dfrac{dy}{dx} = 6x - 9$

⇒ gradient of tangent to the curve at $(1, -2)$ is $6 - 9 = -3$
⇒ an equation of the tangent to the curve at $(1, -2)$ is $y - (-2) = -3(x - 1) \Leftrightarrow y + 3x = 1$
If m is the gradient of the normal to the curve at $(1, -2)$, then $m(-3) = -1 \Rightarrow m = \frac{1}{3}$
⇒ an equation of the normal to the curve at $(1, -2)$ is $y - (-2) = \frac{1}{3}(x - 1) \Leftrightarrow 3y = x - 7$
   The procedure is the same whether the equation of the curve is given either by an implicit function of x or in parametric form.

Worked example

The coordinates of a point on a curve are given in terms of a parameter θ by $x = a\left[\ln\left(\cot\dfrac{\theta}{2}\right) - \cos\theta\right]$, $y = a\sin\theta$, where a is a positive constant and $0 < \theta < \dfrac{\pi}{2}$.
Show that $\dfrac{dy}{dx} = -\tan\theta$
The tangent at the point P of the curve meets the x-axis at T. Prove that the length of PT is a.

$x = a\left[\ln\left(\cot\dfrac{\theta}{2}\right) - \cos\theta\right] = a\left[-\ln\left(\tan\dfrac{\theta}{2}\right) - \cos\theta\right]$

$\Rightarrow \dfrac{dx}{d\theta} = a\left[-\dfrac{\frac{1}{2}\sec^2\frac{\theta}{2}}{\tan\frac{\theta}{2}} + \sin\theta\right] = -a\left[\dfrac{1}{2\sin\frac{\theta}{2}\cos\frac{\theta}{2}} - \sin\theta\right]$

$= -a\left[\dfrac{1}{\sin\theta} - \sin\theta\right] = -a\left[\dfrac{1 - \sin^2\theta}{\sin\theta}\right] = -\dfrac{a\cos^2\theta}{\sin\theta}$

$y = a\sin\theta \Rightarrow \dfrac{dy}{d\theta} = a\cos\theta$

$\Rightarrow \dfrac{dy}{dx} = \dfrac{dy}{d\theta}\cdot\dfrac{d\theta}{dx} = \left[\dfrac{a\cos\theta}{\left(-\frac{a\cos^2\theta}{\sin\theta}\right)}\right] = -\dfrac{\sin\theta}{\cos\theta} = -\tan\theta$

Hence, an equation of the tangent at P is
$y - a\sin\theta = -\tan\theta\left[x - a\ln\left(\cot\dfrac{\theta}{2}\right) + a\cos\theta\right]$

This tangent meets the x-axis at T where $y = 0$
and
$-a\sin\theta = -\tan\theta\left(x - a\ln\left(\cot\dfrac{\theta}{2}\right) + a\cos\theta\right)$

$\Rightarrow x = a\ln\left(\cot\dfrac{\theta}{2}\right)$

$\Rightarrow PT^2 = \left[a\ln\left(\cot\dfrac{\theta}{2}\right) - a\cos\theta - a\ln\left(\cot\dfrac{\theta}{2}\right)\right]^2 + \left(a\sin\theta - 0\right)^2$
$= a^2(\cos^2\theta + \sin^2\theta) = a^2 \Rightarrow PT = a$

INCREASING AND DECREASING FUNCTIONS

If the tangent to the curve with equation $y = f(x)$ at any point (x,y) on it makes an angle ψ with the positive direction of the x-axis then $\frac{dy}{dx} = \tan \psi$. If the curve is as shown in Fig. 7.2a) then as x increases y increases and $0 < \psi < \frac{\pi}{2} \Rightarrow \tan \psi > 0$, i.e. $\frac{dy}{dx}$ is positive and y is said to be an increasing function of x.

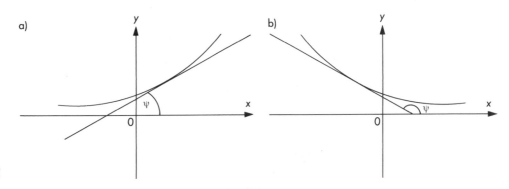

Fig. 7.2

If the curve $y = f(x)$ is as shown in Fig. 7.2b), then $\frac{\pi}{2} < \psi < \pi \Rightarrow \tan \psi < 0$,

i.e. $\frac{dy}{dx}$ is negative and y is said to be a decreasing function of x.

Hence $f(x)$ increases as x increases if $f'(x) > 0$

$f(x)$ decreases as x increases if $f'(x) < 0$

Worked example

Show that, for $x > 0$, the function $f(x) = 8x - \sin 2x + 8\cos x$ increases as x increases: hence show that $f(x) > 0$ for $x \geq 0$

Let $y = 8x - \sin 2x + 8\cos x \Rightarrow \frac{dy}{dx} = 8 - 2\cos 2x - 8\sin x$.

To show that $f(x)$ is an increasing function we need to show that $\frac{dy}{dx}$ is positive. This can be done by incorporating the negative terms in a perfect square.

Using $\cos 2x = 1 - 2\sin^2 x \Rightarrow \frac{dy}{dx} = 8 - 2(1 - 2\sin^2 x) - 8\sin x$

$$\Rightarrow \frac{dy}{dx} = 4(\sin^2 x - 2\sin x) + 6 = 4(\sin x - 1)^2 + 2$$

i.e. $\frac{dy}{dx} > 0 \Rightarrow f(x)$ increases as x increases

Further when $x = 0$, $f(x) = 8$, $\Rightarrow f(x) > 0$ for $x \geq 0$

Worked example

Show that, for $x > 0$, $x > \sin x$

Consider $f(x) = x - \sin x \Rightarrow f'(x) = 1 - \cos x$

$\Rightarrow f'(x) > 0$ for all $x > 0$, except when $x = 2r\pi$, $r\in\mathbb{Z}^+$, and then $f'(x) = 0$.

Hence $f(x)$ is an increasing function for $x > 0$ except at the isolated values $x = 2r\pi$, $r\in\mathbb{Z}^+$, and at these points $f(x) = 2r\pi \Rightarrow f(x) > 0$

Hence, $f(x) = x - \sin x > 0$ for $x > 0$

$\Rightarrow \quad x > \sin x$ for $x > 0$

STATIONARY POINTS

Consider the curve $y = f(x)$ shown in Fig. 7.3 overleaf.

Points such as A, B and C at which a tangent to the curve $y = f(x)$ is parallel to the x-axis are called **stationary points** and at these points $\frac{dy}{dx} = 0$. At A the value of y is greater than the value of y at any point in the immediate vicinity of $A \Rightarrow$ there is a *local maximum* at A. Similarly there is a *local minimum* at B. At C, $\frac{dy}{dx} = 0$, but the value of y is neither greater than, nor less than the value of y in the immediate vicinity of C. We refer to C as a *point of inflexion with zero slope*.

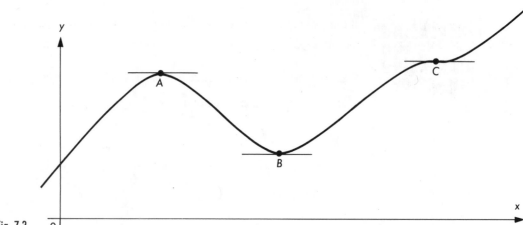

Fig. 7.3

SUMMARISING

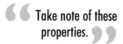
Take note of these
properties.

■ **Stationary points:** $\dfrac{dy}{dx} = f'(x) = 0$

■ **Local maximum:** $\dfrac{dy}{dx} = f'(x) = 0$, and $\dfrac{dy}{dx} = f'(x)$ changes from +ve to −ve as x increases through the point, (or $\dfrac{d^2y}{dx^2} = -ve$)

■ **Local minimum:** $\dfrac{dy}{dx} = f'(x) = 0$ and $\dfrac{dy}{dx} = f'(x)$ changes from −ve to +ve as x increases through the point, (or $\dfrac{d^2y}{dx^2} = +ve$)

■ **Inflexion with horizontal tangent:** $\dfrac{dy}{dx} = f'(x) = 0$, but $\dfrac{dy}{dx} = f'(x)$ does not change sign as x increases through the point.
(If the second derivatives are zero then further investigation is needed).

Worked example

Find the stationary points of the function $f(x) = 3x^4 - 4x^3 - 12x^2 + 10$ and determine the nature of each. Sketch the curve with equation
$y = 3x^4 - 4x^3 - 12x^2 + 10$
$\dfrac{dy}{dx} = 12x^3 - 12x^2 - 24x = 12x(x^2 - x - 2) = 12x(x + 1)(x - 2)$
For stationary point $\dfrac{dy}{dx} = 0 \Rightarrow 12x(x + 1)(x - 2) = 0 \Leftrightarrow x = 0, -1, 2$

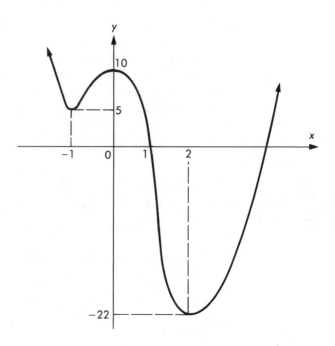

Fig. 7.4

Tests: as the differentiation is simple consider the sign of the second derivative

$$\Rightarrow \frac{d^2y}{dx^2} = 12(3x^2 - 2x - 2)$$

When $x = 0$, $\frac{d^2y}{dx^2} = -24 < 0 \Rightarrow$ local maximum when $x = 0$, $y = 10$

When $x = -1$, $\frac{d^2y}{dx^2} = 36 > 0 \Rightarrow$ local minimum when $x = -1$, $y = 5$

When $x = 2$, $\frac{d^2y}{dx^2} = 72 > 0 \Rightarrow$ local minimum when $x = 2$, $y = -22$

To sketch the curve, first mark in the local maximum and minimum points. Join them by a smooth curve and finally show what happens when $x \to \pm \infty$ (see Fig. 7.4)

Worked example

Show that the function $f(x) = \frac{1}{x}\ln x$, $x \in \mathbb{R}^+$ has a maximum value $\frac{1}{e}$

$$f(x) = \frac{1}{x}\ln x \Rightarrow f'(x) = \frac{1}{x}\left(\frac{1}{x}\right) + \left(-\frac{1}{x^2}\right)\ln x = \frac{1}{x^2}(1 - \ln x)$$

For a stationary point $f'(x) = 0 \to \frac{1}{x^2}(1 - \ln x) = 0 \Rightarrow \ln x = 1 \Leftrightarrow x = e$

When $x = e$ $f(e) = \frac{1}{e}\ln e = \frac{1}{e}$

Test: whilst it would not be difficult to find $f''(x)$ it is easier to test the sign of the first derivative.

$$f'(x) = \frac{1}{x^2}(1 - \ln x) \Rightarrow \text{if } x < e, \; f'(x) > 0$$
$$\text{if } x > e, \; f'(x) < 0$$

$\Rightarrow f'(x)$ changes sign from +ve to −ve as x passes through e

$\Leftrightarrow$ function f has a maximum value of $\frac{1}{e}$

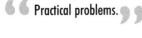

Practical problems.

The method used to maximise or minimise a function can be applied to solve practical problems. Normally, such questions involve the formation of the function to be maximised or minimised. If this is found to be a function of more than one variable, then it is necessary to eliminate all but one of the variables by using all of the information given in the problem. Once this has been done the procedure is the same.

Worked example

A rectangular garden consists of a rectangular lawn, of area $72\,\text{m}^2$ surrounded by a concrete path. The path is 2 m wide at two opposite edges of the garden and 1 m wide along each of the other two edges. Find the dimensions of the garden of smallest area satisfying these requirements.

Let the dimensions be as shown in Fig. 7.5.

Area $A\,\text{m}^2$ of garden is given by $A = (x + 2)(y + 4)$

But area of lawn $= 72\,\text{m}^2 \Rightarrow 72 = xy$

$$\Rightarrow y = \frac{72}{x}$$

$$\Rightarrow A = (x + 2)\left(\frac{72}{x} + 4\right) = 4x + \frac{144}{x} + 80$$

$$\Rightarrow \frac{dA}{dx} = 4 - \frac{144}{x^2}$$

For a stationary value $\frac{dA}{dx} = 0 \Rightarrow 4 - \frac{144}{x^2} = 0 \Leftrightarrow x^2 = 36 \Leftrightarrow x = 6$

Test: $\frac{d^2A}{dx^2} = \frac{288}{x^3} = $ positive when $x = 6 \Rightarrow$ minimum when $x = 6$ and $y = \frac{72}{6} = 12$

Dimensions of the garden are a lawn $6\,\text{m} \times 12\,\text{m}$ within a total area of $8\,\text{m} \times 16\,\text{m} = 128\,\text{m}^2$

Worked example

A right circular cone of radius r and height h has a total surface area S and volume V. Show that $9V^2 = r^2(S^2 - 2\pi r^2 S)$

Hence, or otherwise, show that, for a fixed surface area S, the maximum volume of the cone occurs when its semi-vertical angle θ is given by

$$\tan \theta = \frac{1}{2\sqrt{2}}$$

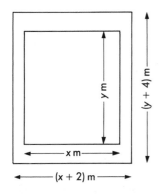

Fig. 7.5

Area of base of cone $= \pi r^2$
Area of curved surface of cone $= \pi r(r^2 + h^2)^{\frac{1}{2}}$
$\Rightarrow S = \pi r^2 + \pi r(r^2 + h^2)^{\frac{1}{2}}$
Volume of cone $V = \frac{1}{3}\pi r^2 h$

Eliminating h between S and $V \Rightarrow S = \pi r^2 + \pi r\left[r^2 + \left(\dfrac{3V}{\pi r^2}\right)^2\right]^{\frac{1}{2}}$

$\Rightarrow S = \pi r^2 + \dfrac{1}{r}(\pi^2 r^6 + 9V^2)^{\frac{1}{2}} \Rightarrow 9V^2 = r^2(S^2 - 2\pi r^2 S)$

If V is to be a maximum, then V^2 will be a maximum so test for V^2 rather than V. This eases the working considerably.

$\dfrac{\mathrm{d}}{\mathrm{d}r}(9V^2) = 2rS^2 - 8\pi r^3 S$

For a stationary value $\dfrac{\mathrm{d}(9V^2)}{\mathrm{d}r} = 0 \Rightarrow 2rS(S - 4\pi r^2) = 0$

$\Rightarrow r = 0$ or $S = 0$ or $S - 4\pi r^2 = 0$
$\Rightarrow S - 4\pi r^2 = 0$ and since $S = \pi r^2 + \pi r(r^2 + h^2)^{\frac{1}{2}}$
$4\pi r^2 = \pi r^2 + \pi r(r^2 + h^2)^{\frac{1}{2}} \Rightarrow 3r = (r^2 + h^2)^{\frac{1}{2}}$

$\qquad$ or $8r^2 = h^2 \Rightarrow \dfrac{r}{h} = \dfrac{1}{2\sqrt{2}} \Rightarrow \tan\theta = \dfrac{1}{2\sqrt{2}}$

Test: $\dfrac{\mathrm{d}^2}{\mathrm{d}r^2}(9V^2) = 2S^2 - 24\pi r^2 S = 2S(S - 12\pi r^2) < 0$ since $S = 4\pi r^2$

$\qquad \Rightarrow$ maximum volume when $\tan\theta = \dfrac{1}{2\sqrt{2}}$

GENERAL POINTS OF INFLEXION

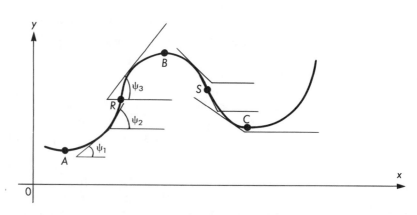

Fig. 7.6

Consider the curve with equation $y = f(x)$ as shown in Fig. 7.6

Between A and R as x increases, ψ increases, $(\psi_2 > \psi_1)$
$\quad \Rightarrow$ as x increases, $\dfrac{\mathrm{d}y}{\mathrm{d}x}$ increases $\Rightarrow \dfrac{\mathrm{d}^2 y}{\mathrm{d}x^2}$ is positive
Between R and B as x increases, ψ decreases, $(\psi_3 < \psi_2)$
$\quad \Rightarrow$ as x increases, $\dfrac{\mathrm{d}y}{\mathrm{d}x}$ decreases $\Rightarrow \dfrac{\mathrm{d}^2 y}{\mathrm{d}x^2}$ is negative

The point R where $\dfrac{\mathrm{d}y}{\mathrm{d}x}$ stops increasing and begins decreasing i.e. where $\dfrac{\mathrm{d}^2 y}{\mathrm{d}x^2} = 0$, is called a **point of inflexion**. Similarly the point S where $\dfrac{\mathrm{d}y}{\mathrm{d}x}$ stops decreasing and begins increasing is also called a point of inflexion.

> The conditions for a point of inflexion.

Thus points of inflexion are to be found where the gradient has a maximum or minimum value $\Rightarrow$ where $\dfrac{\mathrm{d}^2 y}{\mathrm{d}x^2} = 0$ and where $\dfrac{\mathrm{d}^2 y}{\mathrm{d}x^2}$ changes sign from +ve to −ve or −ve to +ve.
(Alternatively where $\dfrac{\mathrm{d}^2 y}{\mathrm{d}x^2} = 0$ and $\dfrac{\mathrm{d}^3 y}{\mathrm{d}x^3} \neq 0$).

Worked example

Find the coordinates of the point of inflexion on the curve with equation
$$y = x^2 + \frac{8}{1-x}, \; x \neq 1$$

$$y = x^2 + \frac{8}{1-x} \Rightarrow \frac{dy}{dx} = 2x + \frac{8}{(1-x)^2}$$

$$\frac{d^2y}{dx^2} = 2 + \frac{8.2}{(1-x)^3}$$

For a point of inflexion $\frac{d^2y}{dx^2} = 0 \Rightarrow 2 + \frac{16}{(1-x)^3} = 0$

$\Rightarrow (1-x)^3 + 8 = 0 \Rightarrow 1-x = -2 \Leftrightarrow x = 3$

Test: for $1 < x < 3$, $\frac{d^2y}{dx^2} < 0$, for $x > 3$, $\frac{d^2y}{dx^2} > 0$

[Alternatively $\frac{d^3y}{dx^3} = \frac{48}{(1-x)^4} \neq 0$ when $x = 3$]

$\Rightarrow$ point of inflexion when $x = 3$, $y = 5$

NEWTON–RAPHSON FORMULA, ROOTS OF AN EQUATION

The tangent to a curve can be used to obtain improved approximations to a root of an equation f(x) = 0. If the curve with equation $y = f(x)$ cuts the x-axis at A then $x = OA$ is a root of the equation f(x) = 0.

However, if $x = x_1 = OB$ is a first approximation to this root, then provided B is sufficiently close to A, the tangent to the curve at P will, in general, cut the x-axis at a point T closer to A than B. Hence $x = x_2 = OT$ is, in general, a second and better approximation to the root OA than is $x = x_1$.

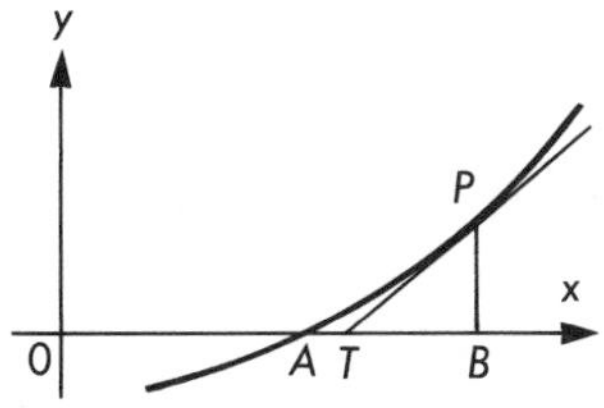

Since $PB = f(x_1)$ and $\tan PTB = f'(x_1)$ then

$$TB = \frac{PB}{\tan PTB} = \frac{f(x_1)}{f'(x_1)}$$

$$\Rightarrow x_2 = OB - TB = x_1 - \frac{f(x_1)}{f'(x_1)}$$

Similarly, repeating the process, the third approximation x_3 is given by

$$x_3 = x_2 - \frac{f(x_2)}{f'(x_2)} \qquad \Rightarrow x_n = x_{n-1} - \frac{f(x_{n-1})}{f'(x_{n-1})}$$

Fig. 7.7

the process being used until the required degree of accuracy is obtained.

Worked example

Show that $x^3 + 2x - 4 = 0$ has a root in the vicinity of $x = 1 \cdot 2$. Use the Newton–Raphson method to obtain this root correct to *five* decimal places.

In order to show that the equation $x^3 + 2x - 4 = 0$ has a root near $x = 1 \cdot 2$ it is necessary to show that the curve $y = x^3 + 2x - 4$ crosses the x-axis near $x = 1 \cdot 2$, i.e. if $f(x) = x^3 + 2x - 4$, then f(x) is continuous and f(x) must be shown to change sign as x takes values on either side of $x = 1 \cdot 2$

$f(1 \cdot 1) = 1 \cdot 1^3 + 2.1 \cdot 1 - 4 = -0 \cdot 469$

$f(1 \cdot 3) = 1 \cdot 3^3 + 2.1 \cdot 3 - 4 = +0 \cdot 797$

$\Rightarrow f(x) = 0$ has a root near $x = 1 \cdot 2$

Note: many students believe that to show that an equation f(x) = 0 has a root near $x = a$, it is only necessary to verify that $f(a) \approx 0$. Such an argument proves nothing: you must either sketch the curve $y = f(x)$, or show that f(x) changes sign as x passes through $x = a$

" A common error. "

$f(x) = x^3 + 2x - 4 \Rightarrow f'(x) = 3x^2 + 2$

n	x_n	$f(x) = x^3 + 2x - 4$	$f'(x) = 3x^2 + 2$	$x_{n+1} = x_n - \frac{f(x_n)}{f'(x_n)}$
1	$1 \cdot 2$	$1 \cdot 728 - 2 \cdot 4 - 4 = 0 \cdot 128$	$6 \cdot 32$	$= 1 \cdot 2 - \frac{0 \cdot 128}{6 \cdot 32} = 1 \cdot 1797 \approx 1 \cdot 18$
2	$1 \cdot 18$	$0 \cdot 0030$	$6 \cdot 1772$	$= 1 \cdot 17951$
3	$1 \cdot 179$	$-0 \cdot 00314$	$6 \cdot 17012$	$= 1 \cdot 17951$

$\Rightarrow$ root $= 1 \cdot 17951$ (to 5 decimal places)

EXERCISES 7.2

1 Find an equation of the tangent to the curve with equation
   $y^3 + y^2 - x^4 = 1$ at the point $(1,1)$

2 The parametric equations of a curve are $x = 2t + 3t^2$, $y = 1 - 2t^2$. Find an equation of the tangent to the curve which is parallel to the line $2y + x = 0$

3   Find an equation of the normal to the curve with equation $y = x \sin 2x$ at the point where $x = \dfrac{\pi}{3}$

4   Find the stationary points and the point of inflexion on the curve with equation $y = 2x^3 - 5x^2 + 4x - 1$. Sketch the curve

5   Find the stationary points and points of inflexion on the curve with equation $y = \dfrac{x^2}{1 + x^2}$. Sketch the curve.

6   Given that $y = 6x^2 + x - 12$ find a) the values of x for which $y = 0$, b) the minimum value of y. Sketch the curves with equations c) $y = 6x^2 + x - 12$, d) $y = \dfrac{1}{6x^2 + x - 12}$

Deduce that there are four values of x for which $(6x^2 + x - 12)^2 = 1$ and find these values, each correct to two decimal places

7   An open rectangular tank is manufactured so as to have width 4 m and volume 50 m³. Find the length and depth of the tank given that the minimum area of sheet metal is used

8   Find the maximum volume of a cone whose slant height is 3 m

9   Show that the equation $x^3 + x^2 + 5x - 1 = 0$ has a root near $x = 0.2$ and find its value correct to three decimal places

10   By sketching the graphs $y = e^{-x}$, $y = \ln(1 + x) - 1$, $x > -1$ show that the equation $\ln(1 + x) = e^{-x} + 1$ has a root near $x = 2$. Find the value of the root correct to one decimal place

EXAMINATION TYPE QUESTIONS WITH MARKING SCHEME AND GUIDED SOLUTION

1   Sketch the graphs $y = e^x - 3$, $y = \ln(x + 3)$ to show that the equation $e^x - \ln(x + 3) = 3$ has two real roots. Use the Newton–Raphson method once to improve the approximation to the positive root, taking $x = 1.5$ as the initial value of x *(12 marks)*

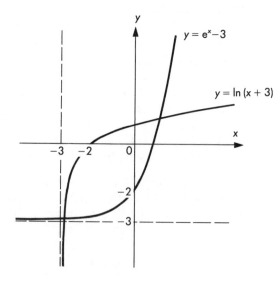

Sketches
Shapes 1, 1
Asymptotes 1, 1
Intersection on axes 1, 1

Fig. 7.8

$$f(x) = e^x - \ln(x + 3) - 3 = 0 \Rightarrow f'(x) = e^x - \frac{1}{x + 3}$$ B1   B1

$$f(1.5) = e^{1.5} - \ln 4.5 - 3 = -0.022$$ B1

$$f'(1.5) = e^{1.5} - \frac{1}{4.5} = 4.259$$ B1

$$x_2 = 1.5 - \frac{(-0.022)}{4.259} = 1.5 + 0.0052$$ M1

$$= 1.505$$ A1

2   A right circular cone of height h is inscribed in a sphere of radius R. Show that the volume V of the cone is given by $V = \dfrac{\pi}{2}(2Rh^2 - h^3)$. Hence find the maximum value of V *(12 marks)*

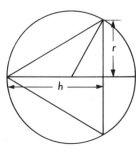

Fig. 7.9

Let the cone have base radius r and height h

$\Rightarrow R^2 = r^2 + (h-R)^2$ M1

$\Rightarrow r^2 = 2Rh - h^2$ A1

Volume of the cone $= V = \dfrac{1}{3}\pi r^2 h$ M1

$\Rightarrow V = \dfrac{1}{3}\pi(2Rh - h^2) \cdot h = \dfrac{\pi}{3}(2Rh^2 - h^3)$ A1

$\dfrac{\mathrm{d}V}{\mathrm{d}h} = \dfrac{\pi}{3}(2R \cdot 2h - 3h^2) = \dfrac{\pi}{3}h(4R - 3h)$ M1 A1

For maximum value $\dfrac{\mathrm{d}V}{\mathrm{d}h} = 0 \Rightarrow h(4R - 3h) = 0 \Rightarrow h = 0$ or $\dfrac{4R}{3}$ A1 M1

But $h \neq 0 \Rightarrow h = \dfrac{4R}{3}$

Test: $\dfrac{\mathrm{d}^2V}{\mathrm{d}h^2} = \dfrac{\pi}{3}[4R - 6h]$ and when $h = \dfrac{4R}{3}$, $\dfrac{\mathrm{d}^2V}{\mathrm{d}h^2} = \dfrac{\pi}{3}[4R - 8R] < 0$

$\Rightarrow$ maximum   M1 A1

Dimensions: $h = \dfrac{4R}{3}$, $r^2 = 2Rh - h^2 = \dfrac{8R^2}{3} - \dfrac{16R^2}{9} = \dfrac{8R^2}{9}$,

$r = \dfrac{2\sqrt{2}.R}{3}$

$\Rightarrow h = \dfrac{4R}{3}$, $r = \dfrac{2\sqrt{2}.R}{3}$ M1 A1

EXAMINATION QUESTIONS

1 i) Differentiate with respect to x

   a) $e^{3x} \sin(\pi x)$,

   b) $\ln\left(\dfrac{x^2 + 1}{\sqrt{x}}\right)$.

  ii) A coloured liquid is poured on to a large flat cloth and forms a circular stain, the area of which grows at a steady rate of $3\,\mathrm{cm^2\,s^{-1}}$. Calculate, in terms of π,

   a) the radius, in cm, of the stain $6\,\mathrm{s}$ after the stain commences,

   b) the rate, in $\mathrm{cm\,s^{-1}}$, of increase of the radius of the stain at this instant.   (L)

2 Sketch the curve $y = (2x - 1)^2(x + 1)$, showing the coordinates of

   a) the points where it meets the axes,

   b) the turning points,

   c) the point of inflexion

  By using your sketch, or otherwise, sketch the graph of

   $$y = \dfrac{1}{(2x - 1)^2(x + 1)}$$

  Show clearly the coordinates of any turning points.   (L)

3 The diagram shows a house at A, a school at D and a straight canal BC, where $ABCD$ is a rectangle with $AB = 2\,\mathrm{km}$ and $BC = 6\,\mathrm{km}$.

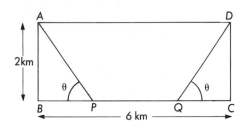

During the winter, when the canal freezes over, a boy travels from A to D by walking to a point P on the canal, skating along the canal to a point Q and then walking from Q

to D. The points P and Q being chosen so that the angles APB and DQC are both equal to θ.

Given that the boy walks at a constant speed of $4\,\text{km}\,\text{h}^{-1}$ and skates at a constant speed of $8\,\text{km}\,\text{h}^{-1}$, show that the time, T minutes, taken for the boy to go from A to D along this route is given by

$$T = 15\left(3 + \frac{4}{\sin\theta} - \frac{2}{\tan\theta}\right).$$

Show that, as θ varies, the minimum time for the journey is approximately 97 minutes.

(L)

4   The tangent to the curve $y = \dfrac{3\tan x}{1 + \sin x}$ at the point on the curve where $x = \pi/6$ cuts the x-axis at the point T. Prove that T is situated at a distance $\dfrac{1}{6}(2\sqrt{3} - \pi)$ from the origin O.

(AEB 1987)

5   A sector S of a circle, of radius R, whose angle at the centre of the circle is ϕ radians, is rolled up to form the curved surface of a right cone standing on a circular base. The semi-vertical angle of this cone is θ radians. Express ϕ in terms of $\sin\theta$ and show that the volume V of the cone is given by
$$3V = \pi R^3 \sin^2\theta \cos\theta$$
If R is constant and θ varies, find the positive value of $\tan\theta$ for which $\dfrac{dV}{d\theta} = 0$. Show further that when this value of $\tan\theta$ is taken, the maximum value of V is obtained.

Hence show that the maximum value of V is $\dfrac{2\pi R^3\sqrt{3}}{27}$ and find, in terms of R, the area of the sector S in this case.

(AEB 1988)

6   a) Find the coordinates of the turning points on the curve with equation
$$y = \frac{x^2}{1 + x^4}$$
Sketch the curve.

   b) A curve has parametric equations $x = 5a\sec\theta$, $y = 3a\tan\theta$, where $-\frac{1}{2}\pi < \theta < \frac{1}{2}\pi$ and a is a positive constant. Find the coordinates of the point on the curve at which the normal is parallel to the line $y = x$.

(C)

7   a) The points A, B have coordinates $(h, 0)$, (h, R) respectively and O is the origin. The triangular region OAB is rotated through four right angles about the x-axis. Show, by integration, that the volume of the right circular cone so formed is $\frac{1}{3}\pi R^2 h$.

   b) A right circular cone of height $a + x$, where $-a < x < a$, is inscribed in a sphere of fixed radius a, so that the vertex and all points of the circumference of the base are on the surface of the sphere. Show that the volume V of the cone is given by
$$V = \tfrac{1}{3}\pi(a - x)(a + x)^2$$

   In either order,
   i) find the maximum value of V as x varies
   ii) sketch the graph of V against x for $-a < x < a$.

(C)

8   A straight metal bar, of square cross-section, is expanding due to heating. After t seconds the bar has dimensions x cm by x cm by $10x$ cm. Given that the area of the cross-section is increasing at $0 \cdot 024\,\text{cm}^2\,\text{s}^{-1}$ when $x = 6$, find the rate of increase of the side of the cross-section at this instant. Find also the rate of increase of the volume when $x = 6$.

(JMB)

9   a) Given that $y = 3^{-x}$, show that
$$\frac{dy}{dx} = -3^{-x}\ln 3.$$

   b) On the same diagram sketch the curves
$$y = 4 - x^2 \text{ and } y = 3^{-x}.$$

   c) Verify that the curves intersect at the point $A\,(-1, 3)$. The curves also intersect at

the point B in the first quadrant whose x-coordinate α is the positive root of the equation
$$x^2 + 3^{-x} - 4 = 0.$$

d) Verify, by calculation, that $1 < \alpha < 2$.

e) By taking 2 as a first approximation to α, use the Newton–Raphson procedure once to find a second approximation, giving your answer to two decimal places.

(L)

10 Given that $f(x) \equiv 3 + 4x - x^4$, show that the equation $f(x) = 0$ has a root $x = a$, where a is in the interval $1 \leqslant a \leqslant 2$.

It may be assumed that if x_n is an approximation to a, then a better approximation is given by x_{n+1}, where
$$x_{n+1} = (3 + 4x_n)^{\frac{1}{4}}.$$
Starting with $x_0 = 1 \cdot 75$, use this result twice to obtain the value of a to 2 decimal places.

(L)

11 i) Differentiate 2^x with respect to x.

ii) Draw the graph of $y = 2^x$, for $-3 \leqslant x \leqslant 3$, by plotting those points whose x values are integral.
   Use your graph to estimate the roots of the equation
   $$x = \log_2 (x + 3).$$

iii) Taking your positive estimate, use Newton's method of successive approximations (applied twice) to obtain the positive root to three places of decimals. (NI)

12 A particle P moves in a straight line so that its displacement s metres, at time t seconds, from a point O on the line is given by
$$s = 2 + e^t \cos t.$$
Find $\dfrac{ds}{dt}$ and calculate the speed of P, in $\mathrm{m\,s}^{-1}$ to 2 significant figures, when $t = 0 \cdot 5$.

(L)

13 A particle moves along the x-axis such that at time t seconds, its displacement x metres from O is given by
$$x = e^{-t} \cos t, \quad t \geqslant 0.$$

a) Find $\dfrac{dx}{dt}$, and hence determine the initial speed, in $\mathrm{m\,s}^{-1}$, of the particle.

b) Find the values of t, in the interval $0 \leqslant t \leqslant 2\pi$, when the particle is instantaneously at rest.

c) Obtain an expression for the acceleration of the particle at time t seconds.

d) Verify that $\quad \dfrac{d^2x}{dt^2} + 2\dfrac{dx}{dt} + 2x = 0.$

(L)

14 A particle moves along a straight line in such a way that at time t seconds the particle has a velocity $v \,\mathrm{ms}^{-1}$, where
$$v = \frac{40 \cos t}{5 - 3 \sin t}, \quad 0 \leqslant t \leqslant \pi.$$
Show that the acceleration is zero when $\sin t = \dfrac{3}{5}$ and find, to two decimal places, the two values of t for which this occurs. Find the exact values of the velocity of the particle for these values of t. Sketch the graph of v.
On a separate diagram, sketch the graph of the speed of the particle. Calculate the exact value of the total distance the particle travels between $t = 0$ and $t = \pi$.   (JMB)

15 A particle moves along the x-axis. At time t seconds its displacement from the origin O is x metres, where
$$x = 2 \sin^2 t - 2 \sin t + 1.$$
Find, in terms of π, the four values of t in the interval $0 \leqslant t \leqslant 2\pi$ at which the particle is stationary. Calculate the exact values of the displacement and of the acceleration for these values of t.
Find also the total distance travelled by the particle during the interval $0 \leqslant t \leqslant 2\pi$.

(JMB)

ANSWERS TO EXERCISES

EXERCISES 7.1

1    $15\,\mathrm{m\,s}^{-1}$, $1{\cdot}84\,\mathrm{m}$, $16{\cdot}6\,\mathrm{m\,s}^{-2}$
2    $0{\cdot}1\,\mathrm{m}^3\,\mathrm{s}^{-1}$

EXERCISES 7.2

1    $4x - 5y + 1 = 0$
2    $2y + x = 3$
3    $y - \dfrac{\pi}{2\sqrt{3}} = \left(\dfrac{6}{2\pi - 3\sqrt{3}}\right)\left(x - \dfrac{\pi}{3}\right)$
4   Max $\left(\dfrac{2}{3}, \dfrac{1}{27}\right)$, Min $(1,0)$, Inflexion $\left(\dfrac{5}{6}, \dfrac{1}{54}\right)$
5   Min $(0, 0)$, Inflexions $\left(\pm\dfrac{1}{\sqrt{3}}, \dfrac{1}{4}\right)$
6    $-\dfrac{3}{2}, \dfrac{4}{3}$; $-11\dfrac{7}{8}$; $1{\cdot}56, -1{\cdot}44, 1{\cdot}27, 1{\cdot}39$
7    $5, 2\frac{1}{2}$
8    $2\pi\sqrt{3}$
9    $0{\cdot}191$
10    $2{\cdot}1$

ANSWERS TO EXAMINATION QUESTIONS

1   i) a) $\mathrm{e}^{3x}(3\sin \pi x + \pi\cos \pi x)$; b) $\dfrac{2x}{x^2 + 1} - \dfrac{1}{2x}$;
     ii) $\dfrac{3\sqrt{2}}{10}, \dfrac{1}{\sqrt{8\pi}}$
6   a) $(0,0)$, $\left(1, \dfrac{1}{2}\right)$, $\left(-1, \dfrac{1}{2}\right)$;
     b) $\left(4a, -\dfrac{9a}{4}\right)$
7    $\dfrac{32}{81}\pi a^3$
8    $0{\cdot}002\,\mathrm{cm\,s}^{-1}, 2{\cdot}16$
9   e) $1{\cdot}97$
10    $1{\cdot}76$
11   i) $2^x \ln 2$; ii) $-2{\cdot}82, 2{\cdot}45$; iii) $2{\cdot}445$
12    $\mathrm{e}^t(\cos t - \sin t), 0{\cdot}66$
13   a) $-\mathrm{e}^{-t}(\sin t + \cos t), -1$; b) $t = \dfrac{3\pi}{4}, \dfrac{7\pi}{4}$;
     c) $2\mathrm{e}^{-t}\sin t$
14    $0{\cdot}64, 2{\cdot}50, \pm 10, \dfrac{80}{3}\ln\left(\dfrac{5}{2}\right)$
15    $\dfrac{\pi}{6}, \dfrac{\pi}{2}, \dfrac{5\pi}{6}, \dfrac{3\pi}{2}$; $\dfrac{1}{2}, 1, \dfrac{1}{2}, 5$; $3, -2, 3, -6$; 10

ELEMENTARY COORDINATE GEOMETRY

GETTING STARTED

For all GCE A-level syllabuses you are expected to be familiar with the various forms for the equation of a straight line, a circle, and the simple forms for the parabola, ellipse and hyperbola. You must be able to find the equations of tangents and normals to curves and to deal with simple loci problems. The formulae you should understand and be able to use are:

Straight line

$ax + by + c = 0$ general equation of straight line

$y = mx + c$ straight line, gradient m, intercept on y-axis $(0,c)$

$\frac{x}{a} + \frac{y}{b} = 1$ straight line, intercept on x-axis $(a,0)$, y-axis $(0,b)$

$y - y_1 = m(x - x_1)$ straight line, gradient m, passes through point (x_1, y_1)

$\frac{y - y_1}{y_2 - y_1} = \frac{x - x_1}{x_2 - x_1}$ straight line, passes through the points (x_1, y_1), (x_2, y_2)

Two lines are parallel if their gradients m_1 and m_2 are equal $\Rightarrow m_1 = m_2$
Two lines are perpendicular if their gradients m_1 and m_2 are such that $m_1 m_2 = -1$.

The perpendicular distance of the point (x_1, y_1) from the straight line $ax + by + c = 0$ is $\frac{|ax_1 + by_1 + c|}{\sqrt{(a^2 + b^2)}}$

Circle

The general equation of the circle is $x^2 + y^2 + 2gx + 2fy + c = 0$, a second degree equation in which there is no term in xy and coefficient of x^2 = coefficient of y^2

$x^2 + y^2 = r^2$, circle, centre origin, radius r.
$(x - a)^2 + (y - b)^2 = r^2$, circle, centre (a,b), radius r.
$x^2 + y^2 + 2gx + 2fy + c = 0 \Rightarrow (x + g)^2 + (y + f)^2 = g^2 + f^2 - c$

i.e. circle centre $(-g, -f)$, radius $\sqrt{(g^2 + f^2 - c)}$.
$(x - x_1)(x - x_2) + (y - y_1)(y - y_2) = 0$, circle having (x_1, y_1) and (x_2, y_2) as coordinates of the extremities of a diameter.

■ Parabola

$y^2 = 4ax$, parametric equations, $x = at^2$, $y = 2at$

■ Ellipse

$\frac{x^2}{a^2} + \frac{y^2}{b^2} = 1$, parametric equations, $x = a\cos\theta$, $y = b\sin\theta$

■ Hyperbola

$\frac{x^2}{a^2} - \frac{y^2}{b^2} = 1$, parametric equations, $x = a\sec\theta$, $y = b\tan\theta$

■ Rectangular hyperbola

$xy = c^2$, parametric equations, $x = ct$, $y = \frac{c}{t}$

ESSENTIAL PRINCIPLES

SOLUTIONS TO STRAIGHT LINE AND CIRCLE QUESTIONS

In solving questions on the straight line and circle it is usually advisable to draw a reasonably accurate diagram. Diagrams often indicate the way forward. For the parabola, ellipse and hyperbola a diagram is not usually so important, but it is still advisable to draw one.

❝ An accurate diagram can help in finding solutions. ❞

Worked example

A variable line passes through the fixed point $(6,3)$ and meets the x-axis at A and the y-axis at B. Find the equation of the locus of N, the mid-point of AB

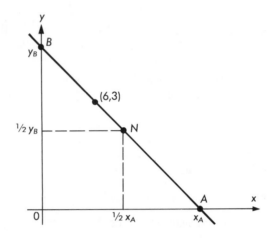

Fig. 8.1

Equation of AB is $y - 3 = m(x - 6)$ where m is the gradient

At A, $y = 0 \Rightarrow x_A = \dfrac{1}{m}(6m - 3)$

At B, $x = 0 \Rightarrow y_B = 3 - 6m$

Let the coordinates of the mid-point of AB be (X, Y)

$\Rightarrow X = \dfrac{1}{2m}(6m - 3)$ and $Y = \dfrac{1}{2}(3 - 6m)$

The gradient m of the line is variable and, in order to obtain the locus of N, it is necessary to eliminate m between the two equations for X and Y

$Y = \dfrac{1}{2}(3 - 6m) \Rightarrow m = \dfrac{1}{6}(3 - 2Y)$

Substituting in $X = \dfrac{1}{2m}(6m - 3) \Rightarrow X = \dfrac{(3 - 2Y - 3)}{\left(\dfrac{3 - 2Y}{3}\right)}$

$\Rightarrow X = \dfrac{6Y}{2Y - 3}$

$\Rightarrow$ locus of mid-point of AB is the curve with equation $x(2y - 3) = 6y$

Worked example

Prove that the points $A(-3, 4)$; $B(1, -4)$; and $C(3, 7)$ in Fig. 8.2 are the vertices of a right-angled triangle. Find the equation of the line passing through the mid-point of the hypotenuse and perpendicular to the hypotenuse

$\left.\begin{array}{l} AB^2 = [-3 - 1]^2 + [4 - (-4)]^2 = 16 + 64 = 80 \\ BC^2 = [1 - 3]^2 + [-4 - 7]^2 = 4 + 121 = 125 \\ CA^2 = [3 - (-3)^2)] + [7 - 4]^2 = 36 + 9 = 45 \end{array}\right\} \Rightarrow BC^2 = AB^2 + CA^2$

$\Rightarrow \triangle ABC$ is right-angled at A

Mid-point of $BC = \left[\dfrac{1}{2}(3 + 1), \dfrac{1}{2}(7 - 4)\right] = \left(2, \dfrac{3}{2}\right)$

Gradient of $BC = m_1 = \dfrac{7 - (-4)}{3 - 1} = \dfrac{11}{2}$

$\Rightarrow$ gradient of line perpendicular to BC is m_2 where $\dfrac{11}{2} \cdot m_2 = -1$

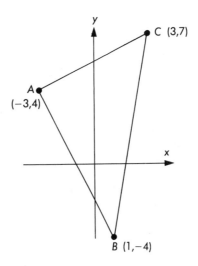

Fig. 8.2

$$\Rightarrow m_2 = -\frac{2}{11}$$

$\Rightarrow$ equation of straight line passing through the mid-point of BC and perpendicular to BC is

$$\left(y - \frac{3}{2}\right) = -\frac{2}{11}(x - 2)$$

$$\Rightarrow 22y + 4x - 41 = 0$$

Worked example

P and Q are the points (x_1, y_1) and (x_2, y_2) respectively. Show that the equation of the circle on PQ as diameter has equation $(x - x_1)(x - x_2) + (y - y_1)(y - y_2) = 0$

Find the centre D of circle C with equation $x^2 + y^2 - 8x - 4y + 10 = 0$. The tangents from the origin O touch circle C at points A and B. Find the equation of circle OAB and determine an equation of the tangent to this circle at the point, other than the origin, where it cuts the y-axis

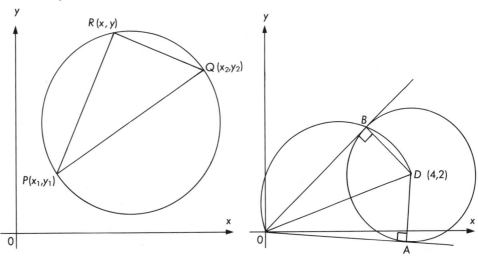

Fig. 8.3

Let $R(x, y)$ be any point on the circle having PQ as diameter

Gradient of $PR = \dfrac{y - y_1}{x - x_1} \equiv m_1$

Gradient of $QR = \dfrac{y - y_2}{x - x_2} \equiv m_2$

But $\angle PRQ = 1$ right angle, angle in semicircle, $\Rightarrow m_1 m_2 = -1$

$$\Rightarrow \frac{y - y_1}{x - x_1} \cdot \frac{y - y_2}{x - x_2} = -1 \Rightarrow (x - x_1)(x - x_2) + (y - y_1)(y - y_2) = 0$$

$x^2 + y^2 - 8x - 4y + 10 = 0 \Rightarrow (x - 4)^2 + (y - 2)^2 = 16 + 4 - 10 = 10$

Centre of circle is at $(4, 2) \equiv D$

Since OA is a tangent to the circle, $O\widehat{A}D = 1$ right angle $= O\widehat{B}D$

$\Rightarrow$ circle OAB is really a circle on OD as diameter

$\Rightarrow$ equation of circle OAB is $x(x - 4) + y(y - 2) = 0$

Circle meets y-axis where $x = 0 \Rightarrow$ i.e. $y = 2$

Equation of the circle $x^2 - 4x + y^2 - 2y = 0 \Rightarrow 2x - 4 + 2y\dfrac{dy}{dx} - 2\dfrac{dy}{dx} = 0$

$\Rightarrow \dfrac{dy}{dx} = \dfrac{2x-4}{2-2y} = 2$ at $(0,2)$

Equation of tangent is $(y-2) = 2(x-0) \Rightarrow y - 2x - 2 = 0$

Note: as the question required the bookwork for the equation of a circle on a given diameter you should expect to use this in the latter part of the question. Consequently look for a right-angle and do not waste time calculating the coordinates of A and B.

Worked example

The points $P(ap^2, 2ap)$, $Q(aq^2, 2aq)$ lie on the parabola $y^2 = 4ax$. The tangents to the parabola at P and Q intersect at R. Show that the coordinates of R are $[apq, a(p+q)]$.

Given that P and Q are variable points on the parabola such that PR is perpendicular to QR, show that this equation of the locus of R is the straight line $x + a = 0$

$y^2 = 4ax \Rightarrow 2y\dfrac{dy}{dx} = 4a \Rightarrow \dfrac{dy}{dx} = \dfrac{2a}{y} = \dfrac{2a}{2ap} = \dfrac{1}{p}$ at $(ap^2, 2ap)$

$\Rightarrow$ equation of tangent to the parabola at P has equation

$y - 2ap = \dfrac{1}{p}(x - ap^2) \Rightarrow py = x + ap^2$

Equation of tangent at $Q = qy = x + aq^2$

Tangents intersect where $(p-q)y = a(p^2 - q^2) \Rightarrow y = \dfrac{a(p^2-q^2)}{(p-q)} = a(p+q)$

and $x = p(a)(p+q) - ap^2 = apq$

$\Rightarrow R \equiv [apq, a(p+q)] \equiv (X, Y)$

$\angle PRQ = 1$ right angle $\Rightarrow \left(\dfrac{1}{p}\right)\left(\dfrac{1}{q}\right) = -1$ or $pq = -1$

$\Rightarrow X = apq = -a \Rightarrow X + a = 0$

i.e. locus of R is the straight line $x + a = 0$

CURVE SKETCHING

A glance at any GCE A-level examination paper will show that candidates are frequently asked to sketch the shape of curves given in cartesian form and in parametric form. One or two boards also require the polar form. Provided you ask yourself a few simple questions, curve sketching should not prove difficult.

CARTESIAN EQUATIONS

The questions are:
  i) is the curve symmetrical about the x-axis? (That is, does the equation contain even powers of y only?)
 ii) is the curve symmetrical about the y-axis? (That is, does the equation contain even powers of x only?)
iii) is the curve symmetrical about the origin? (That is, does the equation stay the same when x is replaced by $(-x)$ and y is replaced by $(-y)$?)

Ask yourself these questions when sketching cartesian equations.

iv) what happens to y if x is made large? what happens to x if y is made large?
 v) Are there any values of x which make y infinite? Are there any values of y which make x infinite?
vi) where does the curve cross the axes?

Finally, if the answers to the questions are not sufficient to sketch the curve, determine the maximum and minimum values of y. Use of the sign of $f'(x)$ also helps to sketch the curve $y = f(x)$

Worked example

Sketch the curve $y^2 = 4(x-1)$
  i) the curve contains even powers of y only $\Rightarrow$ symmetrical about the x-axis
 ii) not symmetrical about y-axis
iii) if $x < 1$, then $4(x-1) < 0 \Rightarrow y^2 < 0$ which is impossible. Hence curve only exists for $x \geq 1$
iv) when $x = 1$, $y = 0 \Rightarrow$ crosses x-axis at $(1,0)$
 v) when x is large, y also is large

vi) $y^2 = 4(x-1) \Rightarrow 2y\dfrac{dy}{dx} = 4 \Rightarrow \dfrac{dy}{dx} = \dfrac{2}{y} \to \infty$ as $y \to 0$

$\Rightarrow$ curve is as shown in Fig. 8.4

Worked example

Sketch the curve $y = \dfrac{3x+2}{x(x-2)}$

i) curve is not symmetrical about either axis

ii) if x is large and positive, y is very small and positive
    if x is large and negative, y is very small and negative

iii) The values $x = 0$ and $x = 2$ make y infinite. If $x = 0+$, i.e. x is positive and just greater
    than 0, then y is negative and very large
    i.e. $x \to 0+$, $y \to -\infty$
    If $x \to 0-$, $y \to +\infty$
    If $x \to 2-$, $y \to -\infty$ and if $x \to 2+$, $y \to +\infty$

iv) curve crosses the x-axis when $x = -\frac{2}{3}$
    $\Rightarrow$ curve is therefore as shown in Fig. 8.5

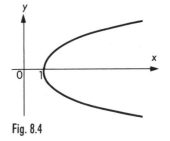

Fig. 8.4

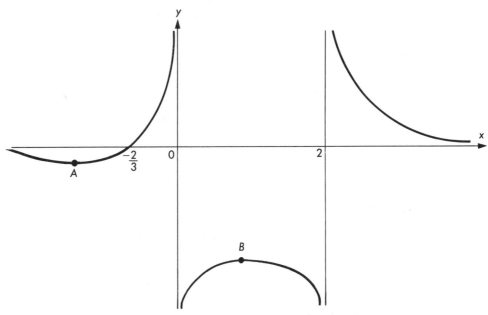

Fig. 8.5

The only question left unresolved by the previous working is the position of the turning
points A and B. These can be determined by differentiation

$y = \dfrac{3x+2}{x^2-2x} \Rightarrow \dfrac{dy}{dx} = \dfrac{(x^2-2x)3 - (3x+2)(2x-2)}{(x^2-2x)}$

$\Rightarrow \dfrac{dy}{dx} = 0 \Rightarrow 3(x^2-2x) - (3x+2)(2x-2) = 0$
$\qquad\qquad \Rightarrow 3x^2 + 4x - 4 = 0 \Leftrightarrow (3x-2)(x+2) = 0$
$\qquad \Rightarrow x = \frac{2}{3}$ or -2

When $x = \frac{2}{3}$, $y = -4\frac{1}{2} \Rightarrow B \equiv (\frac{2}{3}, -4\frac{1}{2})$

When $x = -2$, $y = -\frac{1}{2} \Rightarrow A \equiv (-2, -\frac{1}{2})$

You should note that no part of the curve exists for $-4\frac{1}{2} < y < -\frac{1}{2}$

This can also be shown as follows

$y = \dfrac{3x+2}{x(x-2)} \Rightarrow x^2 y - x(2y+3) - 2 = 0$

$x \in \mathbb{R} \Rightarrow B^2 - 4AC \geq 0 \Rightarrow (2y+3)^2 + 8y \geq 0$
$\qquad \Rightarrow 4y^2 + 20y + 9 \geq 0$
$\qquad \Rightarrow (2y+9)(2y+1) \geq 0$
$\qquad \Rightarrow y \leq -4\frac{1}{2} \cup y \geq -\frac{1}{2}$

PARAMETRIC EQUATIONS

There are no general rules for sketching curves whose equations are given in parametric
form. It is sometimes easy to eliminate the variable parameter and obtain a cartesian
equation from which the curve can be sketched. You can look for symmetry and find

$\dfrac{\mathrm{d}y}{\mathrm{d}x}$ as an aid to sketching the curve, but you may be reduced to giving the parameter

various values and plotting individual points.

Worked example

Sketch the curve whose parametric equations are $x = t^2 + 1$, $y = 2t$

$$y = 2t \Rightarrow t = \frac{y}{2} \Rightarrow x = \left(\frac{y}{2}\right)^2 + 1 \Rightarrow y^2 = 4(x-1)$$

$\Rightarrow$ the curve is the same as the first example shown earlier, i.e. Fig. 8.4.

POLAR EQUATIONS

(You should omit this section if polar coordinates are not contained within your syllabus.)

> **Check whether polar coordinates are in your syllabus.**

A point may be identified in a plane by reference to a fixed line (called the initial line) and a fixed point (called the pole) on the line. Conventionally we take the fixed line as the positive x-axis and the origin O as the pole.

A point P is then identified by the polar coordinates (r, θ), where θ is the angle through which a straight line, coincident with the initial line, must be rotated in an anticlockwise direction in order to pass through P, and r is the distance of P from O.

It is normal practice to define $r > 0$ and θ in $-\pi < \theta \leqslant \pi$

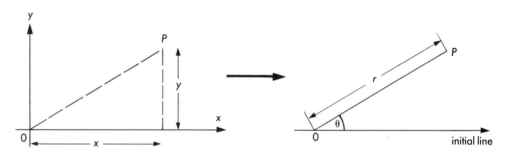

Fig. 8.6

Under these conditions the cartesian coordinates (x, y) and corresponding polar coordinates (r, θ) for a point P are related by $(x, y) \equiv (r\cos\theta, r\sin\theta)$.

When sketching a curve whose equation is given in polar form you should ask the following questions:

> **Ask yourself these questions when sketching polar equations.**

i) is the curve symmetrical about the initial line? (That is, does the equation remain the same if θ is replaced by $-\theta$?)

ii) is the curve symmetrical about a line through the pole and perpendicular to the initial line? (That is, does the equation remain the same if θ is replaced by $(\pi - \theta)$?)

iii) is the curve symmetrical about the origin? (That is, does the equation remain the same if θ is replaced by $(\pi + \theta)$?)

Finally, note the values of θ for which r has special values, i.e. for which $r = 0$, $r = $ maximum value. If the curve $r = f(\theta)$ passes through the pole, it behaves like $f(\theta) = 0$ there. The solution of this equation gives the half-lines which are tangents to the curve at the pole.

Worked example

Sketch the curve $r = 3 + 2\cos\theta$

$f(\theta) = 3 + 2\cos\theta \Rightarrow f(-\theta) = 3 + 2\cos(-\theta) = 3 + 2\cos\theta = f(\theta)$

$\Rightarrow$ curve is symmetrical about the initial line

$f(\pi - \theta) = 3 + 2\cos(\pi - \theta) = 3 - 2\cos\theta \neq f(\theta)$

$\Rightarrow$ curve is not symmetrical about the line $\theta = \dfrac{\pi}{2}$

It is therefore necessary to determine r as θ increases from 0 to π

$$\theta = 0 \Rightarrow r = 5, \quad \theta = \frac{\pi}{2} \Rightarrow r = 3, \quad \theta = \pi \Rightarrow r = 1$$

As θ increases from 0 to π, r steadily decreases from 5 to 1

$\Rightarrow$ curve $r = 3 + 2\cos\theta$ is as shown in Fig. 8.7.

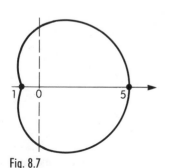

Fig. 8.7

As well as being able to sketch curves it is necessary to be able to interpret the effect of certain simple transformations, namely for $y = f(x)$ state the significance of:

i) $y = af(x)$,   ii) $y = f(x) + a$,   iii) $y = f(x - a)$,   iv) $y = f(ax)$

where a is a known constant. We shall consider a to be a positive constant.

> Learn the impact of these transformations on curves.

i) $y = af(x)$. Multiplication by a positive constant a, $a > 1$, simply increases the y value to a times the original value. The effect then is to stretch the curve out in the direction of the y-axis to give a curve of the form shown in Fig. 8.8.

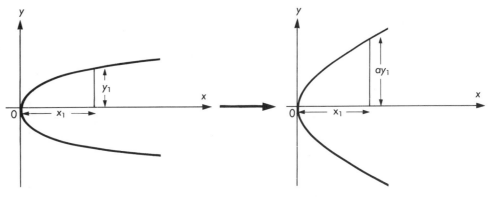

Fig. 8.8

ii) $y = f(x) + a$. This transformation does not affect the shape of the curve. It simply moves it through a distance a in the positive direction of the y-axis to give a curve of the form shown in Fig. 8.9.

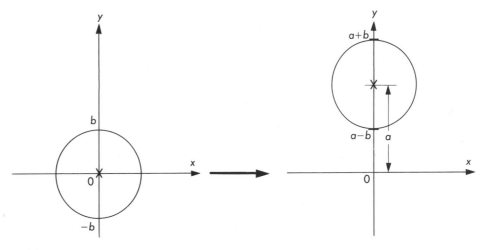

Fig. 8.9

iii) $y = f(x - a)$. This transformation again does not affect the shape of the curve. It simply moves the curve through a distance a in the positive direction of the x-axis to give a curve of the form shown in Fig. 8.10.

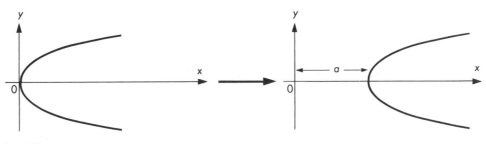

Fig. 8.10

iv) $y = f(ax)$. The effect of this transformation is to extend, (if $a < 1$), or squeeze up, (if $a > 1$), the size of the curve in the x-direction. Thus for $a < 1$, we have the curve in Fig. 8.11, and for $a > 1$, we have the curve in Fig. 8.12.

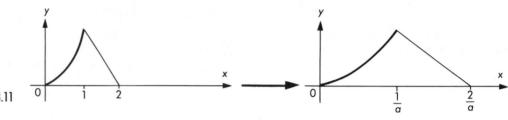

Fig. 8.11

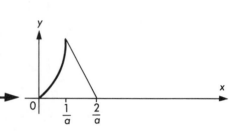

Fig. 8.12

Worked example

The curve with equation $y = f(x)$ exists for $x \in \mathbb{R}$ and that part of the curve for $0 \leqslant x \leqslant 2c$, where c is a positive constant, is shown in Fig. 8.13.

Fig. 8.13

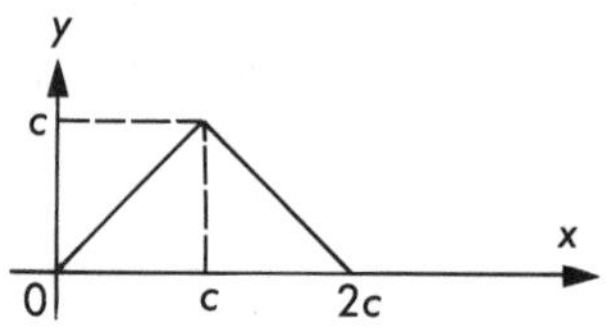

Given that the function f is an odd function of period $4c$, sketch

   i) the curve $y = f(x)$ for $-2c \leqslant x \leqslant 2c$
  ii) the curve $y = f(2x)$ for $-c \leqslant x \leqslant c$
 iii) the curve $y = f(x - c)$ for $0 \leqslant x \leqslant 2c$
 iv) the curve $y = 2f(x)$ for $0 \leqslant x \leqslant 2c$

  i) $y = f(x) \Rightarrow$ ii) $y = f(2c) \Rightarrow$

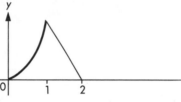

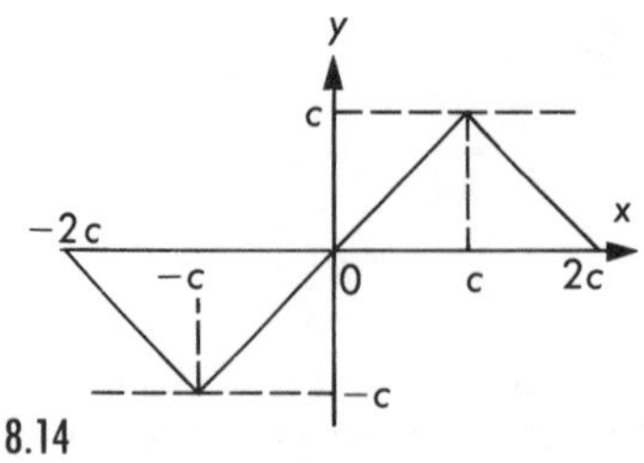

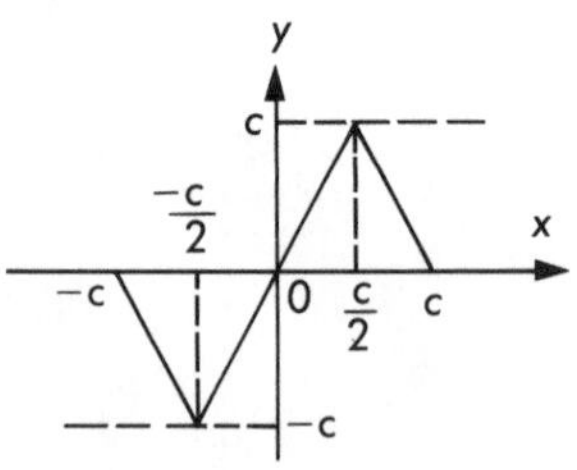

Fig. 8.14   Fig. 8.15

 iii) $y = f(x - c) \Rightarrow$ iv) $y = 2f(x) \Rightarrow$

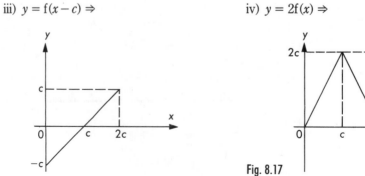

Fig. 8.16   Fig. 8.17

EXERCISES 8.1

1 The sides OA and OC of a parallelogram $OABC$ lie along the lines $3y = x$ and $y = 3x$ respectively, where O is the origin. B is the point $(4, 3)$. Find the equation of the straight line passing through B and parallel to OA. Find also the coordinates of the vertices of the parallelogram

2 Triangle ABC lies entirely in the first quadrant and has an area of $4\frac{1}{2}\,\mathrm{cm}^2$. An equation of one side of the triangle is $2x - 5y + 23 = 0$ and vertices A and B have coordinates $(1, 5)$ and $3, 4)$ respectively. Find a) the coordinates of C, b) equations of the other two sides of the triangle

3 Find an equation of the circle which passes through the points $(1, 1)$, $(1, 7)$ and $(8, 8)$. Find also the equations of the tangents to this circle which pass through the origin

4 A circle touches the straight line $y = x$ at the point $(1, 1)$. Given that the circle passes through the point $(4, 2)$, find the equation of the circle

5 Find the equation of the tangent to the circle $x^2 + y^2 = 5$ at the point $(-2, 1)$. Find also the coordinates of the points P and Q where this tangent meets the circle $x^2 + y^2 - 6x - 12y + 35 = 0$ and show that the tangents to this circle at P and Q are perpendicular to each other

6 Find the equation of the tangent to the curve $y^2 = 12x$ at the point $(3t^2, 6t)$. Prove that the foot of the perpendicular from the point $(3, 0)$ to this tangent lies on the y-axis for all values of t

7 Sketch for positive values of x, the graph of $y = x(x - 2)(x - 4)$

8 Sketch the curve with equation

$$y = \frac{x}{x+1}.$$ Hence, or otherwise, sketch the curve with equation

$$y = \left(\frac{x}{x+1}\right)^2$$

9 Sketch the curve with equation

$$y = \frac{x^2}{1 + x^2}$$

showing clearly the shape of the curve near the origin

10 Sketch the curve with parametric equations $x = t^2$, $y = t^3$. Find an equation of the tangent to the curve at the point $(t^2,\ t^3)$

11 Sketch the curve with parametric equations $x = 1 + 2t$, $y = 9 - t^2$ showing clearly the coordinates of the turning points and the points where the curve crosses the axes. Find an equation of the normal to the curve at the point where $t = 2$

12 In separate diagrams sketch the curves whose equations in polar coordinates are:

 a) $r = 2\sin\theta,\ 0 \leqslant \theta \leqslant \pi$,   b) $r = 1 + \theta,\ 0 < \theta \leqslant \pi$ c) $r = \cos 2\theta,\ -\pi < \theta \leqslant \pi$

13 Sketch the curve $y = \sin x$ for $0 \leqslant x \leqslant \pi$. Show on separate diagrams sketches of the curves with equations
 a) $y = \sin 2x$, $0 \leqslant x \leqslant \pi$
 b) $y = 3\sin x$, $0 \leqslant x \leqslant \pi$

 c) $y = \sin\left(x + \dfrac{\pi}{2}\right)$, $0 \leqslant x \leqslant \dfrac{\pi}{2}$

 d) $y = 2\sin\left(x + \dfrac{\pi}{4}\right)$, $0 \leqslant x \leqslant \dfrac{\pi}{2}$

14 The curve with equation $y = f(x)$ exists for $x \in \mathbb{R}$ and that part of the curve for $0 \leqslant x \leqslant 2T$, where T is a positive constant, is shown in Fig. 8.18.

 Given that the function f is an even function of period $4T$, sketch the curves with equation:
 a) $y = f(x)$ $0 \leqslant x \leqslant 4T$
 b) $y = 2f(x)$, $-2T \leqslant x \leqslant 2T$
 c) $y = f(x - T)$, $-T \leqslant x \leqslant 3T$
 d) $y = f(2x)$ $-T \leqslant x \leqslant T$

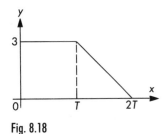

Fig. 8.18

REDUCTION OF EQUATIONS TO LINEAR FORM

The equation $y = mx + c$ is that of a straight line whose gradient is m and whose intercept on the y-axis is c. Many equations, which would normally produce curves (i.e. not straight lines) when values of y are plotted against values of x can be rearranged and new variables chosen so that a straight line graph can be obtained. This exercise is a particularly useful one when it is required to check an experimental law.

Consider the equation $y = ax + bx^2$, where a and b are constants. If a curve is sketched for given values of a and b it would be parabolic. If however the equation is rearranged in the form

$$\frac{y}{x} = a + bx$$

and a new variable Y, where

$$Y = \frac{y}{x}$$

is chosen then the equation becomes

$$Y = a + bx$$

This equation can now be compared with $y = mx + c \Rightarrow Y \equiv y,\ x \equiv x$.

Consequently, a graph of values of Y plotted against values of x yields a straight line from which can be deduced the values of a, (intercept on Y-axis) and b, (the gradient of the straight line).

Further examples of rearrangements to produce straight line graphs are:

			AXES	
GIVEN EQUATION	REARRANGED EQUATION		Y AXIS	X AXIS
$y = ax + \dfrac{b}{x}$	$xy = ax^2 + b$		$Y = xy$	$X = x^2$
$y = \dfrac{1}{ax + b}$	$\dfrac{1}{y} = ax + b$		$Y = \dfrac{1}{y}$	$X = x$
$y = ax^b$	$\lg y = \lg a + b\lg x$		$Y = \lg y$	$X = \lg x$
$y = ab^x$	$\lg y = \lg a + x\lg b$		$Y = \lg y$	$X = x$
$y = ae^{bx}$	$\ln y = \ln a + bx$		$Y = \ln y$	$X = x$
$yx^a = b$	$\lg y = \lg b - a\lg x$		$Y = \lg y$	$X = \lg x$
$y = a(x-2)^b$	$\lg y = \lg a + b\lg(x-2)$		$Y = \lg y$	$X = \lg(x-2)$

> " Useful rearrangements for producing straight line graphs. "

Worked example

The variables x and y below are believed to be related by a law of the form $y = \ln(ax^2 + bx)$ where a and b are constants. By drawing a suitable straight line graph, show that this is so and from your graph estimate values of a and b

x	1	2	3	4	5	6
y	$-1\cdot897$	$0\cdot588$	$1\cdot599$	$2\cdot262$	$2\cdot757$	$3\cdot153$

$$y = \ln(ax^2 + bx) \Rightarrow e^y = ax^2 + bx \Rightarrow \frac{e^y}{x} = ax + b$$

Letting $Y = \dfrac{e^y}{x}$, $\Rightarrow Y = ax + b$ which is of straight line form, a being the gradient and b being the intercept on the Y-axis. Hence plot

$$Y = \frac{e^y}{x} \text{ against } x$$

Intercept on Y-axis $\Rightarrow b = -0{\cdot}575 \approx -0{\cdot}58$

Gradient $= a = \dfrac{2{\cdot}25}{3} = 0{\cdot}75$

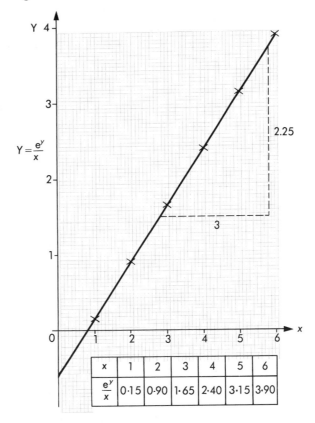

Fig. 8.19

x	1	2	3	4	5	6
$\dfrac{e^y}{x}$	0·15	0·90	1·65	2·40	3·15	3·90

$$\Rightarrow Y = \frac{e^y}{x} = 0{\cdot}75x - 0{\cdot}58$$
$$\Rightarrow y = \ln(0{\cdot}75x^2 - 0{\cdot}58x)$$

An alternative to using the intercept and gradient to find a and b would be to take the coordinates of two points through which the graph passed, substitute them in the equation and solve the simultaneous equations.

EXERCISES 8.2

1 Two variables x and y are connected by a relationship of the form $y = ax^n$ where a and n are constants. By drawing a suitable straight line graph find the values of a and n

x	1·34	3·58	7·60	12·1	14·8
y	208·0	10·9	1.14	0·283	0·155

2 The table gives corresponding values of variables x and y. It is believed that x and y satisfy a relationship of the form $y^2 = Kb^x$, where K and b are constants. Verify graphically that this is so and from your graph obtain approximate values of K and b

x	1	2	3	4	5
y	0·71	0·87	1·06	1·30	1·59

3 Corresponding values of two variables x and y are given below

x	1·0	1·5	2·0	2·5	3·0
y	10·50	5·33	3·56	2·69	2·17

It is believed that x and y are related by an equation of the form $y = \dfrac{a}{x} + \dfrac{b}{x^3}$.

Verify graphically that this is so and determine approximate values of the constants a and b

EXAMINATION TYPE QUESTIONS WITH MARKING SCHEME AND GUIDED SOLUTION

1 A circle touches the line $3x - 4y - 8 = 0$ at the point P $(4, 1)$ and passes through the point Q $(5, 3)$. Find the equation of the circle and show that it touches the y-axis   (*15 marks*)

If the circle touches the line $3x - 4y - 8 = 0$ at $(4, 1)$, the centre of the circle must lie on a line L say, perpendicular to the given line and passing through P $(4, 1)$ M1

$\Rightarrow$ gradient of given line $= \dfrac{3}{4}$ B1

$\Rightarrow$ gradient of line perpendicular to it is $-\dfrac{4}{3}$ B1

$\Rightarrow$ equation of L is $y - 1 = -\dfrac{4}{3}(x - 4) \Longleftrightarrow y = \dfrac{19}{3} - \dfrac{4x}{3}$ M1 A1

Hence if C is the centre of the circle then $C \equiv \left(x, \dfrac{19}{3} - \dfrac{4x}{3}\right)$ M1

But $CP^2 = CQ^2 \Rightarrow (x - 4)^2 + \left(\dfrac{19}{3} - \dfrac{4x}{3} - 1\right)^2 = (x - 5)^2 + \left(\dfrac{19}{3} - \dfrac{4x}{3} - 3\right)^2$ M1 A1

$\Longleftrightarrow x^2 - 8x + 16 + \dfrac{16}{9}(16 - 8x + x^2) = x^2 - 10x + 25 + \dfrac{16}{9}\left(\dfrac{25}{4} - 5x + x^2\right)$

$\Rightarrow \dfrac{10x}{3} = \dfrac{25}{3} \Longleftrightarrow x = \dfrac{5}{2}, y = \dfrac{19}{3} - \dfrac{4}{3}\left(\dfrac{5}{2}\right) = 3$ A1 A1

$\Longleftrightarrow$ Centre of circle is $\left(\dfrac{5}{2}, 3\right)$

Radius of circle is r where $r^2 = CP^2 \Rightarrow \left(4 - \dfrac{5}{2}\right)^2 + (1 - 3)^2 = \dfrac{25}{4} \Rightarrow r = \dfrac{5}{2}$ M1 A1

$\Rightarrow$ equation of circle is $\left(x - \dfrac{5}{2}\right)^2 + (y - 3)^2 = \left(\dfrac{5}{2}\right)^2$ A1

Distance between centre of circle and y-axis $= \dfrac{5}{2} = $ radius   M1

$\Rightarrow$ circle touches the y-axis   A1

2 Sketch the curve with equation $y = \dfrac{x}{2 - x}$ and state

i) the equation of each of the asymptotes
ii) an equation of the tangent at the origin

Show on a separate diagram the curve with equation $y = \left|\dfrac{x}{2 - x}\right|$ (*13 marks*)

$y = \dfrac{x}{2 - x}$

i) Curve is not symmetrical about the x- or the y-axis

ii) $y = \dfrac{x}{2 - x} = \dfrac{1}{\dfrac{2}{x} - 1} \Rightarrow$ when $x \to \pm\infty$, $y \to -1$

iii) When $x = 2+$, $y \to -\infty$, when $x = 2-$, $y \to +\infty$
iv) Curve crosses x axis at $x = 0$

sketch 3 (2,1,0)

$$y = \frac{x}{2-x} \Rightarrow \frac{dy}{dx} = \frac{(2-x)1 - x(-1)}{(2-x)^2}$$ M1

asymptotes $y = -1$ B1
$x = 2$ B1

$$= \frac{2}{(2-x)^2}$$ A1

$$= \tfrac{1}{2} \text{ at } (0,0)$$ A1

equation tangent at origin $y = \frac{1}{2}x$

M1   A1

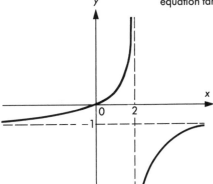

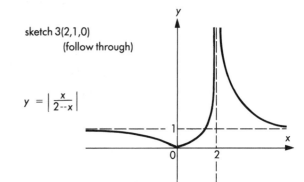

sketch 3(2,1,0)
(follow through)

$$y = \left| \frac{x}{2-x} \right|$$

Fig. 8.20

3   The variables x and y given in the table below are believed to be connected by a relationship of the form $y = Ae^{Bx}$, where A and B are constants. By drawing a suitable graph verify that this is so and determine probable values of A and B

x	0·5	1·0	1·5	2·0	2·5
y	0·663	0·732	0·810	0·895	0·990

(9 marks)

$y = Ae^{Bx} \Rightarrow \ln y = \ln(Ae^{Bx}) = \ln A + Bx$ M1
Plot $\ln y$ against x M1
$\ln y \Rightarrow -0\cdot411, \; -0\cdot312, \; -0\cdot211, \; -0\cdot111, \; -0\cdot010$ B1

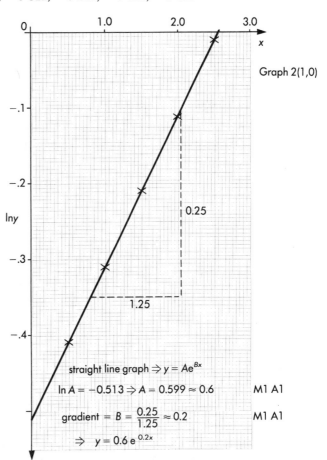

Graph 2(1,0)

straight line graph $\Rightarrow y = Ae^{Bx}$

$\ln A = -0.513 \Rightarrow A = 0.599 \approx 0.6$ M1 A1

gradient $= B = \dfrac{0.25}{1.25} \approx 0.2$ M1 A1

$\Rightarrow y = 0.6\,e^{0.2x}$

Fig. 8.21

ANSWERS TO EXERCISES

EXERCISES 8.1

1 $3y = x + 5$, $\left(\dfrac{5}{8}, \dfrac{15}{8}\right)$; $\left(\dfrac{27}{8}, \dfrac{9}{8}\right)$

2 $(6, 7)$; $2y + x - 11 = 0$; $y - x - 1 = 0$

3 $x^2 + y^2 - 10x - 8y + 16 = 0$; $x = 0$, $40y + 9x = 0$

4 $\left(x - \dfrac{7}{2}\right)^2 + \left(y + \dfrac{3}{2}\right)^2 = \dfrac{25}{2}$

5 $y = 2x + 5$, $(0, 5)$, $(2, 9)$

6 $ty = x + 3t^2$

10 $2y = 3tx - t^3$

11 $2y = x + 5$

EXERCISES 8.2

1 $n = -3$, $a \approx 500$

2 $\frac{1}{3}$, $1 \cdot 5$

3 $6 \cdot 0$, $4 \cdot 5$

CHAPTER 9

BASIC INTEGRATION

BASIC RULES OF INTEGRATION

INTEGRATION BY SUBSTITUTION

INTEGRALS INVOLVING TRIGONOMETRIC FUNCTIONS

INTEGRATION OF RATIONAL ALGEBRAIC FRACTIONS

INTEGRATION BY PARTS

DEFINITE INTEGRALS

GETTING STARTED

In Chapter 6 on Differentiation we listed the basic results for the derivatives of standard functions. We now list those results (in the reverse order) to obtain a table of basic results for the **integrals** of standard functions. As with differentiation it is essential that you know the results and can apply them without difficulty. You are therefore again advised not to become dependent upon the table of results given by the boards in their formulae booklets. The function $f(x)$ to be integrated is known as the **integrand** of the integral.

$f(x)$	$\int f(x)\mathrm{d}x$
$x^n,\ n \neq -1$	$\dfrac{x^{n+1}}{n+1}$
$\dfrac{1}{x}$	$\ln x$
e^x	e^x
$\sin x$	$-\cos x$
$\cos x$	$\sin x$
$\tan x$	$\ln \sec x$
$\sec^2 x$	$\tan x$
$\operatorname{cosec}^2 x$	$-\cot x$

$f(x)$	$\int f(x)\mathrm{d}x$
$\sec x \tan x$	$\sec x$
$\sec x$	$\ln\lvert \sec x + \tan x\rvert$
$\operatorname{cosec} x$	$\ln\lvert \tan \tfrac{1}{2}x\rvert$
$\cot x$	$\ln\lvert \sin x\rvert$
$\dfrac{1}{\sqrt{(1-x^2)}}$	$\arcsin x$
$\dfrac{1}{1+x^2}$	$\arctan x$

You should remember that the results given in the right-hand columns are not unique since, strictly speaking, each should be associated with an arbitrary constant C, say, the constant of integration, i.e.

$$\int x^n \mathrm{d}x = \frac{x^{n+1}}{n+1} + C$$

For convenience we shall drop the arbitrary constant and introduce it only when it is essential to the problem in question, but in any indefinite integral it must be implicit that a **constant of integration** is required. It is advisable to check the results of any integration by differentiating back.

ESSENTIAL PRINCIPLES

BASIC RULES OF INTEGRATION

" Learn these basic rules of integration. "

The basic rules of integration include:

1   $\int cf(x)dx = c\int f(x)dx$ where c is a constant

Worked example

$\int 7\sin x dx = 7\int \sin x dx = -7\cos x$

2   $\int [f(x) \pm g(x)]dx = \int f(x)dx \pm \int g(x)dx$

Worked example

$$\int (7x^2 - 3e^x)dx = 7\int x^2 dx - 3\int e^x dx = \frac{7}{3}x^3 - 3e^x$$

3   If $\int f(x)dx = F(x)$ then $\int f(x+a)dx = F(x+a)$, where a is any constant; i.e. the addition of a constant to the variable makes no difference to the *form* of the integral

Worked example

$$\int x^2 dx = \frac{x^3}{3} \Rightarrow \int (x+4)^2 dx = \frac{(x+4)^3}{3}$$

Worked example

$$\int \frac{1}{x}\ dx = \ln x \Rightarrow \int \frac{1}{x-3}dx = \ln(x-3)$$

You must exercise care in using this rule. It only applies when x is replaced by $(x+a)$ and does not cover the case when the expression inside the brackets is not linear, i.e. first degree in x.

Thus $\int (x^2+3)^2 dx \neq \dfrac{(x^2+3)^3}{3}$

4   If $\displaystyle\int f(x)dx = F(x)$ then $\displaystyle\int f(bx+a)dx = \frac{1}{b}\ F(bx+a)$, where a and b are constants;

i.e. if x is replaced by $bx+a$ then the form of the integral remains the same, but the answer must be divided by b, the coefficient of x.

Worked example

$$\int \sqrt{x}\ dx = \frac{2}{3}x^{\frac{3}{2}} \Rightarrow \int \sqrt{(2x-3)}dx = \frac{1}{2}\cdot\frac{2}{3}(2x-3)^{\frac{3}{2}} = \frac{1}{3}(2x-3)^{\frac{3}{2}}$$

Worked example

$$\int \cos x\ dx = \sin x \Rightarrow \int \cos 4x\ dx = \frac{1}{4}\sin 4x$$

Worked example

$$\int e^x\ dx = e^x \Rightarrow \int e^{2x}\ dx = \frac{1}{2}e^{2x}$$

Worked example

$$\int \frac{1}{x}\ dx = \ln x \Rightarrow \int \frac{1}{4-3x}dx = \frac{1}{(-3)}\ln|4-3x|$$

Worked example

$$\int \frac{1}{1+x^2}\ dx = \arctan x \Rightarrow \int \frac{1}{9+x^2}dx = \frac{1}{9}\int \frac{1}{1+\left(\dfrac{x}{3}\right)^2}dx = 3.\frac{1}{9}\arctan\left(\frac{x}{3}\right)$$

Again you are reminded that the above rule only applies when x is replaced by $bx+a$ and does not apply to integrals of the form

$$\int \frac{1}{\sqrt{(4+3x^2)}}dx$$

You are also warned to look out for the constant b being (-1). A very common mistake is to forget to divide by b when it happens to be -1.

Thus $\int \dfrac{1}{1-x}\,dx = -\ln|1-x|$, not $\ln|1-x|$.

Remember also that to each of the above worked examples should be added a constant of integration.

EXERCISES 9.1

Integrate each of the following functions with respect to the appropriate variable. Check your answers by differentiating and simplifying to obtain the given function.

1 $1 - 2x + 3x^2$

2 $\sqrt{(x+2)}$

3 $\sin 2x$

4 $4e^{1-2x}$

5 $6\sec^2 3t$

6 $\dfrac{1}{e^{4x}}$

7 $\dfrac{2}{\sqrt{(1-x)}}$

8 $\sin(x+\pi)$

9 $\tan 3t$

10 $\dfrac{1}{(1+t)^2}$

11 $\dfrac{1}{9+x^2}$

12 $\dfrac{1}{\sqrt[3]{1+4u}}$

13 $\operatorname{cosec}^2 3u$

14 $(1+e^x)^2$

15 $\tan\left(\dfrac{t}{2}\right)$

16 $(2+7x)^5$

17 $\dfrac{1}{\sqrt{(1-4x^2)}}$

18 $\dfrac{1}{\sqrt{(4-x^2)}}$

It may appear that, using only the small table of standard results and the four basic rules, we have been able to consider a remarkable number of different integrals. We have, however, merely scratched the surface of integration. It must be appreciated that whilst one can almost always be expected to find a derivative of a given function, integration of functions is considerably more demanding and success depends upon being able to recognise a suitable method of approach. This maturity of judgement only comes from experience derived by attempting many different kinds of integrals. We shall outline the standard methods of integration required for the single-subject GCE A-level mathematics examination.

INTEGRATION BY SUBSTITUTION

You are often *given* the required substitution. If you are not, and you find the integral difficult, look to see if one part of the integrand is, apart from a constant factor, the derivative of all or part of the rest of the integrand.

Examples:

$\int x^2(2+3x^3)^4\,dx,$ x^2 is the derivative of $2+3x^3$ divided by 9

$\int \sin^4 x\cos x\,dx,$ $\cos x$ is the derivative of $\sin x$

$\int \dfrac{1}{x}\ln x\,dx,$ $\dfrac{1}{x}$ is the derivative of $\ln x$

$\int \dfrac{e^{2x}}{1+3e^{2x}}\,dx,$ e^{2x} is the derivative of $1+3e^{2x}$ divided by 6

Worked example

$\int x^2(2+3x^3)^4\,dx,$ x^2 is part of the derivative of $2+3x^3$

Let $t = 2+3x^3$, $\dfrac{dt}{dx} = 9x^2 \Rightarrow x^2\,dx \equiv \dfrac{1}{9}\,dt$

$\Rightarrow \displaystyle\int x^2(2+3x^3)^4\,dx = \int (2+3x^3)^4 . x^2\,dx = \int t^4 . \dfrac{1}{9}\,dt$

$= \dfrac{1}{9} \cdot \dfrac{t^5}{5} = \dfrac{(2+3x^3)^5}{45}$

Worked example

$\int \dfrac{1}{x} \ln x \, dx$, $\dfrac{1}{x}$ is the derivative of $\ln x$

Let $t = \ln x$, $\dfrac{dt}{dx} = \dfrac{1}{x} \Rightarrow \dfrac{1}{x} dx \equiv dt$

$\Rightarrow \int \dfrac{1}{x} \ln x \, dx = \int t \, dt = \dfrac{1}{2} t^2 = \dfrac{1}{2} (\ln x)^2$

Worked example

$\int \sin^4 x \cos x \, dx$, $\cos x$ is the derivative of $\sin x$

Let $\sin x = t \Rightarrow \dfrac{dt}{dx} = \cos x \Rightarrow \cos x \, dx \equiv dt$

$\Rightarrow \int \sin^4 x \cos x \, dx = \int t^4 \, dt = \dfrac{1}{5} t^5 = \dfrac{1}{5} \sin^5 x$

Worked example

Use the substitution $x = t^2$ to find $\int \dfrac{1}{2x + \sqrt{x}} \, dx$

$x = t^2$, $\dfrac{dx}{dt} = 2t \Rightarrow dx \equiv 2t \, dt$

$\int \dfrac{1}{2x + \sqrt{x}} \, dx = \int \dfrac{1}{2t^2 + t} \cdot 2t \, dt = \int \dfrac{2}{2t + 1} \, dt = \ln(2t + 1) = \ln(2\sqrt{x} + 1)$

INTEGRALS INVOLVING TRIGONOMETRIC FUNCTIONS

When considering integrals involving trigonometric functions, it is sometimes necessary to seek the aid of the trigonometric identities.

Worked example

$\int \sin 3x \cos x \, dx = \int \dfrac{1}{2} (\sin 4x + \sin 2x) dx$

$= \dfrac{1}{2} \left(-\dfrac{1}{4} \cos 4x - \dfrac{1}{2} \cos 2x \right) = -\dfrac{1}{8} (\cos 4x + 2\cos 2x)$

The product of two sines or two cosines or a sine and a cosine can all be integrated by this method.

> The trigonometric identities can help in finding integrals.

Worked example

$\int \cos^2 3x \, dx = \int \dfrac{1}{2} (\cos 6x + 1) dx = \dfrac{1}{12} \sin 6x + \dfrac{x}{2}$

Here we have used the double-angle formula $\cos 2A \equiv 2\cos^2 A - 1$

Worked example

$\int \sin^5 x \, dx = \int (\sin^2 x)^2 \sin x \, dx = \int (1 - \cos^2 x)^2 \sin x \, dx$
$= \int (1 - 2\cos^2 x + \cos^4 x) \sin x \, dx$

Since $\sin x$ is associated with the derivative of $\cos x$ let $\cos x = t \Rightarrow -\sin x \, dx \equiv dt$

$\Rightarrow \int \sin^5 x \, dx = \int (1 - 2t^2 + t^4)(-dt) = -t + \dfrac{2}{3} t^3 - \dfrac{1}{5} t^5$

$= -\cos x + \dfrac{2}{3} \cos^3 x - \dfrac{1}{5} \cos^5 x$

All odd powers of sines or cosines can be integrated by this method whereas even powers require the use of the double-angle formulae.

Worked example

$\int \tan^3 2x \, dx$. Integrals involving powers of tangents need the identity $\tan^2 A = \sec^2 A - 1$ in conjunction with $\dfrac{d}{dx} (\tan x) = \sec^2 x$

$$\int \tan^3 2x\, dx = \int \tan 2x . \tan^2 2x\, dx = \int \tan 2x (\sec^2 2x - 1) dx$$
$$= \int \tan 2x . \sec^2 2x\, dx - \int \tan 2x\, dx$$
$$= \frac{1}{4} \tan^2 2x - \frac{1}{2} \ln \sec 2x$$

INTEGRATION OF RATIONAL ALGEBRAIC FRACTIONS

Fractions in which the numerator and denominator contain only constants and positive integral powers of the variable can often be integrated using partial fractions and the integral

$$\int \frac{1}{ax+b} dx = \frac{1}{a} \ln(ax+b)$$

If the denominator is of the first degree, then of course, partial fractions are unnecessary, division by the denominator is all that is required.

Worked example
$$\int \frac{6x^2}{3x+1} dx = \int \left(2x - \frac{2}{3} + \frac{2}{3}\frac{1}{3x+1}\right) dx = x^2 - \frac{2}{3}x + \frac{2}{9}\ln(3x+1)$$

Worked example
$$\int \frac{2x+1}{x^2-3x+2} dx = \int \frac{2x+1}{(x-1)(x-2)} dx = \int \left(\frac{5}{x-2} - \frac{3}{x-1}\right) dx = 5\ln(x-2) - 3\ln(x-1)$$

Worked example
$$\int \frac{2x^2+x+5}{(x-3)(x^2+4)} dx = \int \left(\frac{2}{x-3} + \frac{1}{x^2+4}\right) dx = 2\ln(x-3) + \frac{1}{2}\arctan\left(\frac{x}{2}\right)$$

Worked example
$$\int \frac{4x+10}{x^2+5x+2} dx$$

If the denominator does not factorise try differentiating the denominator to see if its derivative is related to the numerator. If this is so then the substitution t = denominator will lead to a solution.
$$t = x^2 + 5x + 2 \Rightarrow dt \equiv (2x+5)dx$$
$$\Rightarrow \int \frac{4x+10}{x^2+5x+2} dx = \int \frac{2}{t} dt = 2\ln t = 2\ln(x^2+5x+2)$$

Worked example
$$\int \frac{1-2x}{x^2+4} dx = \int \left(\frac{1}{x^2+4} - \frac{2x}{x^2+4}\right) dx$$
$$= \frac{1}{2}\arctan\left(\frac{x}{2}\right) - \ln(x^2+4)$$

INTEGRATION BY PARTS

An important method of integration is that of **integration by parts**, derived from the derivative of a product
$$\frac{d}{dx}[u(x).v(x)] = u.\frac{dv}{dx} + v.\frac{du}{dx}$$

Rearranging $\Rightarrow u.\frac{dv}{dx} = \frac{d}{dx}(u.v) - v.\frac{du}{dx}$

Integrating $\Rightarrow \int u.\frac{dv}{dx} dx = uv - \int v.\frac{du}{dx} dx$

Thought must be given as to which part of the integrand shall be equated to u and which part to $\frac{dv}{dx}$. Whilst the following rules do not always work, they are often helpful.

1   If part of the integrand is difficult to integrate, choose this to be u and choose the easier part to integrate to be $\dfrac{dv}{dx}$.

2   If both parts of the integrand are easy to integrate separately, choose that part to be u which when differentiated a sufficient number of times gives zero. If neither parts fall into this category, then it does not matter which is chosen for u or which is chosen for $\dfrac{dv}{dx}$.

Worked example

$\int x \ln x \, dx$

As $\ln x$ is difficult to integrate choose $\ln x \equiv u$

$$\int x \ln x \, dx = \frac{x^2}{2} . \ln x - \int \frac{x^2}{2} . \frac{1}{x} \, dx = \frac{x^2}{2} \ln x - \int \frac{x}{2} dx = \frac{x^2}{2} \ln x - \frac{x^2}{4}$$

Worked example

$\int x \sin 2x \, dx$

Both x and $\sin 2x$ are easy to integrate, but x and not $\sin 2x$ will give zero if differentiated a sufficient number of times $\Rightarrow$ choose $x \equiv u$

$$\int x \sin 2x \, dx = x \left(-\frac{\cos 2x}{2} \right) - \int \left(-\frac{\cos 2x}{2} \right) . 1 \, dx = -\frac{1}{2} x \cos 2x + \frac{1}{4} \sin 2x$$

Worked example

$\int \arctan x \, dx$

Although not a product it is sometimes possible to integrate an inverse trigonometric function such as arctan x, arcsin x or a logarithmic function such as ln x or ln $(1+x)$ by parts.

$\int \arctan x \, dx = \int 1 . \arctan x \, dx$ and taking $u \equiv \arctan x$

$$\Rightarrow x . \arctan x - \int x . \frac{1}{1+x^2} dx$$

$$\Rightarrow x . \arctan x - \frac{1}{2} \ln(1+x^2)$$

Given that $\displaystyle\int f(x) dx = F(x)$ then an integral of the form

$$\int_a^b f(x) dx = \left[F(x) \right]_a^b = F(b) - F(a)$$

is known as a **definite integral**; a and b are called the limits of the integral.

Worked example

$$\int_1^2 (x^2 - 1) dx = \left[\frac{1}{3} x^3 - x \right]_1^2 = \left[\frac{1}{3} 2^3 - 2 \right] - \left[\frac{1}{3} 1^3 - 1 \right] = \frac{8}{3} - 2 - \frac{1}{3} + 1 = 1\frac{1}{3}$$

When the integral is a definite integral care must be taken to ensure that should a change of variable be necessary in order to integrate, then either
i) the limits must be changed to accommodate the new variable or
ii) the integral must be treated as an indefinite integral and the limits inserted at the end of the integration after the function has been expressed in terms of the original variable.

Worked example

Find $\displaystyle\int_0^3 \sqrt{(9-x^2)} \, dx$

Let $x = 3\sin\theta \Rightarrow dx \equiv 3\cos\theta \, d\theta$

When $x = 3$, $3 = 3\sin\theta \Rightarrow \theta = \dfrac{\pi}{2}$

When $x = 0$, $0 = 3\sin\theta \Rightarrow \theta = 0$

$$\int_0^3 \sqrt{(9-x^2)}\,dx = \int_0^{\frac{\pi}{2}} \sqrt{(9-9\sin^2\theta)}.3\cos\theta\,d\theta = \int_0^{\frac{\pi}{2}} 9\cos^2\theta\,d\theta$$

$$= \int_0^{\frac{\pi}{2}} \frac{9}{2}(\cos 2\theta + 1)d\theta = \left[\frac{9}{2}\left(\frac{1}{2}\sin 2\theta + \theta\right)\right]_0^{\frac{\pi}{2}}$$

$$= \frac{9}{2}\left(\frac{1}{2}\sin\pi + \frac{\pi}{2}\right) - \frac{9}{2}\left(\frac{1}{2}\sin 0 + 0\right) = \frac{9}{4}\pi$$

Alternatively

$$\int_0^3 \sqrt{(9-x^2)}\,dx = \frac{9}{2}\left(\frac{1}{2}\sin 2\theta + \theta\right) \text{ where } x = 3\sin\theta$$

$$= \frac{9}{2}\left(\sin\theta\,\cos\theta + \theta\right) = \frac{9}{2}\left[\frac{x}{3}\sqrt{\left(1 - \frac{x^2}{9}\right)} + \arcsin\left(\frac{x}{3}\right)\right]$$

$$\Rightarrow \int_0^3 \sqrt{(9-x^2)}\,dx = \left[\frac{9}{2}\frac{x}{3}\sqrt{\left(1 - \frac{x^2}{9}\right)} + \frac{9}{2}\arcsin\left(\frac{x}{3}\right)\right]_0^3$$

$$= \left[\frac{9}{2}.\frac{3}{3}\sqrt{(1-1)} + \frac{9}{2}\arcsin(1)\right] - \left[0 + \frac{9}{2}\arcsin 0\right]$$

$$= \frac{9}{2}\arcsin(1) = \frac{9}{2}.\frac{\pi}{2} = \frac{9\pi}{4}$$

EXERCISES 9.2

1 Integrate with respect to the appropriate variable

a) $\cos^3 x \sin x$

b) $\dfrac{e^x}{1+e^x}$

c) $\tan\theta\sec^2\theta$

d) $\dfrac{t}{\sqrt{(1-4t^2)}}$

e) $\sin\theta\,e^{\cos\theta}$

f) $x^2\sin(3x^3)$

g) $\dfrac{x}{\sqrt{(x-2)}}$

h) $\dfrac{t^2}{4+t^6}$

i) $e^{2x}\sqrt{(1-e^{2x})}$

2 Find

a) $\int\sin^3 2\theta\,d\theta$

b) $\int\sec^4 x\,dx$

c) $\int\cos 4x\cos 2x\,dx$

d) $\int\sqrt{(1+\cos 3x)}\,dx$

e) $\int\cos^5 x\,dx$

f) $\int\tan^4 x\,dx$

g) $\int\cos^4\theta\,d\theta$

h) $\int\sin 3x\cos 4x\,dx$

i) $\int\sin^3 x\cos^2 x\,dx$

3 Find

a) $\displaystyle\int\frac{dx}{(x+2)(x+3)}$

b) $\displaystyle\int\frac{x+3}{2x+1}\,dx$

c) $\displaystyle\int\frac{1+x}{2+2x+x^2}\,dx$

d) $\displaystyle\int\frac{1+x}{(1-x)(1+x^2)}\,dx$

e) $\displaystyle\int\frac{3-x}{(3+x)^2}\,dx$

f) $\displaystyle\int\frac{2x+1}{x^2(x+1)}\,dx$

4 Integrate with respect to the appropriate variable

a) $\theta\cos 2\theta$

b) x^2e^{3x}

c) $\dfrac{1}{x^2}\ln x$

d) $x\arctan x$

e) $\ln(1+t)$

f) $\arcsin\theta$

5 Evaluate

a) $\displaystyle\int_1^2 \ln x\,dx$

b) $\displaystyle\int_0^1 \frac{1}{2x+1}\,dx$

c) $\displaystyle\int_0^{\frac{\pi}{6}}\cos\theta\sqrt{(1+\sin\theta)}\,d\theta$

d) $\displaystyle\int_0^2 \frac{x}{\sqrt{(4-x^2)}}\,dx$

e) $\displaystyle\int_0^1 \frac{8}{\sqrt{(3+4x)}}\,dx$

f) $\displaystyle\int_1^2 \left(\frac{1}{x} - \sqrt{x}\right)^2 dx$

ANSWERS TO EXERCISES

EXERCISES 9.1

1 $x - x^2 + x^3$

2 $\dfrac{2}{3}(x+2)^{\frac{3}{2}}$

3 $-\dfrac{1}{2}\cos 2x$

4 $-2\,e^{1-2x}$

5 $2\tan 3t$

6 $-\dfrac{1}{4}e^{-4x}$

7 $-4\sqrt{(1-x)}$

8 $-\cos(x+\pi)$

9 $\dfrac{1}{3}\ln(\sec 3t)$

10 $-\dfrac{1}{1+t}$

11 $\dfrac{1}{3}\arctan\left(\dfrac{x}{3}\right)$

12 $\dfrac{3}{8}(1+4u)^{\frac{2}{3}}$

13 $-\dfrac{1}{3}\cot 3u$

14 $x + 2e^x + \dfrac{1}{2}e^{2x}$

15 $-2\ln\cos\left(\dfrac{t}{2}\right)$

16 $\dfrac{1}{42}(2+7x)^6$

17 $\dfrac{1}{2}\arcsin(2x)$

18 $\arcsin\left(\dfrac{x}{2}\right)$

EXERCISES 9.2

1
a) $-\dfrac{1}{4}\cos^4 x$

b) $\ln(1+e^x)$

c) $\dfrac{1}{2}\tan^2\theta$

d) $-\dfrac{1}{4}\sqrt{(1-4t^2)}$

e) $-e^{\cos x}$

f) $-\dfrac{1}{9}\cos(3x^3)$

g) $\dfrac{2}{3}(x+4)\sqrt{(x-2)}$

h) $\dfrac{1}{6}\arctan\left(\dfrac{t^3}{2}\right)$

i) $-\dfrac{1}{3}(1-e^{2x})^{\frac{3}{2}}$

2
a) $-\dfrac{1}{2}\cos 2\theta + \dfrac{1}{6}\cos^3 2\theta$

b) $\tan x + \dfrac{1}{3}\tan^3 x$

c) $\dfrac{1}{12}\sin 6x + \dfrac{1}{4}\sin 2x$

d) $\dfrac{2\sqrt{2}}{3}\sin\left(\dfrac{3x}{2}\right)$

e) $\sin x - \dfrac{2}{3}\sin^3 x + \dfrac{1}{5}\sin^5 x$

f) $\dfrac{1}{3}\tan^3 x - \tan x + x$

g) $\dfrac{1}{32}\sin 4\theta + \dfrac{1}{4}\sin 2\theta + \dfrac{3\theta}{8}$

h) $\dfrac{1}{2}\cos x - \dfrac{1}{14}\cos 7x$

i) $\dfrac{1}{5}\cos^5 x - \dfrac{1}{3}\cos^3 x$

3
a) $\ln\left(\dfrac{x+2}{x+3}\right)$

b) $\dfrac{1}{2}x + \dfrac{5}{4}\ln(2x+1)$

c) $\dfrac{1}{2}\ln(x^2+2x+2)$

d) $\dfrac{1}{2}\ln(1+x^2) - \ln(1-x)$

e) $-\dfrac{6}{x+3} - \ln(x+3)$

f) $\ln\left(\dfrac{x}{x+1}\right) - \dfrac{1}{x}$

4
a) $\dfrac{1}{2}\theta\sin 2\theta + \dfrac{1}{4}\cos 2\theta$

b) $\dfrac{1}{27}e^{3x}(9x^2-6x+2)$

c) $-\dfrac{1}{x}(1+\ln x)$

d) $\dfrac{1}{2}(1+x^2)\arctan x - \dfrac{1}{2}x$

e) $(1+t)\ln(1+t) - t$

f) $\theta\arcsin\theta + \sqrt{(1-\theta^2)}$

5
a) $2\ln 2 - 1$

b) $\dfrac{1}{2}\ln 3$

c) $\sqrt{\left(\dfrac{3}{2}\right)} - \dfrac{2}{3}$

d) 2

e) $4(\sqrt{7} - \sqrt{3})$

f) $6 - 4\sqrt{2}$

APPLICATIONS OF INTEGRATION

GETTING STARTED

Whilst the applications of integration are numerous, those required for GCE A-level single-subject mathematics are normally limited to the formation and solution of simple differential equations, and the evaluation of plane areas, volumes of revolution and centroids. All depend upon the formation and evaluation of an integral, as will be seen in the following examples.

Worked example

Find the equation of a curve which passes through the point $(-1, 4)$ and whose gradient at any point (x, y) on it is $(2x - 1)$

The gradient of the curve at any point (x, y) on it is given by $\dfrac{dy}{dx}$

$$\Rightarrow \frac{dy}{dx} = 2x - 1$$

$$\Rightarrow \int \frac{dy}{dx}\,dx = \int (2x - 1)\,dx \Rightarrow y = x^2 - x + C$$

This is the general equation of all curves having the given gradient. To find the specific curve which also passes through the point $(-1, 4)$ it is necessary to substitute $x = -1$, $y = 4$ in this equation.

$$\Rightarrow 4 = (-1)^2 - (-1) + C \Leftrightarrow C = 2$$
$$\Rightarrow \text{the required equation is } y = x^2 - x + 2$$

Worked example

A point P moves in a straight line so that, at time ts, its speed $V\,\mathrm{m\,s^{-1}}$ is given by

$$V = \frac{1}{t^2 + 3t + 2}$$

Find the distance S m travelled by P during the time between the instant when $t = 1$ and $t = 4$

$$V = \frac{dS}{dt} = \frac{1}{t^2 + 3t + 2} \Rightarrow S = \int \frac{1}{t^2 + 3t + 2}\,dt = \int \frac{1}{(t+1)(t+2)}\,dt$$

$$\Rightarrow \text{Distance } (S \text{ m}) \text{ travelled} = \int_1^4 \frac{1}{(t+1)(t+2)}\,dt = \int_1^4 \left(\frac{1}{t+1} - \frac{1}{t+2} \right)\,dt$$

$$\Rightarrow S = \Big[\ln(t+1) - \ln(t+2) \Big]_1^4 = (\ln 5 - \ln 6) - (\ln 2 - \ln 3) = \ln \frac{5}{4}$$

ESSENTIAL PRINCIPLES

AREA UNDER A CURVE

For questions involving the area of the region between a curve, the x- or y-axis and the appropriate ordinates or abscissae, it is necessary to commit to memory the appropriate integrals. They can however be quite easily built up by considering the area of an appropriate elementary strip, regarding the integral sign as standing for the summation of the areas as designated by the limits of the integral.

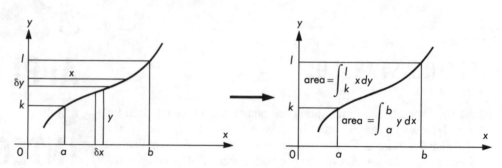

Fig. 10.1

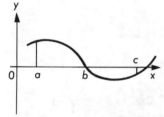

Fig. 10.2

The area of the region between two curves can be dealt with in similar manner.
In calculating such areas it is advisable to sketch the curves, so that mistakes are not made should part of the area be below the x-axis and therefore counted as negative, and so a difference of two areas is obtained rather than their sum. Take a curve with equation $y = f(x)$ which crosses the x-axis between $x = a$ and $x = c$, at $x = b$, say, as shown in Fig. 10.2.

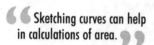
> Sketching curves can help in calculations of area.

Then the area of the region bounded by the curve $y = f(x)$, the x-axis and the ordinates at $x = a$, $x = c$ is given by:

$$\int_a^b y\,dx - \int_b^c y\,dx = \int_a^b f(x)\,dx + \left| \int_b^c f(x)\,dx \right|$$

Worked example

Sketch the arc of the curve with equation $y = 2x - x^2$ for which y is positive. Find the area of the finite region which lies between this arc and the x-axis

You should be able to recognise the equation as that of a parabola. By completing the square of the x terms and writing the equation in the form $y = 1 - (x-1)^2$ it can be seen that the maximum value of y is 1 and occurs when $x = 1$. Further, the curve crosses the x-axis where $x(2-x) = 0 \Rightarrow x = 0$ or 2

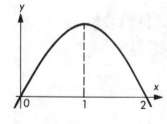

Fig. 10.3

$$\Rightarrow \text{Required area} \Rightarrow \int_0^2 y\,dx = \int_0^2 (2x - x^2)\,dx = \left[x^2 - \frac{x^3}{3} \right]_0^2$$

$$= 4 - \frac{8}{3} = \frac{4}{3}$$

Worked example

Find the area of the finite region bounded by the curves with equation
$y = x^2$, $y^2 = x$
For such a question it is absolutely essential to sketch the curves, as in Fig. 10.4.

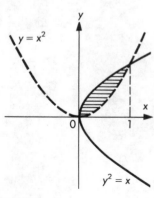

Fig. 10.4

The curves meet where
$y^2 = x = (x^2)^2 \Rightarrow x(x^3 - 1) = 0 \Longleftrightarrow x = 0$ or 1

Required area = area of shaded region = $\displaystyle\int_0^1 y_1\,dx - \int_0^1 y_2\,dx$

where $y_1 = \sqrt{x}$ and $y_2 = x^2$

$$\Rightarrow \text{area} = \int_0^1 (\sqrt{x} - x^2)\,dx = \left[\frac{2}{3}x^{\frac{3}{2}} - \frac{1}{3}x^3 \right]_0^1 = \frac{2}{3} - \frac{1}{3} = \frac{1}{3}$$

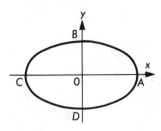

Fig. 10.5

Worked example

Show that the area enclosed by the ellipse with equation $\dfrac{x^2}{a^2}+\dfrac{y^2}{b^2}=1$ is πab

The parametric equations of this ellipse are $x=a\cos\theta$, $y=b\sin\theta$

At A $\theta=0$. At B $\theta=\dfrac{\pi}{2}$ and for points on the ellipse between A

and B $0\leqslant\theta\leqslant\dfrac{\pi}{2}$

Area enclosed by the ellipse $=4$ Area OAB

$$=4\int_0^a y\,dx = 4\int_{\frac{\pi}{2}}^0 b\sin\theta.\,(-a\sin\theta\,d\theta)\text{ since}$$

$y=b\sin\theta$ and $dx\equiv -a\sin\theta\,d\theta$

$$\Rightarrow \text{Area}=-4ab\int_{\frac{\pi}{2}}^0\sin^2\theta\,d\theta=-2ab\int_{\frac{\pi}{2}}^0(1-\cos2\theta)\,d\theta$$

$$=-2ab\left[\theta-\frac{1}{2}\sin2\theta\right]_{\frac{\pi}{2}}^0=-2ab\left[0-0-\frac{\pi}{2}+\frac{1}{2}\sin\pi\right]=\pi ab$$

Note: the limits to the integral in terms of θ are $\dfrac{\pi}{2}$ and 0 since when

$x=0$, $\theta=\dfrac{\pi}{2}$ and when $x=a$, $\theta=0$.

VOLUMES OF REVOLUTION

As with the area under a curve, you can also build up integral formulae for volumes of revolution about the coordinate axes by considering the volume of revolution of an elementary strip.

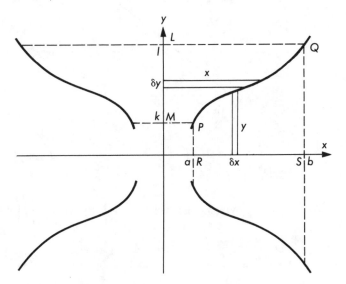

Fig. 10.6

Thus, the strip of area $y\delta x$ rotated about the x-axis produces a disc of volume $\pi y^2\delta x$. Summing all such elementary discs from $x=a$ to $x=b$, the volume of revolution V_{0x} of area $PRSQ$ about the x-axis is given by:

$$V_{0x}=\pi\int_a^b y^2\,dx$$

Similarly the volume of revolution V_{0y} of area $PQLM$ about the y-axis is given by:

$$V_{0y}=\pi\int_k^l x^2\,dy$$

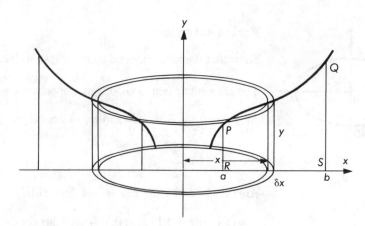

Fig. 10.7

SHELL METHOD

> A useful method for finding the volume of revolution.

When the strip of area $y\,\delta x$ is revolved about the y-axis it produces a cylindrical shell of volume $2\pi xy\,\delta x$. Summing all such elementary shells from $x = a$ to $x = b$ the volume, V, of revolution of area $PRSQ$ about the y-axis is given by:

$$V = 2\pi \int_a^b xy\,dx$$

Worked example

Sketch the curve with equation $y = x - \dfrac{1}{x}$, $x > 0$.

The area bounded by the curve, the x-axis and the lines $x = 2$, $x = 3$ is rotated completely about the x-axis. Calculate the volume of the solid of revolution so formed.

The sketch can easily be obtained by adding the known curves $y = x$ and $y = -\dfrac{1}{x}$

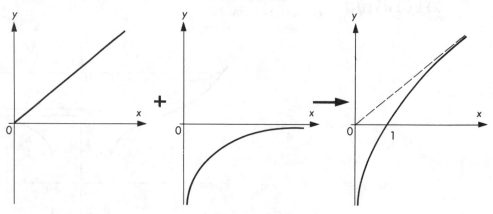

Fig. 10.8

Volume of revolution $\Rightarrow \displaystyle\int_2^3 \pi y^2 dx = \pi \int_2^3 \left(x - \frac{1}{x}\right)^2 dx$

$$\Rightarrow \pi \int_2^3 \left(x^2 - 2 + \frac{1}{x^2}\right)dx = \pi\left[\frac{1}{3}x^3 - 2x - \frac{1}{x}\right]_2^3 = \pi\left(9 - 6 - \frac{1}{3}\right) - \pi\left(\frac{8}{3} - 4 - \frac{1}{2}\right) = 4\frac{1}{2}\pi$$

Worked example

Calculate the volume generated when the finite region enclosed by the curve $y = 1 + 2e^{-x}$ and the lines $x = 0$, $x = 1$, $y = 1$ is revolved completely about the x-axis

The graph of the curve with equation $y = 1 + 2e^{-x}$ is as shown in Fig. 10.9. The region revolved is shown shaded.

The volume of revolution about the x-axis of the region contained between the curve with equation $y = 1 + 2e^{-x}$, the x-axis and the ordinates at $x = 0$, $x = 1$ is given by:

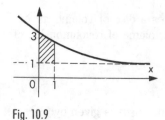

Fig. 10.9

$$\int_0^1 \pi y^2 dx = \pi \int_0^1 (1 + 2e^{-x})^2 dx = \pi \int_0^1 (1 + 4e^{-x} + 4e^{-2x})dx$$

$$= \pi[x - 4e^{-x} - 2e^{-2x}]_0^1 = \pi[(1 - 4e^{-1} - 2e^{-2}) - (0 - 4 - 2)]$$
$$= \pi(7 - 4e^{-1} - 2e^{-2})$$

$\Rightarrow$ Required volume of revolution $= \pi(7 - 4e^{-1} - 2e^{-2})$ – the volume of a cylinder of radius 1 and height 1
$$= \pi(7 - 4e^{-1} - 2e^{-2}) - \pi 1^2 . 1 = \pi(6 - 4e^{-1} - 2e^{-2})$$

Worked example

Find the volume generated when the finite region enclosed between the curve with equation $y = \ln x$, the x-axis and the ordinate at $x = 2$ is revolved completely about the line $x = -1$

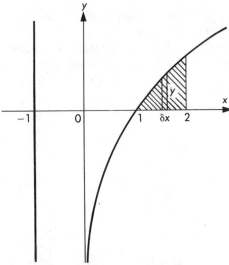

Fig. 10.10

The graph of $y = \ln x$ crosses the x-axis at $x = 1$. The region revolved is the shaded region. The strip of area $y\delta x$ is parallel to and at a distance $(x+1)$ from the line $x = -1$. Consequently, when this strip is revolved completely about the line $x = -1$, a cylindrical shell of volume $2\pi(x+1)y\delta x$ is obtained. Hence, summing all such strips between $x = 1$ and $x = 2$ gives a volume of revolution V, where:

$$V = \int_1^2 2\pi(x+1)y\,dx = \int_1^2 2\pi(x+1)\ln x\,dx$$

Integrating by parts

$$\Rightarrow V = \left[2\pi\left(\frac{1}{2}x^2 + x\right)\ln x\right]_1^2 - \int_1^2 2\pi\left(\frac{1}{2}x^2 + x\right)\frac{1}{x}\,dx$$

$$= 2\pi\left[(2+2)\ln 2 - \left(\frac{1}{2}+1\right)\ln 1\right] - \int_1^2 2\pi\left(\frac{1}{2}x+1\right)dx$$

$$= 2\pi . 4\ln 2 - 2\pi\left[\frac{1}{4}x^2 + x\right]_1^2 = 8\pi\ln 2 - 2\pi\left[(1+2) - \left(\frac{1}{4}+1\right)\right]$$

$$= 8\pi\ln 2 - 3\frac{1}{2}\pi = \pi\left(8\ln 2 - 3\frac{1}{2}\right).$$

MEAN VALUE

The area A contained between the curve $y = f(x)$, the x-axis and the ordinates $x = a$, $x = b$ is given by

$$A = \int_a^b y\,dx = \int_a^b f(x)\,dx$$

When this area is divided by $(b-a)$, the range of integration, the quantity

$$\frac{1}{(b-a)}\int_a^b y\,dx \text{ is known as the mean value of } y \text{ with respect to } x \text{ over the range } a \leqslant x \leqslant b.$$

It represents the mean height of the area A, i.e. the height of the rectangle standing on the same base of length $(b-a)$ and having the same area as A.

Worked example

Calculate the mean value of $y = \cos 2x$ over the interval $-\dfrac{\pi}{4} \leqslant x \leqslant \dfrac{\pi}{4}$

$$\text{Mean value} \Rightarrow \frac{1}{\frac{\pi}{4} - \left(-\frac{\pi}{4}\right)} \int_{-\frac{\pi}{4}}^{\frac{\pi}{4}} \cos 2x \, dx = \frac{1}{\frac{\pi}{2}} \left[\frac{1}{2}\sin 2x\right]_{-\frac{\pi}{4}}^{\frac{\pi}{4}}$$

$$= \frac{2}{\pi}\left[\frac{1}{2}\sin\frac{\pi}{2} - \frac{1}{2}\sin\left(-\frac{\pi}{2}\right)\right] = \frac{2}{\pi}\left[\frac{1}{2} + \frac{1}{2}\right] = \frac{2}{\pi}$$

MOMENTS, CENTROIDS

Integration may also be used to find the moment of an area (or volume) about a fixed axis (or plane).

Consider the curve with equation $y = f(x)$ and the area A contained between the curve, the x-axis and the ordinates $x = a$, $x = b$ as shown in Fig. 10.11

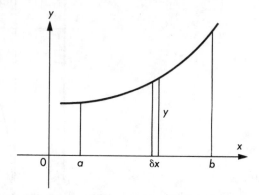

Fig. 10.11

An elementary strip $y\delta x$ has moment $xy\delta x$ about the y-axis. Consequently summing for all such strips the moment of the area A about the y-axis is given by

$$\int_a^b xy \, dx$$

Equating to $A\bar{x} \Rightarrow \bar{x}\int_a^b y \, dx = \int_a^b xy \, dx$

i.e. $\bar{x}$ the x coordinate of the *centroid* of area $A = \int_a^b xy \, dx \bigg/ \int_a^b y \, dx$

Similarly, the moment of the elementary strip about the x-axis is

$\frac{1}{2}y \cdot y\delta x = \frac{1}{2}y^2 \delta x$, giving

$$A\bar{y} = \bar{y}\int_a^b y \, dx = \int_a^b \frac{1}{2}y^2 \, dx \Rightarrow \bar{y} = \int_a^b \frac{1}{2}y^2 \, dx \bigg/ \int_a^b y \, dx$$

Worked example

Find the coordinates of the centroid of the finite area A in the first quadrant bounded by the curve with equation $y^2 = 4x$, the x-axis and the ordinate at $x = 4$

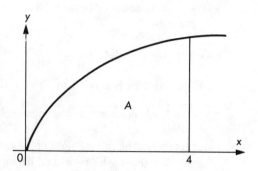

Fig. 10.12

$$\text{Area } A = \int_0^4 y \, dx = \int_0^4 \sqrt{(4x)} \, dx = \int_0^4 2x^{\frac{1}{2}} \, dx = \left[2 \cdot \frac{2}{3} \cdot x^{\frac{3}{2}}\right]_0^4 = \frac{32}{3}$$

$$\Rightarrow \frac{32}{3}\bar{x} = \int_0^4 xy\,dx = \int_0^4 2x^{\frac{3}{2}}\,dx = \left[2.\frac{2}{5}x^{\frac{5}{2}}\right]_0^4 = \frac{128}{5}$$

$$\Rightarrow \bar{x} = \frac{128}{5}.\frac{3}{32} = \frac{12}{5} = 2\frac{2}{5}$$

Further $\dfrac{32}{3}\bar{y} = \displaystyle\int_0^4 \frac{1}{2}y^2\,dx = \int_0^4 \frac{1}{2}.4x\,dx = \int_0^4 2x\,dx = \left[2.\frac{x^2}{2}\right]_0^4 = 16$

$$\Rightarrow \bar{y} = 16.\frac{3}{32} = \frac{3}{2} = 1\frac{1}{2}$$

$$\Rightarrow (\bar{x}, \bar{y}) \equiv \left(2\frac{2}{5}, 1\frac{1}{2}\right)$$

The same procedure can also be used to find the centre of mass of a solid of revolution, except that moments must be taken about a plane rather than an axis.

Worked example

Find the centre of mass of a right circular cone of height h and base radius r

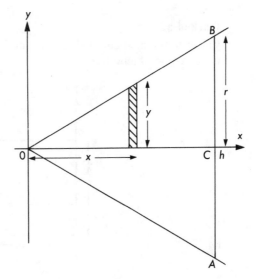

Fig. 10.13

The cone can be regarded as the volume of the solid of revolution of the triangular area OCB about the x-axis.

Volume of revolution $V = \displaystyle\int_0^h \pi y^2\,dx$

From similar triangles $\dfrac{y}{x} = \dfrac{r}{h} \Rightarrow y = \dfrac{r}{h}x$

$$\Rightarrow V = \int_0^h \pi.\frac{r^2}{h^2}x^2\,dx = \pi\frac{r^2}{h^2}\left[\frac{x^3}{3}\right]_0^h = \frac{1}{3}\pi r^2 h$$

Moment of an elementary disc about a plane containing the y-axis and perpendicular to the x-axis $= x.\,\pi y^2 \delta x$

$$\Rightarrow V\bar{x} = \int_0^h \pi x y^2\,dx = \int_0^h \pi x.\left(\frac{r}{h}x\right)^2\,dx = \int_0^h \frac{\pi r^2}{h^2}.x^3\,dx$$

$$= \pi\frac{r^2}{h^2}\left[\frac{x^4}{4}\right]_0^h = \pi\frac{r^2}{h^2}.\frac{h^4}{4} = \frac{1}{4}\pi r^2 h^2$$

$$= \frac{1}{3}\pi r^2 h\bar{x} = \frac{1}{4}\pi r^2 h^2 \Longleftrightarrow \bar{x} = \frac{3}{4}h$$

MOMENT OF INERTIA

(Do not study this section if it is not contained in your syllabus.) Consider a plane lamina of uniform density ρ in the shape of the area A contained between the curve with equation $y = f(x)$, the x-axis and the ordinates $x = a$, $x = b$. Consider further an elementary strip $y\delta x$ parallel to the y-axis as shown in Fig. 10.14.

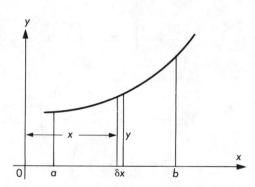

Fig. 10.14

The second moment of the strip about the y-axis is defined as $x^2.\rho y\,\delta x$
i.e. (distance from y-axis)^2. mass of the strip.
Summing all such second moments for the area A gives what is defined as the **moment of inertia** of the lamina about the y-axis.

i.e. moment of inertia of the lamina about the y-axis $= \displaystyle\int_a^b x^2.\rho y\,dx$

Worked example
Calculate the moment of inertia of a uniform rectangle of mass M and sides length a and b about a side of length a

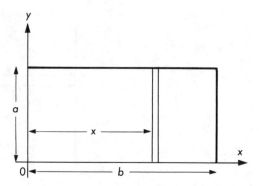

Fig. 10.15

Consider the rectangular lamina to lie in the x-y plane as shown, with a side of length a along the y-axis and a side of length b along the x-axis. Consider a strip parallel to the y-axis of length a and breadth δx, distance x from the y-axis. If ρ is the mass per unit area of the lamina, then the second moment of the strip about the y-axis is $x^2.\rho a\,\delta x$
$\Rightarrow$ Moment of inertia of the lamina about a side of length a is

$$\int_0^b x^2\rho a\,dx = \rho a\left[\frac{1}{3}x^3\right]_0^b = \rho a\frac{1}{3}b^3$$

But $M = ab\rho$

$\Rightarrow$ Moment of inertia of the lamina about a side of length $a = \dfrac{1}{3}Mb^2$

Worked example
The uniform lamina OAB shown in Fig. 10.16 has the shape of the finite area in the first quadrant bounded by the curve with equation $y = 4 - x^2$, the line $y = 3x$ and the y-axis. Find the moment of inertia of the lamina about OB given that the mass of the lamina is M.
The parabola $y = 4 - x^2$ and the straight line $y = 3x$ intersect where
$4 - x^2 = 3x \Rightarrow x^2 + 3x - 4 = 0 \Rightarrow (x-1)(x+4) = 0 \Longleftrightarrow x = 1$ or -4
$\Rightarrow A \equiv (1,3)$
Consider an elementary strip PQ parallel to the y-axis
Length $PQ = (4 - x^2) - 3x$

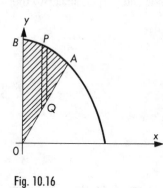

Fig. 10.16

$\Rightarrow$ Area $OAB = \displaystyle\int_0^1 [(4-x^2) - 3x]dx = \left[4x - \frac{1}{3}x^3 - \frac{3}{2}x^2\right]_0^1 = 2\frac{1}{6}$

$\Rightarrow M = 2\dfrac{1}{6}\rho$ where ρ is the mass per unit area of the lamina

Moment of inertia of the lamina about $OB = \int_0^1 x^2 . \rho(4 - x^2 - 3x) dx$

$$= \rho \left[\frac{4}{3} x^3 - \frac{1}{5} x^5 - \frac{3}{4} x^4 \right]_0^1$$

$$= \left(\frac{4}{3} - \frac{1}{5} - \frac{3}{4} \right) \rho = \frac{23}{60} \rho = \frac{23}{60} . \frac{6M}{13} = \frac{23}{130} M$$

EXERCISES 10.1

1   A particle P moves in a straight line Ox so that at time t its velocity v in the direction of increasing x is $wt\sin 2wt$ where w is a constant. Given that P starts from rest at O when $t = 0$ find

   a) the acceleration of P when $t = \dfrac{\pi}{(2w)}$

   b) The distance OP when P comes momentarily to rest for the first time after leaving O

2   Find the equation of the curve passing through the point $\left(0, \dfrac{3}{4} \right)$ and whose gradient at any point (x, y) on it is xe^{2x}

3   Find the area of the finite region bounded by the curve with equation $y = e^{-x}$, the x-axis and the lines $x = 1$, $x = -1$

4   The curve with equation $y = \sin x$ from $x = 0$ to $x = \pi$ cuts off a finite region S with the x-axis. Find:
   a) the area of S
   b) the volume of the solid of revolution when S is rotated completely about the x-axis

5   A finite region S is bounded by the curves with equation $y = x^2$, $y = 2 - x^2$. Calculate
   a) the area of S, b) the volume of the solid generated when S is rotated through π radians about the y-axis

6   Find the area of the finite region bounded by the curves with equation $y^2 = 3x$, $x^2 = 3y$
   Find the volume generated when this region is rotated completely about the x-axis

7   The parametric equations of a curve joining the points $(0, 1)$ and $(0, -1)$ are $x = \sin \theta$, $y = (1 + \sin \theta) \cos \theta$, $0 \leqslant \theta \leqslant \pi$. Sketch the curve.
   The finite region in the first quadrant bounded by the curve and the coordinate axes is rotated completely about the x-axis. Find the volume of the solid of revolution, leaving your answer in terms of π

8   Find the volume of the solid generated by revolving about the line $y = -4$, the finite region bounded by the parabola $y = 1 - x^2$ and the x-axis

9   Find the area of the finite region S bounded by the curve whose equation is $y^2 = 4x$, the x-axis and the lines $x = 1$, $x = 4$. Find the coordinates of the centroid of S

10   Find the centroid of the finite region bounded by the parabola with equation $y = x^2$ and the straight line $y = 2x + 3$.

11   A uniform equilateral triangular lamina has side $2a$ and mass m per unit area. Find the moment of inertia of the lamina about an axis coincident with one of its edges.

APPROXIMATE INTEGRATION

It is not always possible to find an integral of a given function of x, for example

$\int e^{x^2} dx$ since we cannot find a function, say f(x), whose derivative with respect to x is e^{x^2}.

However an *approximate* value of a definite integral such as

$\int_0^1 e^{x^2} dx$ can be found by equating the integrand e^{x^2} to y and regarding the problem of integration as that of finding the area of the region between the curve $y = e^{x^2}$, the x-axis and the ordinates given by the limits of the integral, i.e. $x = 0$ and $x = 1$. Two such methods are of importance at A-level.

1 THE TRAPEZIUM RULE

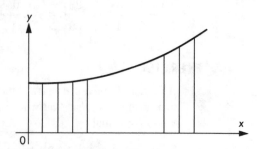

Fig. 10.17

This method divides the area required into a number of parallel strips, each of equal width. Each strip is then regarded as a trapezium whose area is one half the sum of the parallel sides multiplied by the perpendicular distance between them.

Thus, if the area required is divided into n strips by $(n+1)$ equidistant ordinates of length $y_1, y_2, y_3 \ldots y_{n+1}$, each strip being of width $\dfrac{b-a}{n}$

then
$$\int_a^b f(x)dx \approx \left(\frac{b-a}{n}\right)\left[\frac{1}{2}(y_1+y_2)+\frac{1}{2}(y_2+y_3)+\frac{1}{2}(y_3+y_4)+\ldots+\frac{1}{2}(y_n+y_{n+1})\right]$$

$$\approx \left(\frac{b-a}{n}\right)\left[\frac{1}{2}(y_1+y_{n+1})+(y_2+y_3+y_4+\ldots+y_n)\right]$$

This is known as the **trapezium rule**.

In practice we do not construct the trapeziums, we only calculate the width interval and the length of the ordinates as given by the integrand.

Worked example

Use the trapezium rule with a strip width $\dfrac{\pi}{12}$ to obtain an approximate value of

$$\int_0^{\frac{\pi}{3}} \tan x \, dx$$

giving your answer correct to 2 significant figures. Explain with the aid of a sketch, why the trapezium rule gives an estimate which is greater than the correct value

$y_1 = \tan 0 = 0{\cdot}0000$

$y_2 = \tan \dfrac{\pi}{12} = \qquad 0{\cdot}2679$

$y_3 = \tan \dfrac{\pi}{6} = \qquad 0{\cdot}5774$

$y_4 = \tan \dfrac{\pi}{4} = \qquad \underline{1{\cdot}0000}$

$\qquad\qquad\qquad\qquad 1{\cdot}8453$

$y_5 = \tan \dfrac{\pi}{3} = 1{\cdot}7321$

$\qquad\qquad \underline{1{\cdot}7321}$

$$\Rightarrow \int_0^{\frac{\pi}{3}} \tan x \, dx = \frac{\pi}{12}\left[\frac{1{\cdot}7321}{2}+1{\cdot}8453\right] = 0{\cdot}7098 = 0{\cdot}71 \text{ (to 2 significant figures)}$$

Fig. 10.18 shows each of the trapeziums. It can be seen that the curve $y = \tan x$ lies below the sloping upper side of at least three of the trapezia. Consequently, the area of the region between the curve $y = \tan x$, the x-axis and the ordinate at $x = \dfrac{\pi}{3}$ is less than the

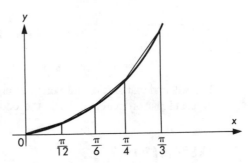

Fig. 10.18

sum of the area of the trapezia. Hence the approximate value given by the trapezium rule is greater than the true value.

2 SIMPSON'S RULE

This method divides the area into an even number of strips and takes points on the curve corresponding to these x-values, drawing through each set of three adjacent points a parabola whose axis is parallel to the y-axis and which, in general, gives a closer approximation to the curve over that section than is obtained by simply connecting the adjacent points by straight lines.

Thus, dividing the area into an even number of strips, say, $2n$ by $2n+1$ equidistant ordinates of lengths $y_1, y_2, y_3 \ldots y_{2n+1}$, each strip being of width h, then the required area is approximately

$$\frac{h}{3}[y_1 + y_{2n+1} + 4(y_2 + y_4 + \ldots + y_{2n}) + 2(y_3 + y_5 + \ldots + y_{2n-1})]$$

or in words,

$$\int_a^b f(x)dx = \text{one third the width interval of a strip} \times [\text{sum of the first and last ordinates}$$

$$+ \ 4 \text{ times the sum of all } even \text{ ordinates} + 2 \text{ times the sum of all the remaining } odd \text{ ordinates.}]$$

Caution: you are warned that the above statement is used with the first ordinate labelled y_1 and *not* y_0 as occurs in some texts. You are strongly recommended to commence labelling your ordinates at y_1 rather than y_0. Many students who commence labelling at y_0 get mixed up with the odd and even ordinates in that they still tend to think of the remaining odd ordinates as being $y_1, y_3, y_5 \ldots$, when in this notation they are in fact $y_2, y_4, y_6 \ldots$

> **It will help to start labelling your ordinates at y_1 rather than y_0.**

Worked example

Use Simpson's rule with six strips to estimate the value of

$$\int_{\frac{1}{2}}^{2} \sqrt{(8 - t^3)} \, dt$$

Range of integration is $2 - \frac{1}{2} = 1\frac{1}{2}$.   Six strips $\Rightarrow$ strip width $= \dfrac{1\frac{1}{2}}{6} = \dfrac{1}{4}$

$t = \dfrac{1}{2}$ $y_1 = \sqrt{7\dfrac{7}{8}} = 2 \cdot 806$

$t = \dfrac{3}{4}$ $y_2 = \sqrt{7\dfrac{37}{64}} =$ $2 \cdot 753$

$t = 1$ $y_3 = \sqrt{7} =$ $2 \cdot 646$

$t = 1\dfrac{1}{4}$ $y_4 = \sqrt{6\dfrac{3}{64}} =$ $2 \cdot 459$

$t = 1\dfrac{1}{2}$ $y_5 = \sqrt{4\dfrac{5}{8}} =$ $\underline{2 \cdot 151}$
$\underline{4 \cdot 797}$

$t = 1\dfrac{3}{4}$ $y_6 = \sqrt{2\dfrac{41}{64}} =$ $\underline{1 \cdot 625}$
$\underline{6 \cdot 837}$

$t = 2$ $y_7 = \quad 0 \quad = \underline{0 \cdot 000}$
$\underline{2 \cdot 806}$

$$\int_{\frac{1}{2}}^{2} \sqrt{(8-t^3)}\, dt = \frac{1}{3} \cdot \frac{1}{4} [2 \cdot 806 + 4 \times 6 \cdot 837 + 2 \times 4 \cdot 797] = 3 \cdot 312$$

$$= 3 \cdot 31 \text{ (to 3 significant figures)}$$

You will find that setting out your working as shown above helps in that it separates out the first and last ordinates and also the odd and even ordinates.

EXERCISES 10.2

1   Use the trapezium rule with a strip width of $0 \cdot 1$ to obtain an approximation of

$$\int_{0}^{0 \cdot 5} \frac{1}{\sqrt{(1-x^2)}}\, dx$$

2   Use the trapezium rule with a strip width of $0 \cdot 2$ to obtain an approximation of

$$\int_{0}^{1} \frac{1}{\sqrt{(1+x^2)}}\, dx$$

Sketch the curve with equation $y = \dfrac{1}{\sqrt{(1+x^2)}}$, $x \geqslant 0$ and hence determine whether your estimate is too large or too small, giving reasons

3   Use Simpson's rule with a strip width of $\dfrac{\pi}{4}$ to find an approximate value of $\displaystyle\int_{0}^{\pi} \sin^{\frac{3}{2}} x\, dx$.

Work to 4 decimal places

4   Show that the formula

$$\int_{0}^{2h} f(x)dx = \frac{h}{3}[f(0) + 4f(h) + f(2h)]$$

is exact when $f(x)$ is a cubic polynomial in x. Hence, or otherwise, deduce Simpson's rule

5   The region bounded by the curve with equation $y = \tan x$, the x-axis and the ordinate at

$x = \dfrac{\pi}{4}$ is rotated completely about the x-axis. Find the volume of the solid of

revolution. Use Simpson's rule with 2 intervals to find an approximate value of this volume and calculate the percentage error in your result

DIFFERENTIAL EQUATIONS

Any equation involving the derivatives of one variable with respect to another variable is called a **differential equation**. In particular if the only derivative present is first order then the equation is said to be a first-order differential equation. An example of such an equation is

$$2\frac{dy}{dx} + 3xy = x \cos x$$

In fact all integrals can be looked upon as the solution of a first order differential equation for given $\dfrac{dy}{dx} = f(x)$ then

$$y = \int f(x)dx$$

Worked example
Solve the first order differential equation

$$\frac{dy}{dx} = 3\tan x + x^2 \quad \Rightarrow \quad y = \int (3\tan x + x^2)dx$$

$$\Rightarrow y = 3\ln \sec x + \frac{1}{3}x^3 + C$$

where C is an arbitrary constant

Such a solution containing an arbitrary constant is said to be the general solution of the equation. If we are further given the value of y for a particular value of x, say, $y = 1$ when $x = 0$, then C can be determined,

i.e. $1 = 3\ln \sec 0 + \dfrac{1}{3}(0)^3 + C \Rightarrow C = 1$

$\Rightarrow y = 3\ln \sec x + \dfrac{1}{3}x^3 + 1$, known as the particular solution of the differential equation.

FIRST ORDER VARIABLES SEPARABLE

Differential equations of the form $\dfrac{dy}{dx} = f(x)g(y)$ which can be rearranged in the form

$\dfrac{1}{g(y)}\dfrac{dy}{dx} = f(x)$ and each side integrated with respect to x to give

$$\int \frac{1}{g(y)}\frac{dy}{dx}dx = \int f(x)dx \Rightarrow \int \frac{1}{g(y)}dy = \int f(x)dx$$

are called **first order variables separable**, i.e. equations in which all the y's can be collected with the dy and all the x's with the dx. Their solution is simply the evaluation of two integrals.

Worked example
Find the general solution of the differential equation

$y\dfrac{dy}{dx} + x = xy^2$

$\Rightarrow y\dfrac{dy}{dx} = x(y^2 - 1) \Longleftrightarrow \displaystyle\int \frac{y}{y^2 - 1}dy = \int x\,dx$

$\Rightarrow \dfrac{1}{2}\ln(y^2 - 1) = \dfrac{1}{2}x^2 + C$

Worked example
Express y in terms of x given that

$\cot x\dfrac{dy}{dx} = 1 - y^2$

and that $y = 0$ when $x = \dfrac{\pi}{4}$

$\cot x\dfrac{dy}{dx} = 1 - y^2 \Rightarrow \displaystyle\int \frac{1}{1-y^2}dy = \int \tan x\,dx$

$\Longleftrightarrow \dfrac{1}{2}\displaystyle\int\left(\frac{1}{1-y} + \frac{1}{1+y}\right)dy = \int \frac{\sin x}{\cos x}dx$

$\Rightarrow -\dfrac{1}{2}\ln(1-y) + \dfrac{1}{2}\ln(1+y) = -\ln\cos x + \ln C$

$\Rightarrow \dfrac{1}{2}\ln\left(\dfrac{1+y}{1-y}\right) = \ln\dfrac{C}{\cos x} \Longleftrightarrow \left(\dfrac{1+y}{1-y}\right)^{\frac{1}{2}} = \dfrac{C}{\cos x}$

$x = \dfrac{\pi}{4}, y = 0 \Rightarrow 1 = \dfrac{C}{\cos\left(\dfrac{\pi}{4}\right)} = \sqrt{2}C \Rightarrow C = \dfrac{1}{\sqrt{2}}$

$\Rightarrow \left(\dfrac{1+y}{1-y}\right)^{\frac{1}{2}} = \dfrac{1}{\sqrt{2}\cos x} \Rightarrow \dfrac{1+y}{1-y} = \dfrac{1}{2\cos^2 x}$

$\Rightarrow y = \dfrac{1 - 2\cos^2 x}{1 + 2\cos^2 x}$

Questions are sometimes worded so that the student is required to form the differential equation and then solve it.

Worked example

The population of Sunville is increasing annually at a rate which is proportional to the population at that instant. Given that the population at the beginning of 1950 was 20 000 and at the beginning of 1980 was 24 000 calculate the expected population at the beginning of the year 2000.

Let the population t years after 1950 be P. Then the increase in population at that instant is

$$\frac{dP}{dt} = kP \text{ where } k \text{ is a constant}$$

$$\Rightarrow \int \frac{dP}{P} = \int k\,dt \iff \ln P = kt + C$$

When $t = 0$, (the beginning of 1950), $P = 20\,000$
$\Rightarrow \ln 20\,000 = C \Rightarrow \ln P = kt + \ln 20\,000$

When $t = 30$, (the beginning of 1980), $P = 24\,000$
$\Rightarrow \ln 24\,000 = 30k + \ln 20\,000$

$$\Rightarrow k = \frac{1}{30} \ln (1 \cdot 2)$$

When $t = 50$, (the beginning of 2000)

$$\ln P = \frac{1}{30} \ln (1 \cdot 2) . 50 + \ln 20\,000$$

$\Rightarrow \ln P = \ln [(20\,000)(1 \cdot 2)^{\frac{5}{3}}] \Rightarrow P \approx 27\,100$
$\Rightarrow$ at the beginning of the year 2000 the expected population of Sunville is approximately 27 000

FIRST-ORDER LINEAR DIFFERENTIAL EQUATIONS

A differential equation of the form

$$f(x)\frac{dy}{dx} + yg(x) = h(x)$$

is said to be a **first-order linear differential equation** since when the equation is rearranged so that the coefficient of $\frac{dy}{dx}$ is one, i.e. $\frac{dy}{dx} + Py = Q$, where P and Q are functions of x only, the only remaining term involving y is y to the power 1.

E.g. $x^2\frac{dy}{dx} + 3xy = 1 - x \Rightarrow \frac{dy}{dx} + \left(\frac{3}{x}\right)y = \left(\frac{1}{x^2} - \frac{1}{x}\right)$, which is of the form $\frac{dy}{dx} + Py = Q$

where $P = \frac{3}{x}$, $Q = \frac{1}{x^2} - \frac{1}{x}$

This equation is easily solved by multiplying the equation

$$\frac{dy}{dx} + \left(\frac{3}{x}\right)y = \frac{1}{x^2} - \frac{1}{x} \text{ throughout by } x^3 \text{ to give}$$

$$x^3\frac{dy}{dx} + 3x^2y = x - x^2$$

A careful look at the L.H.S. of this equation shows that

$$x^3\frac{dy}{dx} + 3x^2y = \frac{d}{dx}(x^3y), \text{ so that the equation can be rewritten}$$

$$\frac{d}{dx}(x^3y) = x - x^2$$

Integrating with respect to x

$$\Rightarrow \int \frac{d}{dx}(x^3y)dx = \int (x - x^2)\,dx$$

$$\Rightarrow x^3 y = \frac{1}{2}x^2 - \frac{1}{3}x^3 + C$$

All first-order linear differential equations can be reduced to integrating in a similar manner, i.e. by multiplying the equation by some function of x, (the integrating factor), so as to make the L.H.S. into the derivative of a product and then integrating both sides with respect to x.

The rules to follow are:

Remember these rules.

1   Rearrange the equation in the form $\frac{dy}{dx} + Py = Q$, i.e. make the coefficient of $\frac{dy}{dx}$ equal to 1

2   Calculate the integrating factor $R = e^{\int P\,dx}$
i.e. $R = e^{\int (\text{coefficient of } y)\,dx}$
making sure the sign of the coefficient is also taken into account

3   Multiply the equation by the integrating factor

4   Integration gives $Ry = \int QR\,dx$

Worked example

Solve the differential equation $\cos x \frac{dy}{dx} + y \sin x = 1$

1   Make the coefficient of $\frac{dy}{dx} = 1$, so that $\frac{dy}{dx} + y \cdot \frac{\sin x}{\cos x} = \frac{1}{\cos x}$

2   Calculate the integrating factor R
$$R = e^{\int \frac{\sin x}{\cos x}\,dx} = e^{-\ln \cos x} = e^{\ln \sec x} = \sec x$$

3   Multiplying the equation by the integrating factor

$$\sec x \frac{dy}{dx} + y \tan x \sec x = \sec^2 x$$

4   Integrate
$$\Rightarrow y \sec x = \int \sec^2 x\,dx = \tan x + C$$
i.e. solution is $y \sec x = \tan x + C$ or $y = \sin x + C \cos x$

Note: in 2 we used $e^{\ln \sec x} = \sec x$. This is a property of the logarithmic function that $e^{\ln f(x)} = f(x)$. It can be easily verified as follows

Let $y = e^{\ln f(x)}$
Taking logarithms of both sides $\Rightarrow \ln y = \ln e^{[\ln f(x)]}$
$\Rightarrow \ln y = \ln f(x) \cdot \ln e$ [since $\ln x^n = n \ln x$]
$\quad\quad\; = \ln f(x) \cdot 1$ [since $\ln e = 1$]
i.e. $\ln y = \ln f(x) \Rightarrow y = f(x)$

Worked example

Solve the differential equation $x\frac{dy}{dx} + 2y = e^{3x}$ given that $y = 0$ when $x = 0$

$$x\frac{dy}{dx} = 2y = e^{3x} \Rightarrow \frac{dy}{dx} + \left(\frac{2}{x}\right)y = \frac{1}{x}e^{3x}$$

Integrating factor $= e^{\int \frac{2}{x}\,dx} = e^{2\ln x} = e^{\ln x^2} = x^2$

Multiplying by the integrating factor $\Rightarrow x^2 \frac{dy}{dx} + 2xy = xe^{3x} \Rightarrow \frac{d}{dx}(x^2 y) = xe^{3x}$

$$\Rightarrow x^2 y = \int xe^{3x}\,dx = x \cdot \frac{1}{3}e^{3x} - \int \frac{1}{3}e^{3x} \cdot 1\,dx = \frac{x}{3}e^{3x} - \frac{1}{9}e^{3x} + C \quad\quad \text{(Integration by parts)}$$

When $x = 0$, $y = 0 \Rightarrow C = \frac{1}{9}$

The solution is therefore

$$x^2 y = \frac{1}{9}(1 + 3xe^{3x} - e^{3x})$$

EXERCISES 10.3

1   Solve the differential equation $2\dfrac{dy}{dx} = 2xe^{-2y} + e^{-2y}$ given that $y = 0$ when $x = 0$

2   Solve the differential equation $x\dfrac{dy}{dx} - y = xy$

3   Solve the differential equation $\dfrac{dy}{dx} = e^{2x+3y}$

4   Express y as a function of x given that

$(1+x^2)\dfrac{dy}{dx} = x(1-y^2)$ and $y = 0$ when $x = 1$

5   Solve the equation $\sin x \dfrac{dy}{dx} + y\cos x = x^2$ given that when $x = \dfrac{\pi}{3}$, $y = 0$

6   Solve the equation $x\dfrac{dy}{dx} + 4y = x$

7   Solve the equation $\dfrac{dy}{dx} + y = e^x$ given that when $x = 0$, $y = 2$

8   Find the general solution of the differential equation $\dfrac{dy}{dx} + 2y\cot x = \cot x$

9   The gradient of the tangent to a curve C at any point (x, y) is $3x^2y^2$. Find the equation of C given that it passes through the point $(2, 1)$.

EXAMINATION QUESTIONS

1   Use the trapezium rule with 4 ordinates to find the value of

$$\int_0^{\frac{\pi}{6}} \ln(1 + \sin x)\, dx,$$

giving your answer to 3 significant figures.
Hence find

$$\int_0^{\frac{\pi}{6}} \ln(1 + \sin x)^{13}\, dx,$$

giving your answer to 2 significant figures.   (L)

2

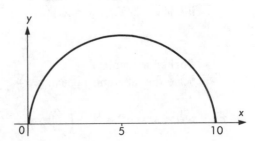

The figure above shows the graph of

$$y = \frac{2}{5}[x(10-x)]^{\frac{1}{2}}, \quad 0 \leqslant x \leqslant 10.$$

An engineer is studying the effect of the shape of the cross-section of a drainage channel on the flow of water along it. One shape she studies is modelled by the above equation, where y metres is the depth of the channel at a distance of x metres perpendicularly across the channel from one side.
Copy and complete the following table for values of y calculated using

$$y = \frac{2}{5}[x(10-x)]^{\frac{1}{2}}.$$

x	0	2·5	5	7·5	10
y					

Hence, by using Simpson's rule with 5 ordinates, obtain an estimate for the cross-sectional area of the channel in m^2, giving your answer to 3 significant figures.
Given that 22 500 litres of water pass through the channel each second, calculate the speed of the water in ms^{-1} to 1 decimal place. (L)

3 a) Sketch the curve $y = 2x^2 - 1$, showing on your sketch the coordinates of the points at which the curve meets the coordinate axes.
The region R is the finite region bounded by the curve and the line $y = 1$.

b) By using $\int x\,dy$, with appropriate limits, find the area of R.
A uniform thin sheet is made in the shape of R.

c) Determine the position of the centre of mass of this sheet. (L)

4 Given that x and y are positive, find the general solution of the differential equation
$$\frac{dy}{dx} = -\frac{y}{x^2}$$
Show that the solution for which $y = e$ when $x = 1$ may be expressed in the form
$$y = e^{1/x}$$
(L)

5 Use the trapezium rule, with six ordinates, to obtain an approximate value of the integral
$$\int_0^{2.5} \frac{x^2}{1+x^3}\,dx.$$
What is the exact value of the integral? (NI 1987)

6 a) Find $\int x \ln x\,dx$.

b) By means of the substitution $t = \tan x$, or otherwise, find
$$\int \frac{1}{1 + \cos^2 x}\,dx$$

c) The region bounded by the curve $y = (1 + \cos x)^{-\frac{1}{2}}$, the x-axis, and the lines $x = 0$ and $x = \frac{1}{2}\pi$, is denoted by R. Use the trapezium rule with ordinates at $x = 0$, $x = \frac{1}{4}\pi$ and $x = \frac{1}{2}\pi$ to estimate the area of R, giving two signficant figures in your answer. (C)

7 a) Find the following integrals:

i) $\displaystyle\int \sin^2\left(\frac{x}{3}\right) dx$

ii) $\displaystyle\int x^2 \cos 2x\,dx$.

b) Express $x(4-x)$ as the difference of two squares and hence, or otherwise, find
$$\int_1^3 \frac{1}{\sqrt{\{x(4-x)\}}}\,dx.$$

8 A curve is given by the equation
$$y = \sin x + \tfrac{1}{2}\sin 2x, \quad 0 \le x \le 2\pi.$$
Find the values of x for which y is zero.
Find the exact coordinates of the stationary points on the curve and sketch the curve.
Find the area of the region bounded by the curve and the x-axis for $0 \le x \le \pi$. Deduce the mean value of y over the interval $0 \le x \le \pi$. (JMB)

9 Show that

a) $\displaystyle\int_0^{\pi/2} \cos^2 x\,dx = \frac{\pi}{4}$,

b) $\displaystyle\int_0^{\pi/2} x\cos x\,dx = \dfrac{\pi}{2} - 1.$

Sketch and label the region R defined by

$$x \geq 0,\ y \geq 0,\ y \leq \cos x,\ x \leq \dfrac{\pi}{2}.$$

Find, in terms of π,

c) the x-coordinate of the centroid of the region R,

d) the y-coordinate of the centroid of the region R.

Find also, in terms of π, the volume obtained when the region R is revolved through 2π about the x-axis. (L)

10 Express $\mathrm{f}(x) = \dfrac{3+x}{(1+x^2)(1+2x)}$ in partial fractions.

a) Prove that the area of the region enclosed by the curve with equation

$y = \mathrm{f}(x)$, the coordinate axes and the line $x = 1$ is $\dfrac{\pi}{4} + \dfrac{1}{2}\ln\left(\dfrac{9}{2}\right).$

b) Obtain the expansion of $\mathrm{f}(x)$ in ascending powers of x up to and including the term in x^3. State the set of values of x for which the expansion is valid.   (AEB 1988)

11 a) Use integration by parts to find

$$\int x^{\frac{1}{2}}\ln x\,dx.$$

b) Find the solution of the differential equation $\dfrac{dy}{dx} = (xy)^{\frac{1}{2}}\ln x,$

for which $y = 1$ when $x = 1$ (AEB 1989)

12 Verify that $\int \ln x\,dx = x\ln x - x + c$, where c is an arbitrary constant.

Use the trapezium rule with trapezia of unit width to find an estimate for $\displaystyle\int_2^5 \ln x\,dx.$

Explain with the aid of a sketch why the trapezium rule underestimates the value of the integral in this case, and calculate correct to one significant figure, the percentage error involved.

By again using trapezia of unit width, show that, when n is a positive integer greater than 1, the trapezium rule approximation to $\displaystyle\int_2^{n+1} \ln x\,dx$ is

$$\ln(n!) + \dfrac{1}{2}\ln\left(\dfrac{n+1}{2}\right).$$
(C)

13 Find the solution of the differential equation $(x^2 - 5)^{\frac{1}{2}}\dfrac{dy}{dx} = 2xy^{\frac{1}{2}},\ x^2 > 5,$

for which $y = 4$ when $x = 3$, expressing y in terms of x.   (JMB)

ANSWERS TO EXERCISES

EXERCISES 10.1

1 $-\pi w,\ \dfrac{\pi}{4w}$

2 $y = \dfrac{1}{2}xe^{2x} - \dfrac{1}{4}e^{2x} + 1$

3 $e - \dfrac{1}{e}$

4 $2,\ \dfrac{1}{2}\pi^2$

5 $2\dfrac{2}{3},\ \pi$

6 $3,\ 8\dfrac{1}{10}\pi$

7   $\dfrac{13}{10}\pi$

8   $11\dfrac{11}{15}\pi$

9   $9\dfrac{1}{3}, \left(2\dfrac{23}{35}, 1\dfrac{17}{28}\right)$

10   $\left(1, 3\dfrac{2}{5}\right)$

11   $\dfrac{1}{4}a^4\sqrt{3}$

EXERCISES 10.2

1   0·525
2   1·1484
3   1·769
5   0·6742, 0·6935, 2·86%

EXERCISES 10.3

1   $y = -\ln(1 - \tan x)$
2   $\ln y = x + \ln x + C$
3   $2e^{-3y} + 3e^{2x} = C$
4   $2(1 + y) = (1 - y)(1 + x^2)$
5   $y\sin x = \dfrac{1}{3}x^3 - \dfrac{\pi^3}{81}$
6   $x^4 y = \dfrac{1}{5}x^5 + C$
7   $2ye^x = e^{2x} + 3$
8   $y\sin^2 x = \dfrac{1}{2}\sin^2 x + C$
9   $\dfrac{1}{y} = \dfrac{3}{2} - x^3$

ANSWERS TO EXAMINATION QUESTIONS

1   0·115, 1·5
2   14·9, 1·5
3   b) $\dfrac{8}{3}$;   c) $(0, 0·2)$
4   $\ln y = \dfrac{1}{x} + c.$
5   $0·9345, \dfrac{1}{3}\ln\left(\dfrac{133}{8}\right).$
6   a) $\dfrac{x^2}{2}\ln x - \dfrac{x^2}{4} + C,$

   b) $\dfrac{1}{\sqrt{2}}\arctan\left(\dfrac{1}{\sqrt{2}}\tan x\right),$ c) 1·3.

7   a) i) $\dfrac{1}{2}\left(x - \dfrac{3}{2}\sin\dfrac{2x}{3}\right) + C,$

   ii) $\dfrac{x^2}{2}\sin 2x + \dfrac{x}{2}\cos 2x - \dfrac{1}{4}\sin 2x + C$

   b) $\dfrac{\pi}{3}$

8   $0, \pi, 2\pi; \left(\dfrac{\pi}{3}, \dfrac{3\sqrt{3}}{2}\right),$

   $(\pi, 0), \left(\dfrac{5}{3}\pi - \dfrac{3\sqrt{3}}{2}\right); 2, \dfrac{2}{\pi}$

9   $\dfrac{\pi}{2} - 1, \dfrac{\pi}{8}, \dfrac{\pi^2}{4}.$

10   $\dfrac{1-x}{1+x^2} + \dfrac{2}{1+2x},$

   b) $3 - 5x + 7x^2 - 15x^3, |x| < \dfrac{1}{2}.$

11   $\dfrac{2}{3}x^{\frac{3}{2}}\ln x - \dfrac{4}{9}x^{\frac{3}{2}},$

   $2y^{\frac{1}{2}} = \dfrac{2}{3}x^{\frac{3}{2}}\ln x - \dfrac{4}{9}x^{\frac{3}{2}} + \dfrac{22}{9}$

12   3·636, 0·7%

13   $y^{\frac{1}{2}} = (x^2 - 5)^{\frac{1}{2}}$

COMPLEX NUMBERS

GETTING STARTED

We have seen that the roots of the equation $Ax^2 + Bx + C = 0$ where A, B and $C \in \mathbb{R}$, $A \neq 0$ are

$$\frac{-B \pm \sqrt{(B^2 - 4AC)}}{2A}$$

We noted that, since there is no real number whose square is negative, then when $B^2 - 4AC < 0$, the roots of the equation are not real and we therefore referred to them as being imaginary (or complex). We must now consider such roots in more detail and introduce a new set of numbers $\mathbb{C}$, the set of complex numbers.

COMPLEX NUMBERS

All negative numbers can be written as the product of a positive number and (-1). Thus $-8 = 8(-1)$ or generally $-n^2 = n^2(-1)$. If we denote $\sqrt{(-1)}$ by i (or j) so that i^2 (or j^2) $= -1$ then $-n^2 = i^2 n^2$ and we can write $\sqrt{(-n^2)} = \sqrt{(i^2 n^2)} = in$, where $i = \sqrt{(-1)}$

Such a number is said to be an **imaginary number** and when associated with a real number by the operation of addition or subtraction then the resultant number is said to be a **complex number**.
E.g. $2 + 3i$, $-1 - i$, $2 - 3i$ where $i = \sqrt{(-1)}$ all belong to the set $\mathbb{C}$ of complex numbers.

Thus the roots of the equation $x^2 - 6x + 13 = 0$ are

$$x = \frac{-6 \pm \sqrt{(36 - 52)}}{2} = 3 \pm \sqrt{(-4)} = 3 \pm 2i$$

The equation is said to have complex roots.

ESSENTIAL PRINCIPLES

DEFINITION

Generally a **complex number** is denoted by the letter z and is written in the form

$z = x + iy$ where $i = \sqrt{(-1)}$ and $x, y \in \mathbb{R}$
x is called the *real part* of $z \Rightarrow \text{Re}(z) = x$
y is called the *imaginary part* of $z \Rightarrow \text{Im}(z) = y$

Both x and y can take zero values, so that the set of real numbers and the set of imaginary numbers $\in \mathbb{C}$.

EQUALITY OF COMPLEX NUMBERS

Given that $z_1 = x_1 + iy_1$, $z_2 = x_2 + iy_2$, then $z_1 = z_2$ implies
$$x_1 + iy_1 = x_2 + iy_2 \text{ and } x_1 = x_2,\ y_1 = y_2$$
i.e. two complex numbers are equal provided their real parts are equal **and** provided their imaginary parts are equal. Thus, a single equation between complex numbers is equivalent to two relations between real numbers; one derived by equating the real parts of the two sides of the equation and the other derived by equating the imaginary parts of the two sides of the equation.

> **A useful property of equations involving complex numbers.**

Consequently, in any equation involving complex numbers, the equality sign can still be retained when i is replaced by $-i$, a property which often proves invaluable when working with complex numbers.

Worked example
Find x and y given that $x + 3iy + i(2x + iy) = 5$
$x + 3iy + 2ix + i^2y = 5 \Rightarrow x + 3iy + 2ix - y = 5$, since $i^2 = -1$
Consequently $x - 3iy - 2ix - y = 5$
Adding $2x - 2y = 10$
Subtracting $6iy + 4ix = 0 \Rightarrow 2i(3y + 2x) = 0 \Rightarrow 3y + 2x = 0$
Solving $5y = -10$, $y = -2$, $x = 3$

ELEMENTARY RULES IN CARTESIAN FORM

When you have to combine complex numbers, treat the numbers as ordinary algebraic expressions and use the ordinary laws of algebra, replacing i^2 whenever it occurs by -1.

Let $z_1 = a + ib$, $z_2 = c + id$
Addition: $z_1 + z_2 = (a + ib) + (c + id) = (a + c) + i(b + d)$
Subtraction: $z_1 - z_2 = (a + ib) - (c + id) = (a - c) + i(b - d)$
i.e. in each case, compound the real parts, compound the imaginary parts
Multiplication: $z_1 z_2 = (a + ib)(c + id) = ac + i^2bd + ibc + iad$
$$= (ac - bd) + i(bc + ad)$$

You should note $(a + ib)(a - ib) = a^2 + b^2$, a real number, i.e. the product of two complex numbers whose real parts are equal and whose imaginary parts are equal in magnitude, but opposite in sign, is a real number.

Such complex numbers are said to be **conjugate complex numbers**, i.e. $a - ib$ is the conjugate of $a + ib$, (likewise $a + ib$ is the conjugate of $a - ib$).
If $z = x + iy$ then we denote the conjugate of z by $z^* \Rightarrow z^* = x - iy$

$$\Rightarrow zz^* = (x + iy)(x - iy) = x^2 + y^2$$

Division: $\dfrac{z_1}{z_2} = \dfrac{a + ib}{c + id} = \dfrac{(a + ib)}{(c + id)} \cdot \dfrac{(c - id)}{(c - id)} = \dfrac{(ac + bd)}{(c^2 + d^2)} + \dfrac{i(bc - ad)}{(c^2 + d^2)}$

i.e. to separate out the real and imaginary parts, make the denominator real by multiplying numerator and denominator by the conjugate of the denominator

Worked example

$z_1 = 10 + 2i$, $z_2 = 2 - 3i$. Find in $x + iy$ form   a) $z_1 z_2$,   b) $\dfrac{z_1}{z_2}$

$z_1 z_2 = (10 + 2i)(2 - 3i) = 20 - 30i + 4i - 6i^2 = 26 - 26i$
$\dfrac{z_1}{z_2} = \dfrac{(10 + 2i)}{(2 - 3i)} \cdot \dfrac{(2 + 3i)}{(2 + 3i)} = \dfrac{(20 + 6i^2 + 4i + 30i)}{2^2 + 3^2} = \dfrac{14 + 34i}{13}$

Worked example

Find z in the form $x+iy$ where x and y are real given that $(z+1)(2-i) = 3-4i$

$$z+1 = \frac{3-4i}{2-i} = \frac{(3-4i)}{(2-i)} \cdot \frac{(2+i)}{(2+i)} = \frac{6-4i^2-8i+3i}{2^2+1} = \frac{10-5i}{5} = 2-i$$

$$\Rightarrow z = 2-i-1 = 1-i$$

Worked example

Find, in the form $a+ib$, a, $b \in \mathbb{R}$ the values of $\sqrt{(5+12i)}$

$a+ib = \sqrt{(5+12i)} \Rightarrow (a+ib)^2 = 5+12i$

$\Longleftrightarrow a^2 - b^2 + 2iab = 5 + 12i$

$$\Rightarrow a^2 - b^2 = 5 \text{ and } 2ab = 12 \text{ or } a = \frac{6}{b}$$

$$a^2 - b^2 = 5 \Rightarrow \left(\frac{6}{b}\right)^2 - b^2 = 5 \Longleftrightarrow b^4 + 5b^2 - 36 = 0$$

$$\Longleftrightarrow (b^2 - 4)(b^2 + 9) = 0 \Longleftrightarrow b^2 = 4 \text{ or } -9$$

But $b \in \mathbb{R} \Rightarrow b^2 = 4$, $b = \pm 2$

$$a = \frac{6}{b} \Rightarrow a = 3 \text{ when } b = 2, \; a = -3 \text{ when } b = -2$$

$$\sqrt{(5+12i)} = \pm(3+2i)$$

Worked example

Given that $x+iy = (1+i)^n$, where $n \in \mathbb{Z}^+$ prove that $x^2 + y^2 = 2^n$.

$x+iy = (1+i)^n, \Rightarrow x - iy = (1-i)^n$

Hence $(x+iy)(x-iy) = (1+i)^n(1-i)^n \Rightarrow x^2 + y^2 = [(1+i)(1-i)]^n = 2^n$

Note: success in the solution of this question depends upon recognising　i) $x^2 + y^2$ is the real number obtained by multiplying $x+iy$ by its conjugate $x-iy$　ii) $(1-i)^n$ is the conjugate of $(1+i)^n$　iii) $(1+i)^n(1-i)^n = [(1+i)(1-i)]^n = (1+1)^n = 2^n$

ARGAND DIAGRAM

> *Make use of the Argand diagram.*

The ordinary two-dimensional cartesian diagram is used to represent a complex number, the real part x being measured along the x-axis or real axis; the imaginary part y being measured along the y-axis or imaginary axis. It is then called an **Argand diagram**.

The complex number $z_1 = 4 + 2i$ is therefore represented by the point A in Fig. 11.1. B represents $-2-3i$, C represents $-4+3i$, D represents $2+i0$ and E represents i.

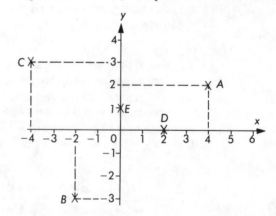

Fig. 11.1

MODULUS ARGUMENT

If P represents the complex number $z = x + iy$ then the length of OP, usually denoted by r, $|z|$ or $|x+iy|$, is called the **modulus** of z. Clearly $|z| = r = \sqrt{(x^2+y^2)}$, $r > 0 \Rightarrow$ the modulus of a complex number can always be found by multiplying the number by its conjugate, since if $z = x + iy$, and taking the positive square root
$$zz^* = (x+iy)(x-iy) = x^2 + y^2 = |z|^2 \Rightarrow |z| = \sqrt{zz^*}.$$
The angle $\theta = P\hat{O}X$ is called the **argument** of z and is written as

$$\arg z = \arg(x+iy) = \theta = \arctan\left(\frac{y}{x}\right).$$

The examining boards have agreed to use the convention that $-\pi < \arg z \leqslant \pi$.

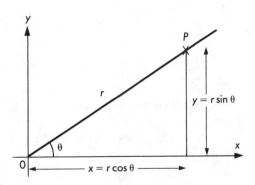

Fig. 11.2

Warning: in finding θ, you must plot the complex number on an Argand diagram since there is an infinite number of angles whose tangent is $\left(\dfrac{y}{x}\right)$; two of these lie in $(-\pi, \pi)$, but only one of these corresponds to the unique position of P. The value of θ in $(-\pi, \pi)$ is known as the **principal value** of the argument.

Thus $z_1 = 2 + 3i \Rightarrow \arg(2 + 3i) = \arctan\left(\dfrac{3}{2}\right) = \theta_1$, say,

$$z_2 = -2 - 3i \Rightarrow \arg(-2 - 3i) = \arctan\left(\dfrac{-3}{-2}\right) = \arctan\left(\dfrac{3}{2}\right) = \theta_2, \text{ say.}$$

$$\theta_1 = 56 \cdot 3° \text{ or } 0 \cdot 982^c, \quad \theta_2 = -123 \cdot 7° = -2 \cdot 16^c$$

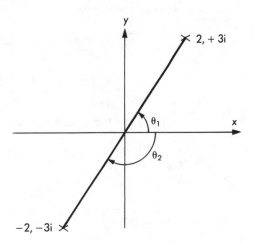

Fig. 11.3

POLAR FORM	The representation of a complex number in terms of its modulus r and argument θ is often referred to as the **polar form** of the complex number.

$$\Rightarrow z = x + iy = r\cos\theta + ir\sin\theta \equiv r \angle \theta$$

Note: $z_1 = r_1 \angle \theta_1$, $z_2 = r_2 \angle \theta_2$ and $z_1 = z_2 \Leftrightarrow r_1 = r_2$ and $\theta_1 = \theta_2$.

Worked example

Find the modulus and argument of the complex number $z = -\dfrac{1}{2} + i\dfrac{\sqrt{3}}{2}$

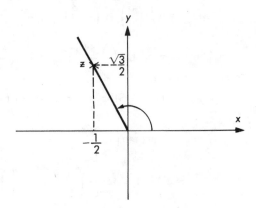

Fig. 11.4

$$z = -\frac{1}{2} + i\frac{\sqrt{3}}{2} = x + iy \Rightarrow x = -\frac{1}{2}, \ y = \frac{\sqrt{3}}{2}$$

$$|z| = \sqrt{(x^2+y^2)} = \sqrt{\left[\left(-\frac{1}{2}\right)^2 + \left(\frac{\sqrt{3}}{2}\right)^2\right]} = \sqrt{\left(\frac{1}{4} + \frac{3}{4}\right)} = 1$$

$$\arg z = \arctan\left(\frac{y}{x}\right) = \arctan\left[\left(\frac{\sqrt{3}}{2}\right)\Big/\left(-\frac{1}{2}\right)\right] = \arctan(-\sqrt{3}) = 180° - 60° = 120°$$

since z lies in the second quadrant of the Argand diagram.

PRODUCT AND QUOTIENT IN POLAR FORM

1 Product $z_1 z_2 = r_1 \angle \theta_1 \cdot r_2 \angle \theta_2$
$$= r_1(\cos\theta_1 + i\sin\theta_1) \cdot r_2(\cos\theta_2 + i\sin\theta_2)$$
$$= r_1 r_2[(\cos\theta_1\cos\theta_2 - \sin\theta_1\sin\theta_2) + i(\sin\theta_1\cos\theta_2 + \cos\theta_1\sin\theta_2)]$$
$$= r_1 r_2[\cos(\theta_1 + \theta_2) + i\sin(\theta_1 + \theta_2)] \equiv (r_1 r_2) \angle (\theta_1 + \theta_2).$$
$$\text{i.e. we multiply the moduli and add the arguments.}$$

Worked example

$z_1 \equiv 2\angle 135°, \ z_2 \equiv 3\angle 63°$

$z_1 z_2 \equiv (2 \times 3) \angle (135° + 63°) = 6\angle 198° \equiv 6\angle(-162°)$; $-162°$ being the principal value of the argument

2 Quotient $\dfrac{z_1}{z_2} = \dfrac{r_1 \angle \theta_1}{r_2 \angle \theta_2} = \dfrac{r_1}{r_2} \dfrac{(\cos\theta_1 + i\sin\theta_1)}{(\cos\theta_2 + i\sin\theta_2)} \cdot \dfrac{(\cos\theta_2 - i\sin\theta_2)}{(\cos\theta_2 - i\sin\theta_2)}$

$$= \frac{r_1}{r_2} \frac{(\cos\theta_1\cos\theta_2 + \sin\theta_1\sin\theta_2) + i(\sin\theta_1\cos\theta_2 - \cos\theta_1\sin\theta_2)}{(\cos^2\theta_2 + \sin^2\theta_2)}$$

$$= \frac{r_1}{r_2}\left[\frac{\cos(\theta_1 - \theta_2) + i\sin(\theta_1 - \theta_2)}{1}\right] \equiv \frac{r_1}{r_2} \angle (\theta_1 - \theta_2)$$

i.e. we divide the moduli and subtract the arguments

Worked example

$z_1 = 3\angle 124°, \ z_2 = 4\angle(-16°)$

$$\frac{z_1}{z_2} = \frac{3\angle 124°}{4\angle(-16°)} = 0{\cdot}75 \angle [(124° - (-16°)] = 0{\cdot}75 \angle 140°$$

Worked example

Show that $z_1 = \dfrac{9 + i\sqrt{3}}{5 - i\sqrt{3}}$ has modulus $\sqrt{3}$ and argument $\dfrac{\pi}{6}$

Find the polar form of $z_2 = \dfrac{3}{2}(1 - i\sqrt{3})$. Hence or otherwise show that $z_1{}^4 + z_2{}^2$ is real

$$z_1 = \frac{9 + i\sqrt{3}}{5 - i\sqrt{3}} \cdot \frac{5 + i\sqrt{3}}{5 + i\sqrt{3}} = \frac{42 + i14\sqrt{3}}{28} = \frac{3}{2} + i\frac{\sqrt{3}}{2}$$

$$|z_1| = \sqrt{\left[\left(\frac{3}{2}\right)^2 + \left(\frac{\sqrt{3}}{2}\right)^2\right]} = \frac{\sqrt{12}}{2} = \sqrt{3}$$

$$\arg z_1 = \arctan\left(\frac{\sqrt{3}}{2}\Big/\frac{3}{2}\right) = \arctan\left(\frac{1}{\sqrt{3}}\right) = \frac{\pi}{6}$$

$$z_2 = \frac{3}{2}(1 - i\sqrt{3}) \Rightarrow |z_2| = \frac{3}{2}\sqrt{[1^2 + (-\sqrt{3})^2]} = 3$$

$$\arg z_2 = \arctan\left(\frac{-\sqrt{3}}{1}\right) = -\frac{\pi}{3}$$

$$z_1{}^4 + z_2{}^2 = \left(\sqrt{3}\angle\frac{\pi}{6}\right)^4 + \left(3\angle - \frac{\pi}{3}\right)^2 = 9\angle\frac{2\pi}{3} + 9\angle - \frac{2\pi}{3}$$

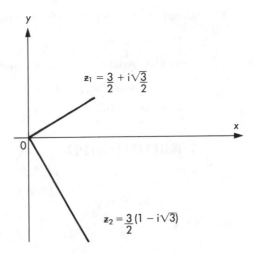

Fig. 11.5

$$= 9\left(\cos\frac{2\pi}{3}+i\sin\frac{2\pi}{3}\right) + 9\left[\cos\left(-\frac{2\pi}{3}\right)+i\sin\left(-\frac{2\pi}{3}\right)\right] = 18\cos\left(\frac{2\pi}{3}\right) = -9$$

That is, $z_1{}^4 + z_2{}^2$ is real.

GEOMETRICAL REPRESENTATION OF OPERATIONS

1 ADDITION AND SUBTRACTION

Let $z_1 = x_1 + iy_1$ and $z_2 = x_2 + iy_2$ and let P and Q be their representations respectively on an Argand diagram.

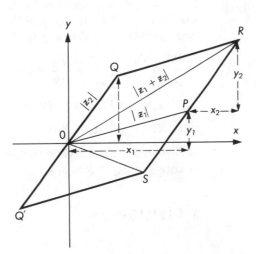

Fig. 11.6

It can be seen quite easily by algebraic and geometrical considerations that the sum $z_1 + z_2$ is represented on the Argand diagram by R, the vertex of the parallelogram $OPRQ$ where OP and OQ are two adjacent sides of the parallelogram. Thus complex numbers follow the laws of addition for ordinary vectors.

Similarly, subtraction can be regarded as addition
$z_1 - z_2 = z_1 + (-z_2)$ and if Q' represents $(-z_2)$ then the vertex S of parallelogram $OQ'SP$ represents $z_1 - z_2$.

You must be able to interpret distances and angles on an Argand diagram in terms of moduli and arguments.

Thus $OP = |z_1|$, $OQ = |z_2|$, $OR = |z_1 + z_2|$, $OS = |z_1 - z_2| = QP$ (by elementary geometry)

$\angle POx = \arg z_1$, $\angle QOx = \arg z_2$, $\angle ROx = \arg(z_1 + z_2)$, $\angle SOx = \arg(z_1 - z_2)$.

Further, since the sum of the lengths of any two sides of a triangle is greater than the length of the third side, we have from $\triangle OPR$.

$$OP + PR > OR \Rightarrow |z_1| + |z_2| > |z_1 + z_2|$$

However if $\arg z_1 = \arg z_2$, then OP and OQ would be parallel and in this case
$OP + OQ = OP + PR = OR$ or $|z_1| + |z_2| = |z_1 + z_2|$

Consequently, in general for two complex numbers z_1 and z_2,

$$|z_1| + |z_2| \geqslant |z_1 + z_2|$$

Note also by writing $z_1 - z_2$ for z_1 we obtained

$$|z_1 - z_2| + |z_2| \geqslant |(z_1 - z_2) + z_2| = |z_1|$$
$$\Rightarrow |z_1 - z_2| \geqslant |z_1| - |z_2|$$

2 MULTIPLICATION

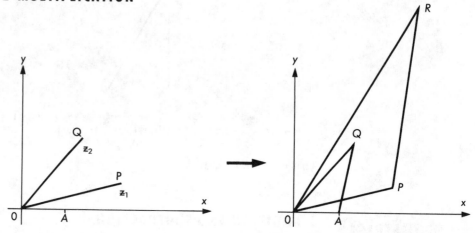

Fig. 11.7

Let $P \equiv z_1$, $Q \equiv z_2$ and $A \equiv 1 + i0$.

Construct $\triangle OPR$, as shown in Fig. 11.7, similar to $\triangle OAQ$, then $R \equiv z_1 z_2$

Since $\triangle OAQ$ is similar to $\triangle OPR$

$$\frac{OA}{OQ} = \frac{OP}{OR} \Rightarrow \frac{1}{|z_2|} = \frac{|z_1|}{OR} \text{ or } OR = |z_1| \cdot |z_2|$$

and $\angle AOQ = \angle POR \Rightarrow \angle POR = \arg z_2$ and hence

$$\angle ROx = \angle POx + \angle POR = \arg z_1 + \arg z_2$$

$\Rightarrow R$ represents a complex number whose modulus is $|z_1| \cdot |z_2|$ and whose argument is $\arg z_1 + \arg z_2 \Rightarrow R \equiv z_1 z_2$

3 DIVISION

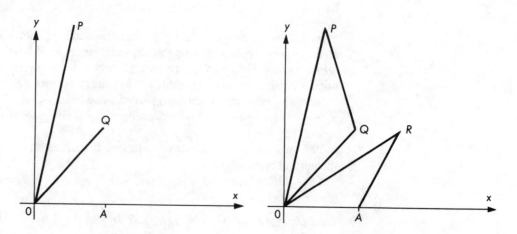

Fig. 11.8

Let $P \equiv z_1$, $Q \equiv z_2$, $A \equiv 1 + i0$

Construct $\triangle OAR$, as shown in Fig. 11.8, similar to $\triangle OQP$, then $R \equiv \dfrac{z_1}{z_2}$

Since $\triangle OAR$ is similar to $\triangle OQP$

$$\frac{OR}{OA} = \frac{OP}{OQ} \Rightarrow \frac{OR}{1} = \frac{|z_1|}{|z_2|} \text{ or } OR = \frac{|z_1|}{|z_2|}$$

and $\angle AOR = \angle QOP = \angle POx - \angle QOx = \arg z_1 - \arg z_2$

Hence R represents a complex number whose modulus is $\dfrac{z_1}{z_2}$

and whose argument is $\arg z_1 - \arg z_2 \Rightarrow R \equiv \dfrac{z_1}{z_2}$

4 MODULI AND ANGLES

Two extremely useful results which students tend to neglect concern the modulus and arguments involved in the complex numbers z_1 and z_2. If P and Q represent the complex numbers z_1 and z_2 respectively in an Argand diagram then if PV and QW are straight lines parallel to the real axis, as shown in Fig. 11.9,

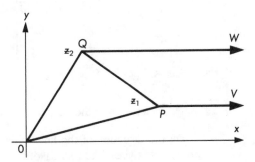

Fig. 11.9

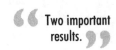
Two important results.

i) $PQ = |z_1 - z_2| = |z_2 - z_1|$

ii) $\angle QPV = \arg(z_2 - z_1)$, $-\angle WQP = \arg(z_1 - z_2)$

Worked example

Sketch the locus of a point $P \equiv z = x + iy$ on an Argand diagram when
i) $|z| = 3$,   ii) $|z - 2 - i| = 3$

i) If O is the origin and $P \equiv z$ then $OP = |z|$.
   $\Rightarrow OP = |z| = 3$
   $\Rightarrow$ distance between the origin O and a variable point P is always 3
   $\Rightarrow P$ must lie on a circle, centre the origin, radius 3 (see Fig. 11.10).

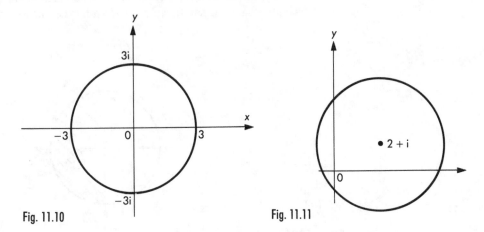

Fig. 11.10   Fig. 11.11

ii) $|z - 2 - i| = 3 \Rightarrow |z - (2 + i)| = 3$
   $\Rightarrow$ distance between a fixed point $2 + i$ and a variable point
      $P \equiv z$ is always 3
   $\Rightarrow P$ must lie on a circle, centre $2 + i$, radius 3 (see Fig. 11.11).

Worked example

Shade on separate Argand diagrams the regions for which

i) $\dfrac{2}{3}\pi > \arg(z-1) > -\dfrac{\pi}{2}$,   ii) $|z-1| < 2$ iii) $\dfrac{2\pi}{3} > \arg(z-1) > -\dfrac{\pi}{2}$ and $|z-1| < 2$

i) The argument of $(z-1)$ is the angle contained between the straight line joining $1 \equiv (1+i0)$ to the point z and the line through 1 parallel to the positive x (or real) axis.

Thus if $A \equiv 1$, $P_1 \equiv z$ then $\arg(z-1) = \dfrac{2\pi}{3} \Rightarrow \angle P_1 Ax = \dfrac{2\pi}{3}$ (see Fig. 11.12)

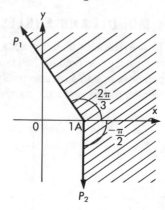

Fig. 11.12

Similarly for $P_2 \equiv z$ then $\arg(z-1) = -\dfrac{\pi}{2} \Rightarrow \angle P_2 Ax = -\dfrac{\pi}{2}$

Note: the direction of the lines must always be outwards from the right-hand term in the bracket, i.e. 1 in this case.

ii) $|z-1| < 2$ represents the inside of the circle centre $z = 1$ and having radius 2 (see Fig. 11.13).

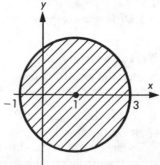

Fig. 11.13

iii) The final part of the question requires those parts of the shaded region in i) and ii) which are common, and these are shown by the shaded region in Fig. 11.14.

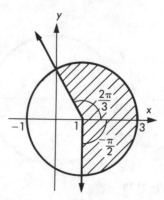

Fig. 11.14

EXERCISES 11.1

1 Find in the form $a + ib$ i) $(1+2i)(3-4i)$ ii] $\dfrac{1+2i}{3-4i}$

Hence express in similar form iii) $(1-2i)(3+4i)$ iv) $\dfrac{1-2i}{3+4i}$

2 Given that $z_1 = 3 + 4i$, $z_2 = \dfrac{5}{2} + \dfrac{5}{2}i$ and $z_3 = \dfrac{1}{z_1} + \dfrac{1}{z_2}$ find z_3 in the form

i) $(x + iy)$ ii) $r\angle\theta$

3 Express in the form $a + ib$ i) $(-7 + 4i)(3 - 4i)$ ii) $\dfrac{1}{5 - 3i} - \dfrac{1}{5 + 3i}$

iii) $\left(\cos\dfrac{\pi}{3} + i\sin\dfrac{\pi}{3}\right)\left(\cos\dfrac{\pi}{6} + i\sin\dfrac{\pi}{6}\right)$ iv) $(-1 + i)^3$

4 Given that $z_1 = \dfrac{2 - i}{2 + i}$ and $z_2 = \dfrac{2i - 1}{1 - i}$ find z_1 and z_2 in the form $x + iy$. Show on an

Argand diagram the points P and Q representing the complex numbers $5z_1 + 2z_2$, $5z_1 - 2z_2$ respectively

5 Find the modulus and argument of each root of the equation $z^2 - 4z + 8 = 0$

6 Given that $z_1 = -1 + i\sqrt{3}$, $z_2 = \sqrt{3} + i$, find z_1 and z_2 in polar form and show on an

Argand diagram points representing the complex numbers a) $z_1 z_2$, b) $\dfrac{z_1}{z_2}$

7 Find z in the form $x + iy$ given that z satisfies the equation $\dfrac{2z - 3}{1 - 4i} = \dfrac{3}{1 + i}$

8 Given that $|z| = 2$ and $\arg z = -\dfrac{2\pi}{3}$ find the values of a) z^2, b) $|z^2 + z|$, c) $\arg z^2$,

d) $\arg(z^2 + z)$

9 Given that the complex number z has modulus 5 and argument $\dfrac{\pi}{2}$

and is given by $z = \dfrac{a}{1 + 2i} + \dfrac{b}{1 - 3i}$ find the value of the real constants a and b

10 Solve the simultaneous equations

$|z + 1| + |z - 1| = 4$, $\arg(2z) = \dfrac{\pi}{2}$

11 Represent on an Argand diagram the loci given by:
a) $|z - 4| = 4$, b) $|z| = |z - 2|$. Determine the complex numbers corresponding to the points of intersection of these loci

12 Sketch the loci in an Argand diagram given by the equations
a) $|z + 1| = 2$, b) $|z - 1 - i| = |z|$

13 Show in separate diagrams the regions for which

a) $0 < \arg(z - 2) < \dfrac{\pi}{3}$, b) $|z - 2| < |z - 2i|$

14 Use modulus notation to express the fact that a point P, representing the complex number z in an Argand diagram lies inside the circle centre $(-4, 3)$, radius 5. If further the modulus of z is less than 5 show on an Argand diagram the region in which z lies

EXAMINATION TYPE QUESTIONS WITH MARKING SCHEME AND GUIDED SOLUTION

1 z is the point $x + iy$ in an Argand diagram and $\left|\dfrac{z}{z - 3}\right| = \dfrac{1}{2}$

Find the cartesian equation of the locus of z (6 marks)

$\left|\dfrac{z}{z - 3}\right| = \dfrac{1}{2} \Rightarrow \dfrac{|z|}{|z - 3|} = \dfrac{1}{2} \Rightarrow \dfrac{|x + iy|}{|x + iy - 3|} = \dfrac{1}{2}$ M1   M1

$$\Rightarrow \frac{|x+iy|^2}{|(x-3)+iy|^2} = \left(\frac{1}{2}\right)^2 \qquad \text{A1}$$

$$\Rightarrow \frac{x^2+y^2}{(x-3)^2+y^2} = \frac{1}{4} \qquad \begin{array}{l}\text{A1}\\\text{A1}\end{array} \quad \text{A1}$$

$$\Rightarrow 4(x^2+y^2) = (x-3)^2+y^2$$

Note: i) a considerable amount of algebraic manipulation is avoided by using the fact that the modulus of a quotient is the quotient of the individual moduli

ii) it is not necessary to simplify the answers beyond that shown unless you are specifically asked to express it in a particular form

2 Given that $z = 3-4i$ find in the form $x+iy$, $x,y \in \mathbb{R}$, i) z^2, ii) $\frac{1}{z}$.

Given that $w^2 = z$ find the values of w in cartesian form. *(12 marks)*

$z = 3-4i \Rightarrow z^2 = (3-4i)^2 = 9-24i+16i^2 = -7-24i$ B1

$\frac{1}{z} = \frac{1}{3-4i} = \frac{1}{3-4i}\cdot\frac{3+4i}{3+4i} = \frac{3+4i}{3^2+4^2} = \frac{3}{25}+i\frac{4}{25}$ M1   A1

Let $w = u+iv \Rightarrow (u+iv)^2 = 3-4i$
$\Rightarrow (u^2-v^2)+2uvi = 3-4i$ M1

Equating real parts and equating imaginary parts
$u^2-v^2 = 3, \ 2uv = -4$ M1   A1   A1

Substituting $v = -\frac{2}{u}$ in $u^2-v^2 = 3 \Rightarrow u^2-\left(-\frac{2}{u}\right)^2 = 3$ M1

$\Rightarrow u^4-3u^2-4 = 0 \Rightarrow (u^2-4)(u^2+1) = 0, \ u^2 = 4 \text{ or } -1$ A1   A1
But u^2 cannot be negative $\Rightarrow u^2 = 4$ or $u = \pm 2$
$u = 2, \ v = -1$ or $u = -2, \ v = 1$ M1
$\Rightarrow w = 2-i$ or $-2+i$ A1

RECENT BOARD EXAMINATION QUESTIONS

1 Given that $z = 3+4i$, express in the form $p+qi$, where p and $q \in \mathbb{R}$,

a) $\frac{1}{z}$, b) $\frac{1}{z^2}$.

Find the argument of $\frac{1}{z^2}$, giving your answer in degrees to 1 decimal place.   (L)

2 Given that $z_1 = -1+i\sqrt{3}$ and $z_2 = \sqrt{3}+i$, find arg z_1 and arg z_2.
Express z_1/z_2 in the form $a+ib$, where a and b are real, and hence find $\arg(z_1/z_2)$.
Verify that $\arg(z_1/z_2) = \arg z_1 - \arg z_2$.   (L)

3 Given that the real and imaginary parts of the complex number $z = x+iy$ satisfy the equation
$(2-i)x - (1+3i)y - 7 = 0,$
find x and y.
State the values of
a) $|z|$, b) $\arg z$.   (L)

4 Given that $z = a+ib$, where a and b are real, and that z satisfies the equation
$(2+i)(z+3i) = (7i-6),$
find the value of a and the value of b.

Show z on an Argand diagram, stating the modulus and the argument of z.

(AEB 1988)

5 Given that p and q are real and that $1+2i$ is a root of the equation
$z^2+(p+5i)z+q(2-i)=0$
determine   a) the values of p and q,
            b) the other root of the equation.

(AEB 1987)

6 Express the complex number

$$z = \frac{2+9j}{4-2j}$$

in the form $x+yj$, and calculate its modulus r and argumentθ.
Express the moduli and arguments of $-z, z^{-1}, z^2$ and $\bar{z}$ (the conjugate of z), each in terms of r and θ.

(NI 1987)

7 The complex numbers u, v and w are related by

$$\frac{1}{u} = \frac{1}{v} + \frac{1}{w}.$$

Given that $v = 3+4i$ and $w = 4-3i$, find u in the form $x+iy$.

(WJEC)

8 The equation
$z^2+iz+p+10i=0$
has a complex root $2+iy$, where y is real. Given that p is real, find y and p.

(JMB)

9 Indicate on an Argand diagram the set S of complex numbers z which satisfy the inequality
$|z-(8+6i)| \leqslant 3$.
Find the least value of $|z|$ for $z \in S$. Calculate correct to three significant figures the corresponding value, θ, of arg z, where $-\pi < \theta \leqslant \pi$.
Mark on your diagram the least value, α, of arg z for $z \in S$.

(JMB)

10 The complex number z has modulus 4 and argument $\dfrac{\pi}{6}$.

The complex number w has modulus 2 and argument $-\dfrac{2\pi}{3}$.

i) Express z and w in the form $a+ib$, where a and b are integers or surds.
ii) Mark in an Argand diagram the points P and Q which represent z and w, respectively. Calculate PQ^2, giving your answer in the form $r+s\sqrt{3}$, where r and s are integers.

iii) Find the modulus and the argument of $\dfrac{z}{w}$ and w^2,

giving each argument between $-\pi$ and π.

(JMB)

ANSWERS TO EXERCISES

EXERCISES 11.1

1 i) $11+2i$; (ii) $\frac{1}{5}(-1+2i)$;

   iii) $11-2i$; iv) $\frac{1}{5}(-1-2i)$

2 i) $\frac{1}{25}(8-9i)$; ii) $0\cdot482 \angle(-48\cdot4°)$

3 i) $-5+40i$; ii) $\frac{3}{17}i$; iii) i; iv) $2+2i$

4 $\frac{1}{5}(3-4i), \frac{1}{2}(-3+i)$

5 $2\sqrt{2}, \angle\frac{\pi}{4}$; $2\sqrt{2}, \angle-\frac{\pi}{4}$

6 a) $2, \angle\frac{2\pi}{3}$; b) $2, \angle\frac{\pi}{6}$

7 $\frac{3}{4}(-1-5i)$

8 a) $-2+2\sqrt{3}i$; b) $2\sqrt{3}$;

   c) $\dfrac{2\pi}{3}$; d) $\dfrac{5\pi}{6}$

9 $-5, 10$

10 $i\sqrt{3}$

11 $1\pm i\sqrt{7}$

14 $|z+4-3i|<5$

ANSWERS TO EXAMINATION QUESTIONS

1 a) $\dfrac{3-4i}{25}$; b) $\dfrac{-7-24i}{625}$, $-106\cdot3°$

2 $\dfrac{2\pi}{3}, \dfrac{\pi}{6}, i$

3 a) $\sqrt{10}$; b) $-18\cdot4°$

4 $a=-1, b=1, \left(2^{\frac{1}{2}}, \dfrac{3\pi}{4}\right)$

5 a) $-1, 7$; b) $-7i$

6 $-\dfrac{1}{2}+2i$; $\dfrac{1}{2}\sqrt{17}$, $104°$; $r, 180°+\theta°$;

    $\dfrac{1}{r}, -\theta°; r^2, 2\theta°; r, -\theta°$

7 $\dfrac{7}{2}+\dfrac{1}{2}i$

8 $y=-3, p=2$

9 $7, 0\cdot644$

10 i) $z=2\sqrt{3}+2i; w=-1-\sqrt{3}i$;
   ii) $PQ^2 = 20+8\sqrt{3}$;

   iii) $\dfrac{z}{w} \equiv 2, \dfrac{5\pi}{6}; w^2 \equiv 4, \dfrac{2\pi}{3}$

VECTORS

GETTING STARTED

Scalars: a quantity which has magnitude and no direction is referred to as a scalar quantity. It is specified by a real number when appropriate units have been chosen, e.g. length in metres, temperature in degrees centigrade.

Vectors: you should remember that a non-zero vector is a combination of (i) a positive real number, i.e. the magnitude or length of the vector and (ii) a direction in space. A vector is often represented by a segment of a straight line, the length of the line segment indicating the magnitude of the vector, and the direction of the line (indicated by an arrow), the direction of the vector. Consequently, such vectors are often denoted by $\overrightarrow{PQ}$, where P and Q are the end points of the segment and the arrow indicates the direction. An alternative notation is to denote the vector by a single bold letter say **F**. The magnitude of the vector is usually denoted by $|\overrightarrow{PQ}|$ or PQ and $|\mathbf{F}|$ or F.

PROPERTIES OF VECTORS

❝ **Eight important properties of vectors.** ❞

ESSENTIAL PRINCIPLES

Equality

$\mathbf{a} = \mathbf{b}$ provided i) $|\mathbf{a}| = |\mathbf{b}|$ and ii) the direction of $\mathbf{a}$ is the same as the direction of $\mathbf{b}$. Consequently, the actual position in space of the line segment representing a vector is immaterial so long as it has the correct length and direction.

Further, $\overrightarrow{PQ} = -\overrightarrow{QP}$

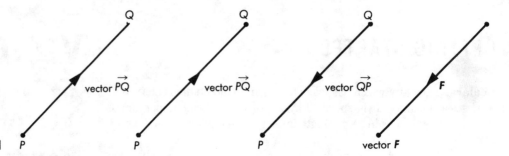

Fig. 12.1

Zero or null vector

Any vector of zero magnitude. It is denoted by $\mathbf{0}$, $\Rightarrow \overrightarrow{PQ} + \overrightarrow{QP} = \mathbf{0}$ or $\mathbf{a} + (-\mathbf{a}) = \mathbf{0}$

Multiplication by a scalar

Multiplication of a vector $\mathbf{a}$ by a positive scalar quantity, say λ, changes the magnitude of the vector to $\lambda|\mathbf{a}|$, but leaves its direction unaltered.

Unit vector

$\mathbf{a}$ is defined to be a **unit vector** when $|\mathbf{a}| = 1$.
It is often denoted by $\hat{\mathbf{a}}$.

Note: $\dfrac{\mathbf{b}}{|\mathbf{b}|} = \hat{\mathbf{b}}$, a unit vector in the direction of $\mathbf{b}$.

Addition or sum of vectors

The addition or sum of two vectors $\mathbf{a}$ and $\mathbf{b}$ is defined by the so-called **triangle** or **parallelogram law**.

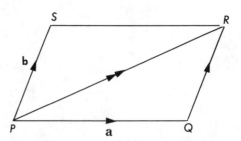

Fig. 12.2

For $\mathbf{a} = \overrightarrow{PQ}$, $\mathbf{b} = \overrightarrow{PS}$ then $\mathbf{a} + \mathbf{b} = \overrightarrow{PQ} + \overrightarrow{PS} = \overrightarrow{PQ} + \overrightarrow{QR} = \overrightarrow{PR}$, where $PQRS$ is a parallelogram.

$\overrightarrow{PR}$ can be looked upon as either the vector represented by the diagonal of the parallelogram $PQRS$ formed by $\mathbf{a} = \overrightarrow{PQ}$ and $\mathbf{b} = \overrightarrow{PS}$ or, as the third side of triangle PQR whose other two sides $\overrightarrow{PQ} = \mathbf{a}$ and $\overrightarrow{QR} = (\overrightarrow{PS}) = \mathbf{b}$ represent the given vectors whose sum is required, drawn in the order which ensures continuity of the direction of the vectors.

Repeated application of the triangle law leads to the polygon of vectors, i.e. the sum of any number of vectors can be obtained by joining them end to end so that their directions are in the same sense. The **resultant** is then given in both magnitude and direction by the join of the initial point to the final point.

$\mathbf{a} + \mathbf{b} + \mathbf{c} + \mathbf{d} + \mathbf{e} = \overrightarrow{OR} = \mathbf{r}$ as shown in Fig. 12.3

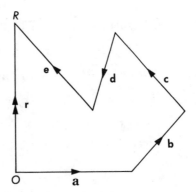

Fig. 12.3

From the polygon of vectors we can deduce

i) $\mathbf{a}+\mathbf{b}=\mathbf{b}+\mathbf{a}$ (commutative law)
ii) $(\mathbf{a}+\mathbf{b})+\mathbf{c}=\mathbf{a}+(\mathbf{b}+\mathbf{c})$ (associative law)

Difference of two vectors
The difference between two vectors, say $\mathbf{a}-\mathbf{b}$ is defined as $\mathbf{a}+(-\mathbf{b})$ i.e. as the sum of vectors $\mathbf{a}$ and $(-\mathbf{b})$.

Position vector

Given O, a fixed point, say the origin, then $\mathbf{r}=\overrightarrow{OP}$ is said to be the **position vector** of the point P relative to O.

Ratio formula
Given that A and B have position vectors $\mathbf{a}$ and $\mathbf{b}$ respectively, with respect to a fixed origin O, and R a point on AB such that $AR:RB=m:n$ then the position vector $\mathbf{r}$ of R with respect to O is given by:

$$\mathbf{r}=\frac{n\mathbf{a}+m\mathbf{b}}{m+n}$$

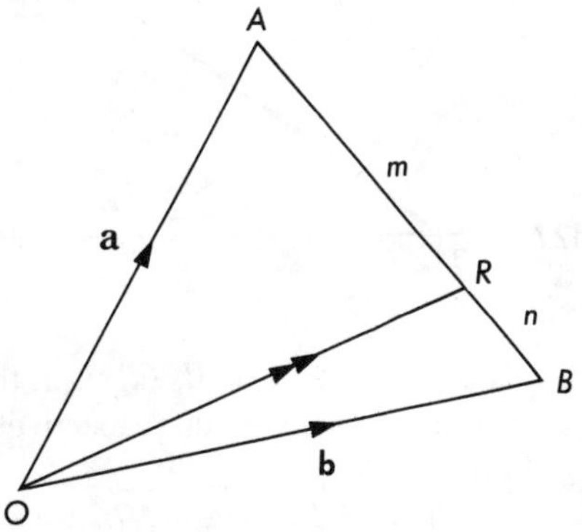

Fig. 12.4

Proof: $\overrightarrow{OA}+\overrightarrow{AB}=\overrightarrow{OB}\Rightarrow\overrightarrow{AB}=\overrightarrow{OB}-\overrightarrow{OA}=\mathbf{b}-\mathbf{a}$

$$\Rightarrow\overrightarrow{AR}=\frac{m}{m+n}\overrightarrow{AB}=\frac{m}{m+n}(\mathbf{b}-\mathbf{a})$$

$$\Rightarrow\mathbf{r}=\overrightarrow{OR}=\overrightarrow{OA}+\overrightarrow{AR}=\mathbf{a}+\frac{m}{m+n}(\mathbf{b}-\mathbf{a})=\frac{n\mathbf{a}+m\mathbf{b}}{m+n}$$

Worked example

Given $\triangle ABC$ such that $\overrightarrow{AB}=\mathbf{a}$, $\overrightarrow{AC}=\mathbf{b}$, find the position vector with respect to A of G, the intersection of the medians.

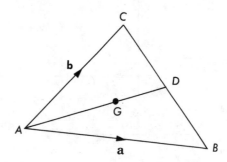

Fig. 12.5

If D is the mid-point of BC, then G lies on AD such that $AG:GD = 2:1$ (see Fig. 12.5).

$$\mathbf{a} + \overrightarrow{BC} = \mathbf{b} \Rightarrow \overrightarrow{BC} = \mathbf{b} - \mathbf{a} \Rightarrow \overrightarrow{BD} = \tfrac{1}{2}(\mathbf{b} - \mathbf{a})$$

$$\overrightarrow{AD} = \overrightarrow{AB} + \overrightarrow{BD} = \mathbf{a} + \tfrac{1}{2}(\mathbf{b} - \mathbf{a}) = \tfrac{1}{2}(\mathbf{a} + \mathbf{b})$$

$$\Rightarrow \overrightarrow{AG} = \tfrac{2}{3}\overrightarrow{AD} = \tfrac{2}{3}.\tfrac{1}{2}(\mathbf{a} + \mathbf{b}) = \tfrac{1}{3}(\mathbf{a} + \mathbf{b})$$

Worked example

Given that $|\mathbf{a}| = 7$, $|\mathbf{b}| = 3$ and $|\mathbf{a} - \mathbf{b}| = 6$ find the angle between vectors $\mathbf{a}$ and $\mathbf{b}$.

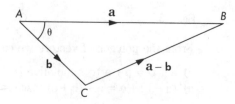

Fig. 12.6

$$\overrightarrow{AC} + \overrightarrow{CB} = \overrightarrow{AB} \qquad \Rightarrow \overrightarrow{CB} = \overrightarrow{AB} - \overrightarrow{AC} = \mathbf{a} - \mathbf{b}$$

$$AB = |\overrightarrow{AB}| = |\mathbf{a}| = 7,\ AC = |\overrightarrow{AC}| = |\mathbf{b}| = 3,\ CB = |\overrightarrow{CB}| = |\mathbf{a} - \mathbf{b}| = 6$$

$$\Rightarrow \text{for } \angle CAB = \theta \qquad \Rightarrow \cos\theta = \frac{7^2 + 3^2 - 6^2}{2.7.3} = \frac{49 + 9 - 36}{42} = \frac{11}{21}$$

$$\Rightarrow \theta = 58\cdot4°$$

COMPONENTS OF A VECTOR

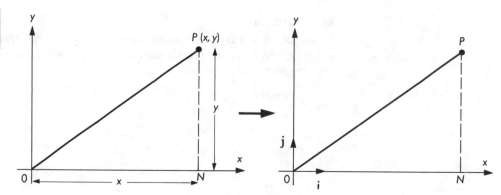

Fig. 12.7

Consider a point P, (x, y), in the xy plane and unit vectors $\mathbf{i}$ and $\mathbf{j}$ in the direction of the x-axis and the y-axis respectively. Let N be the foot of the ordinate from P (see Fig. 12.7).

$$\text{Then } \overrightarrow{OP} = \overrightarrow{ON} + \overrightarrow{NP} \Rightarrow \mathbf{r} = x\mathbf{i} + y\mathbf{j}$$

and x and y are said to be the components of $\mathbf{r}$ in the directions of Ox (i.e. the direction of $\mathbf{i}$) and Oy, (i.e. the direction of $\mathbf{j}$) respectively.

Similarly for three-dimensional space and a point $P(x, y, z)$ and unit vectors $\mathbf{i}$, $\mathbf{j}$ and $\mathbf{k}$ in the directions of the x-axis, the y-axis and the z-axis respectively

$$\overrightarrow{OP} = \mathbf{r} = x\mathbf{i} + y\mathbf{j} + z\mathbf{k}$$

Further $OP = |\overrightarrow{OP}| = \sqrt{(x^2 + y^2 + z^2)}$

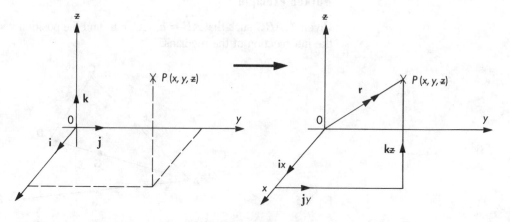

Fig. 12.8

OPERATIONAL RULES FOR VECTORS IN COMPONENT FORM

Let $\mathbf{a} = x_1\mathbf{i} + y_1\mathbf{j} + z_1\mathbf{k}$ and $\mathbf{b} = x_2\mathbf{i} + y_2\mathbf{j} + z_2\mathbf{k}$, then

1 $\mathbf{a} = \mathbf{b} \Rightarrow x_1 = x_2,\ y_1 = y_2,\ z_1 = z_2$

2 $\mathbf{a} \pm \mathbf{b} = (x_1 \pm x_2)\mathbf{i} + (y_1 \pm y_2)\mathbf{j} + (z_1 \pm z_2)\mathbf{k}$

3 $\lambda\mathbf{a} = \lambda x_1\mathbf{i} + \lambda y_1\mathbf{j} + \lambda z_1\mathbf{k}$ where λ is a scalar quantity

4 $|\mathbf{a}| = (x_1^2 + y_1^2 + z_1^2)^{\frac{1}{2}},\ |\mathbf{b}| = (x_2^2 + y_2^2 + z_2^2)^{\frac{1}{2}}$

5 If point A has position vector $\mathbf{a}$ and point B has position vector $\mathbf{b}$, then

$$AB = |\overrightarrow{AB}| = |\mathbf{b} - \mathbf{a}| = [(x_2 - x_1)^2 + (y_2 - y_1)^2 + (z_2 - z_1)^2]^{\frac{1}{2}}$$

EQUATIONS OF A STRAIGHT LINE

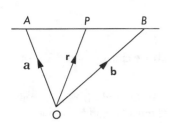

Fig. 12.9

1 Equation of a straight line passing through a fixed point A, position vector $\mathbf{a}$ and parallel to the vector $\mathbf{u}$.
Let P, position vector $\mathbf{r}$ be any point on the straight line.

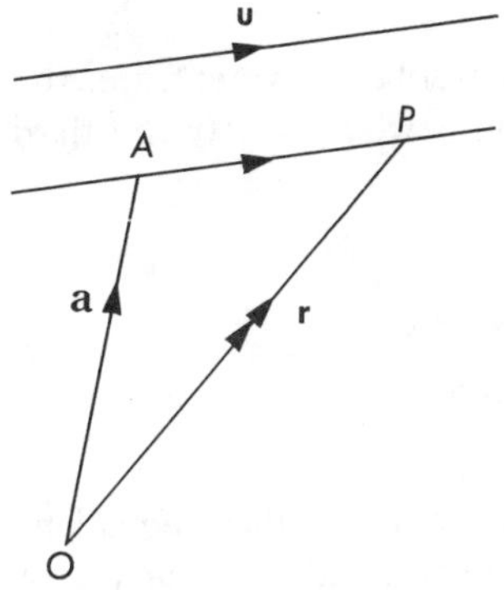

Then $\overrightarrow{AP}$ is parallel to $\mathbf{u} \Rightarrow \overrightarrow{AP} = \lambda\mathbf{u}$ where λ is a scalar variable dependent upon the position of P.
$\overrightarrow{OP} = \mathbf{r} = \overrightarrow{OA} + \overrightarrow{AP} = \mathbf{a} + \lambda\mathbf{u}$
$\Rightarrow \mathbf{r} = \mathbf{a} + \lambda\mathbf{u}$ is the vector equation of the required line.

If $P \equiv (x, y, z)$, $A \equiv (x_1, y_1, z_1)$ and $\mathbf{u} = l\mathbf{i} + m\mathbf{j} + n\mathbf{k}$ then

$\mathbf{r} = \mathbf{a} + \lambda\mathbf{u} \Rightarrow x\mathbf{i} + y\mathbf{j} + z\mathbf{k} = x_1\mathbf{i} + y_1\mathbf{j} + z_1\mathbf{k} + \lambda(l\mathbf{i} + m\mathbf{j} + n\mathbf{k})$

and comparing coefficients of $\mathbf{i}$, $\mathbf{j}$ and $\mathbf{k}$ $\Rightarrow$
$x = x_1 + \lambda l,\ y = y_1 + \lambda m,\ z = z_1 + \lambda n$

or $\dfrac{x - x_1}{l} = \dfrac{y - y_1}{m} = \dfrac{z - z_1}{n}\ (= \lambda)$

which is the cartesian equation of the line passing through the point (x_1, y_1, z_1) and parallel to the vector $\mathbf{u} = l\mathbf{i} + m\mathbf{j} + n\mathbf{k}$.
(l, m, n) are known as the direction ratios of the line.

2 Equation of a straight line passing through two fixed points A and B having position vectors $\mathbf{a}$ and $\mathbf{b}$ respectively relative to an origin O.

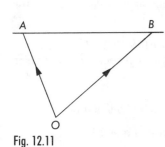

Fig. 12.10

$\overrightarrow{OA} + \overrightarrow{AB} = \overrightarrow{OB} \Rightarrow \overrightarrow{AB} = \overrightarrow{OB} - \overrightarrow{OA} = \mathbf{b} - \mathbf{a}$

Let P, position vector $\mathbf{r}$, be any point on the straight line.

$\overrightarrow{AP} = \lambda\overrightarrow{AB} = \lambda(\mathbf{b} - \mathbf{a})$ where λ is a scalar variable dependent upon the position of P relative to A and B.
$\overrightarrow{OP} = \mathbf{r} = \overrightarrow{OA} + \overrightarrow{AP} = \mathbf{a} + \lambda(\mathbf{b} - \mathbf{a})$
$\Rightarrow \mathbf{r} = \mathbf{a} + \lambda(\mathbf{b} - \mathbf{a})$ is the vector equation of the required line.

If $P = (x, y, z)$, $A = (x_1, y_1, z_1)$, $B = (x_2, y_2, z_2)$ then

$\mathbf{r} = \mathbf{a} + \lambda(\mathbf{b} - \mathbf{a}) \Rightarrow x\mathbf{i} + y\mathbf{j} + z\mathbf{k} = x_1\mathbf{i} + y_1\mathbf{j} + z_1\mathbf{k} + \lambda(x_2\mathbf{i} + y_2\mathbf{j} + z_2\mathbf{k} - x_1\mathbf{i} - y_1\mathbf{j} - z_1\mathbf{k})$

and comparing coefficients of $\mathbf{i}$, $\mathbf{j}$, $\mathbf{k}$ $\Rightarrow$
$x = x_1 + \lambda(x_2 - x_1),\ y = y_1 + \lambda(y_2 - y_1),\ z = z_1 + \lambda(z_2 - z_1)$

or $\dfrac{x - x_1}{x_2 - x_1} = \dfrac{y - y_1}{y_2 - y_1} = \dfrac{z - z_1}{z_2 - z_1}\ (= \lambda)$, called cartesian equations of the line.

Worked example

The position vectors of the points A and B are given by

$\overrightarrow{OA} = 2\mathbf{i} + \mathbf{j} - 4\mathbf{k},\ \overrightarrow{OB} = -2\mathbf{i} + 4\mathbf{j} - 2\mathbf{k}$

Find a vector equation of the line passing through A and B.
Find the position vector of the point where this line meets the plane $z = 0$

$\overrightarrow{OA} + \overrightarrow{AB} = \overrightarrow{OB} \Rightarrow \overrightarrow{AB} = \overrightarrow{OB} - \overrightarrow{OA} = (-2\mathbf{i} + 4\mathbf{j} - 2\mathbf{k}) - (2\mathbf{i} + \mathbf{j} - 4\mathbf{k})$
$= -4\mathbf{i} + 3\mathbf{j} + 2\mathbf{k}$

Fig. 12.11

A vector equation of AB is $\mathbf{r} = \overrightarrow{OA} + \lambda(\overrightarrow{AB})$

$\Rightarrow \mathbf{r} = (2\mathbf{i} + \mathbf{j} - 4\mathbf{k}) + \lambda(-4\mathbf{i} + 3\mathbf{j} + 2\mathbf{k})$

Meets plane $z = 0$ where $z = 0 \Rightarrow -4\mathbf{k} + 2\lambda\mathbf{k} = 0, \ \lambda = 2$

$\Rightarrow$ point has position vector $\mathbf{r} = (2 - 8)\mathbf{i} + (1 + 6)\mathbf{j} = -6\mathbf{i} + 7\mathbf{j}$

Worked example

Show that the lines whose vector equations are

$\mathbf{r} = 3\mathbf{i} + \mathbf{j} + \lambda(2\mathbf{j} + \mathbf{k})$ and $\mathbf{r} = 4\mathbf{k} + \mu(\mathbf{i} + \mathbf{j} - \mathbf{k})$ intersect, and find the position vector of their point of intersection

If the lines intersect, the position vector of their point of intersection satisfies the equation of each of the lines so that:

$\mathbf{r} = 3\mathbf{i} + \mathbf{j} + \lambda(2\mathbf{j} + \mathbf{k}) = 4\mathbf{k} + \mu(\mathbf{i} + \mathbf{j} - \mathbf{k})$

$\Rightarrow 3\mathbf{i} + (1 + 2\lambda)\mathbf{j} + \lambda\mathbf{k} = \mu\mathbf{i} + \mu\mathbf{j} + (4 - \mu)\mathbf{k}$

Comparing corresponding terms $\quad \mathbf{i} \Rightarrow 3 = \mu \qquad\quad (1)$

$\mathbf{j} \Rightarrow 1 + 2\lambda = \mu \quad (2)$

$\mathbf{k} \Rightarrow \lambda = 4 - \mu \quad (3)$

If the lines intersect, these three equations in two unknowns will be consistent, i.e. the solution of any two of the equations will give values of λ and μ which satisfy the third equation.

From (1), $\mu = 3$

Substituting in (2) $1 + 2\lambda = 3 \Rightarrow \lambda = 1$

Check in (3), L.H.S. $= \lambda = 1$, R.H.S. $= 4 - \mu = 4 - 3 = 1 = $ L.H.S.

$\Rightarrow \lambda = 1, \ \mu = 3$ satisfies all three equations, i.e. the lines intersect.

They do so where $\mathbf{r} = 3\mathbf{i} + \mathbf{j} + 1(2\mathbf{j} + \mathbf{k}) = 3\mathbf{i} + 3\mathbf{j} + \mathbf{k}$

Note: each vector equation is in fact equivalent to three equations. These three equations are those derived by equating the coefficients of $\mathbf{i}$, $\mathbf{j}$ and $\mathbf{k}$ on both sides of the vector equation.

EXERCISES 12.1

1   Point A has position vector $\mathbf{a}$, point B has position vector $\mathbf{b}$, point C has position vector $3\mathbf{a} + 2\mathbf{b}$ and point D has position vector $2\mathbf{b} - \mathbf{a}$. Express the vectors $\overrightarrow{BA}$, $\overrightarrow{DB}$, $\overrightarrow{CD}$, $\overrightarrow{AB} + 2\overrightarrow{BC}$, $2\overrightarrow{DB} + 3\overrightarrow{CA}$ in terms of $\mathbf{a}$ and $\mathbf{b}$

2   Given that $\mathbf{a} = 2\mathbf{i} + \mathbf{j} - 3\mathbf{k}$, $\mathbf{b} = \mathbf{i} - 2\mathbf{j} + \mathbf{k}$ and $\mathbf{c} = -3\mathbf{i} + 4\mathbf{j}$
find in terms of $\mathbf{i}$, $\mathbf{j}$ and $\mathbf{k}$ \ \ i) $3\mathbf{a} + \mathbf{b}$ \ \ ii) $2\mathbf{a} - 3\mathbf{c}$ \ \ iii) $\mathbf{a} - \mathbf{b} + \mathbf{c}$ and iv) $-2\mathbf{a} + 3\mathbf{b} - 4\mathbf{c}$
Find the unit vector in the direction of $\mathbf{a} + \mathbf{b} + \mathbf{c}$

3   Coplanar points $\mathbf{A}$, $\mathbf{B}$, $\mathbf{C}$ and $\mathbf{D}$ have position vectors $\mathbf{a}$, $\mathbf{b}$, $5\mathbf{a}$ and $3\mathbf{b}$ respectively where $\mathbf{a}$ and $\mathbf{b}$ are non-zero, non-parallel vectors. Find in terms of $\mathbf{a}$ and $\mathbf{b}$ the position vector of the point of intersection of AB and CD

4   Points A, B and C have position vectors $\mathbf{a}$, $\mathbf{b}$ and $\mathbf{c}$ respectively with respect to a fixed origin O. State the position vector of P, the mid-point of AB and find the position vector of the point which divides the line segment PC internally in the ratio $3:4$

5   Points A and B have coordinates $(3, 3, -2)$ and $(5, -1, 4)$ respectively. Find a vector equation of the straight line passing through the mid-point of AB and parallel to the vector $\mathbf{a} = \mathbf{i} - 2\mathbf{j} - 4\mathbf{k}$.
Find the coordinates of the point where this line meets the plane $y = 3$

6   Show that the points whose position vectors are $2\mathbf{i} + 5\mathbf{j} + 6\mathbf{k}$ and $-7\mathbf{i} + 2\mathbf{j} - 9\mathbf{k}$ lie on the line with equation $\mathbf{r} = -\mathbf{i} + 4\mathbf{j} + \mathbf{k} + \lambda(3\mathbf{i} + \mathbf{j} + 5\mathbf{k})$.
Find the coordinates of the point of intersection of this line and the line with vector equation $\mathbf{r} = 2\mathbf{i} - 3\mathbf{j} - 10\mathbf{k} + \mu(-2\mathbf{i} + 2\mathbf{j} + 2\mathbf{k})$

7   Show that the lines with vector equations $\mathbf{r} = 2(1 - \lambda)\mathbf{i} + (3\lambda - 1)\mathbf{j} + (1 + 2\lambda)\mathbf{k}$ and $\mathbf{r} = (10 + 4\mu)\mathbf{i} + (2 - \mu)\mathbf{j} + (8 + \mu)\mathbf{k}$ intersect and find the position vector of their point of intersection.

THE SCALAR OR DOT PRODUCT

The **scalar** or **dot product** of two vectors $\mathbf{a}$ and $\mathbf{b}$, written $\mathbf{a} \cdot \mathbf{b}$ is defined by

$\mathbf{a} \cdot \mathbf{b} = |\mathbf{a}| \cdot |\mathbf{b}| \cos\theta = ab \cos\theta \qquad$ where θ is the angle between $\mathbf{a}$ and $\mathbf{b}$.

It is called the scalar product because the answer is a scalar quantity. The alternative name 'dot product' is derived from the use of the 'dot' to denote multiplication. The dot *must not* be left out; **a b** has no meaning.

Note:

1   $\mathbf{a} \cdot \mathbf{b} = ab\cos\theta = \mathbf{b} \cdot \mathbf{a}$

2   If **a** and **b** are parallel, $\theta = 0$ and $\mathbf{a} \cdot \mathbf{b} = |\mathbf{a}| \cdot |\mathbf{b}| = ab$
   For unit vectors **i**, **j** and **k**, $\mathbf{i} \cdot \mathbf{i} = \mathbf{j} \cdot \mathbf{j} = \mathbf{k} \cdot \mathbf{k} = 1$

3   If **a** and **b** are perpendicular, $\theta = \dfrac{\pi}{2}$ and $\cos\dfrac{\pi}{2} = 0$, so $\mathbf{a} \cdot \mathbf{b} = 0$

   For the unit vectors **i**, **j** and **k**,
   $\mathbf{i} \cdot \mathbf{j} = \mathbf{j} \cdot \mathbf{i} = 0, \ \mathbf{j} \cdot \mathbf{k} = \mathbf{k} \cdot \mathbf{j} = 0, \ \mathbf{k} \cdot \mathbf{i} = \mathbf{i} \cdot \mathbf{k} = 0$

4   For $\mathbf{a} = a_1\mathbf{i} + a_2\mathbf{j} + a_3\mathbf{k}$ and $\mathbf{b} = b_1\mathbf{i} + b_2\mathbf{j} + b_3\mathbf{k}$ then
   $\mathbf{a} \cdot \mathbf{b} = a_1 b_1 + a_2 b_2 + a_3 b_3$ and $|\mathbf{a}|^2 = \mathbf{a} \cdot \mathbf{a} = a_1^2 + a_2^2 + a_3^2$

5   i) $\mathbf{a} \cdot (\mathbf{b} \pm \mathbf{c}) = \mathbf{a} \cdot \mathbf{b} \pm \mathbf{a} \cdot \mathbf{c}$
   ii) $(\mathbf{a} \pm \mathbf{b}) \cdot \mathbf{c} = \mathbf{a} \cdot \mathbf{c} \pm \mathbf{b} \cdot \mathbf{c}$
   iii) $(\lambda\mathbf{a}) \cdot \mathbf{b} = \lambda(\mathbf{a} \cdot \mathbf{b}) = \mathbf{a} \cdot (\lambda\mathbf{b})$

6   The angle θ between two non-zero vectors **a** and **b** is given by $\cos\theta = \dfrac{\mathbf{a} \cdot \mathbf{b}}{|\mathbf{a}| \cdot |\mathbf{b}|}$

Worked example

Given that $\mathbf{a} = 2\mathbf{i} - \mathbf{j} + 3\mathbf{k}$ and $\mathbf{b} = \mathbf{i} - 2\mathbf{k}$, find $\mathbf{a} \cdot \mathbf{b}$ and hence determine the angle between **a** and **b**

$\mathbf{a} \cdot \mathbf{b} = (2\mathbf{i} - \mathbf{j} + 3\mathbf{k}) \cdot (\mathbf{i} - 2\mathbf{k}) = 2 - 6 = -4$

$|\mathbf{a}|^2 = \mathbf{a} \cdot \mathbf{a} = 4 + 1 + 9 = 14, \ |\mathbf{b}|^2 = 1 + 4 = 5$

$$\Rightarrow \cos\theta = \frac{\mathbf{a} \cdot \mathbf{b}}{|\mathbf{a}| \cdot |\mathbf{b}|} = \frac{-4}{\sqrt{14} \cdot \sqrt{5}} = \frac{-4}{\sqrt{70}} \Rightarrow \theta = 118 \cdot 6°$$

$\Rightarrow$ angle between **a** and **b** is $118 \cdot 6°$

Worked example

Given that $\overrightarrow{OA} = 3\mathbf{i} + \mathbf{j} - 2\mathbf{k}$, $\overrightarrow{OB} = \mathbf{i} - 2\mathbf{j} + 3\mathbf{k}$ and $\overrightarrow{OC} = -\mathbf{i} + 4\mathbf{j} + 2\mathbf{k}$, show that $\overrightarrow{OC}$ is perpendicular to $\overrightarrow{AB}$

$\overrightarrow{AB} = \overrightarrow{OB} - \overrightarrow{OA} = (\mathbf{i} - 2\mathbf{j} + 3\mathbf{k}) - (3\mathbf{i} + \mathbf{j} - 2\mathbf{k}) = -2\mathbf{i} - 3\mathbf{j} + 5\mathbf{k}$

$\overrightarrow{OC}$ is perpendicular to $\overrightarrow{AB}$ provided $\overrightarrow{OC} \cdot \overrightarrow{AB} = 0$

But $\overrightarrow{OC} \cdot \overrightarrow{AB} = (-\mathbf{i} + 4\mathbf{j} + 2\mathbf{k}) \cdot (-2\mathbf{i} - 3\mathbf{j} + 5\mathbf{k}) = +2 - 12 + 10 = 0$

$$\Rightarrow \overrightarrow{OC} \text{ is perpendicular to } \overrightarrow{AB}$$

Worked example

Find a vector equation of the straight line passing through the points $A\,(11, 0, -1)$ and $B\,(-9, 4, 5)$ with respect to a fixed origin O. Find the coordinates of a point P on AB such that OP is perpendicular to AB. Hence find the distance of O from AB.

$A \equiv (11, 0, -1), \ B \equiv (-9, 4, 5) \Rightarrow \overrightarrow{AB} = \mathbf{i}(-9 - 11) + \mathbf{j}(4 - 0) + \mathbf{k}(5 + 1)$

$\Rightarrow \overrightarrow{AB} = -20\mathbf{i} + 4\mathbf{j} + 6\mathbf{k}$

$\Rightarrow$ Equation of AB is $\mathbf{r} = (11\mathbf{i} - \mathbf{k}) + \lambda(-20\mathbf{i} + 4\mathbf{j} + 6\mathbf{k})$

   or $\mathbf{r} = (11\mathbf{i} - \mathbf{k}) + \mu(-10\mathbf{i} + 2\mathbf{j} + 3\mathbf{k})$ where $\mu = 2\lambda$

The coordinates of any point on AB are $[(11 - 10\mu), 2\mu, (-1 + 3\mu)]$.

Let these be the coordinates of P, then $\overrightarrow{OP} = (11 - 10\mu)\mathbf{i} + 2\mu\mathbf{j} + (-1 + 3\mu)\mathbf{k}$

But OP is perpendicular to $AB \Rightarrow \overrightarrow{OP} \cdot \overrightarrow{AB} = 0$

$\Rightarrow [(11 - 10\mu)\mathbf{i} + 2\mu\mathbf{j} + (-1 + 3\mu)\mathbf{k}] \cdot [-20\mathbf{i} + 4\mathbf{j} + 6\mathbf{k}] = 0$

or $-20(11 - 10\mu) + 2\mu \cdot 4 + (-1 + 3\mu) \cdot 6 = 0 \Rightarrow 226\mu = 226 \Rightarrow \mu = 1$

$\Rightarrow P \equiv (1, 2, 2)$

Hence distance of O from $AB = |\overrightarrow{OP}| = \sqrt{(1^2 + 2^2 + 2^2)} = 3$

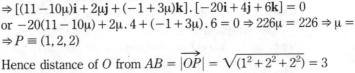

Fig. 12.12

EQUATIONS OF A PLANE

1 An equation of the plane through points A, B and C having position vectors **a**, **b** and **c** respectively with respect to an origin O

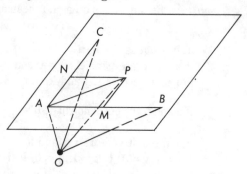

Fig. 12.13

Let P, position vector **r**, be any point in the plane.
Then constructing a parallelogram $AMPN$ such that AP is a diagonal and M and N are on AB and AC (or AB, AC produced if necessary) respectively
$$\Rightarrow \overrightarrow{AP} = \overrightarrow{AM} + \overrightarrow{MP} = \overrightarrow{AM} + \overrightarrow{AN}$$
But $\overrightarrow{AM} = s\overrightarrow{AB}$ and $\overrightarrow{AN} = t\overrightarrow{AC}$ where s and t are scalar variables depending upon the position of P in the plane with respect to A
$$\Rightarrow \overrightarrow{AP} = s\overrightarrow{AB} + t\overrightarrow{AC}$$
$$\Rightarrow \mathbf{r} - \mathbf{a} = s(\mathbf{b} - \mathbf{a}) + t(\mathbf{c} - \mathbf{a})$$
or $\mathbf{r} = (1 - s - t)\mathbf{a} + s\mathbf{b} + t\mathbf{c}$; a vector equation of plane ABC

2 An equation of a plane which passes through a fixed point A, position vector **a** with respect to an origin O and which is perpendicular to the direction given by a unit vector $\hat{\mathbf{n}}$ (or a vector **n**)

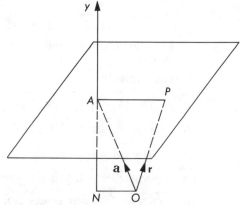

Fig. 12.14

Let P, position vector **r** be any point in the plane
$$\Rightarrow \overrightarrow{AP} = \mathbf{r} - \mathbf{a}$$
But AP is perpendicular to $\hat{\mathbf{n}} \Rightarrow \hat{\mathbf{n}} . (\mathbf{r} - \mathbf{a}) = 0$
$\Rightarrow \hat{\mathbf{n}} . \mathbf{r} = \hat{\mathbf{n}} . \mathbf{a} = 1 . OA\cos \angle NAO = AN$, where N is the foot of the perpendicular from the origin to the normal to the plane through A, i.e. AN = perpendicular distance p, of the origin from the plane.
$\Rightarrow$ vector equation of the plane is $\hat{\mathbf{n}} . \mathbf{r} = p$, where p is the perpendicular distance of the origin from the plane; or $\mathbf{n} . \mathbf{r} = \mathbf{n} . \mathbf{a}$ where **n** is any vector normal to the plane and **a** is the position vector of any point in the plane.

Cartesian form:
If $P \equiv (x, y, z)$, $A \equiv (x_1, y_1, z_1)$ and $\hat{\mathbf{n}} = l\mathbf{i} + m\mathbf{j} + n\mathbf{k}$ then
$\hat{\mathbf{n}} . \mathbf{r} = \hat{\mathbf{n}} . \mathbf{a} \Rightarrow (l\mathbf{i} + m\mathbf{j} + n\mathbf{k}) . (x\mathbf{i} + y\mathbf{j} + z\mathbf{k}) = (l\mathbf{i} + m\mathbf{j} + n\mathbf{k}) . (x_1\mathbf{i} + y_1\mathbf{j} + z_1\mathbf{k})$
$\Rightarrow lx + my + nz = lx_1 + my_1 + nz_1$ (or p)
or $l(x - x_1) + m(y - y_1) + n(z - z_1) = 0$
The general equation of a plane is therefore $Ax + By + Cz = D$
where A, B, C are the direction ratios of the normal to the plane and D is a constant. If the unit normal $\hat{\mathbf{n}}$ is used rather than **n** so that $\sqrt{(A^2 + B^2 + C^2)} = 1$ then A, B, C are called the **direction cosines** of the normal to the plane and D will be the distance of the origin from the plane.

Worked example

Find i) a vector equation, ii) a cartesian equation of the plane which passes through the point A $(1, -2, 3)$ and is normal to the vector $\mathbf{n}$ where $\mathbf{n} = \mathbf{i} - 2\mathbf{j} + 4\mathbf{k}$

Find the distance of the origin from the plane

Vector equation of the plane is $\mathbf{r} . \mathbf{n} = \mathbf{a} . \mathbf{n}$ where

$\mathbf{r} = x\mathbf{i} + y\mathbf{j} + z\mathbf{k}$, $\mathbf{a} = \overrightarrow{OA} = \mathbf{i} - 2\mathbf{j} + 3\mathbf{k}$ and $\mathbf{n} = \mathbf{i} - 2\mathbf{j} + 4\mathbf{k}$

$\Rightarrow (x\mathbf{i} + y\mathbf{j} + z\mathbf{k}) . (\mathbf{i} - 2\mathbf{j} + 4\mathbf{k}) = (\mathbf{i} - 2\mathbf{j} + 3\mathbf{k}) . (\mathbf{i} - 2\mathbf{j} + 4\mathbf{k}) = 1 + 4 + 12 = 17$

Vector equation is $\mathbf{r} . (\mathbf{i} - 2\mathbf{j} + 4\mathbf{k}) = 17$ and Cartesian form $\Rightarrow x - 2y + 4z = 17$

Rearranging so that the sum of the squares of the coefficients of x, y and z is 1

$$\Rightarrow \frac{x - 2y + 4z}{\sqrt{(1^2 + 2^2 + 4^2)}} = \frac{17}{\sqrt{(1^2 + 2^2 + 4^2)}} \Rightarrow \frac{1}{\sqrt{21}} (x - 2y + 4z) = \frac{17}{\sqrt{21}}$$

$\Rightarrow$ distance of the origin from the plane is $\dfrac{17}{\sqrt{21}}$

Worked example

With respect to a fixed origin O, a point P has position vector
$3\mathbf{i} - \mathbf{j} + 2\mathbf{k}$ and a plane Π has equation $\mathbf{r} . (2\mathbf{i} - 4\mathbf{j} - \mathbf{k}) = 8$

Show that P lies in the plane Π

The point Q has position vector $7\mathbf{i} - 9\mathbf{j}$

a) Show that $\overrightarrow{QP}$ is perpendicular to the plane Π
b) Calculate, to the nearest one tenth of a degree, $\angle OQP$

If P lies in the plane Π then the coordinates of P satisfy the equation
of $\Pi \Rightarrow \mathbf{r} . (2\mathbf{i} - 4\mathbf{j} - \mathbf{k}) = (3\mathbf{i} - \mathbf{j} + 2\mathbf{k}) . (2\mathbf{i} - 4\mathbf{j} - \mathbf{k}) = 6 + 4 - 2 = 8$
$\quad \Rightarrow P$ lies in the plane

$\overrightarrow{QP}$ is perpendicular to the plane Π provided $\overrightarrow{QP}$ is parallel to a vector which is normal to

the plane, i.e. provided $\overrightarrow{QP}$ is parallel to the vector $2\mathbf{i} - 4\mathbf{j} - \mathbf{k} = \mathbf{n}$, say

$\overrightarrow{QP} = \overrightarrow{OP} - \overrightarrow{OQ} = (3\mathbf{i} - \mathbf{j} + 2\mathbf{k}) - (7\mathbf{i} - 9\mathbf{j}) = -4\mathbf{i} + 8\mathbf{j} + 2\mathbf{k}$
$\qquad\qquad = -2(2\mathbf{i} - 4\mathbf{j} - \mathbf{k}) = -2\mathbf{n}$

$\Rightarrow \overrightarrow{QP}$ is perpendicular to the plane Π

$\overrightarrow{QO} = -7\mathbf{i} + 9\mathbf{j}$, $\overrightarrow{QP} = -4\mathbf{i} + 8\mathbf{j} + 2\mathbf{k}$

$\overrightarrow{QO} . \overrightarrow{QP} = |\overrightarrow{QO}| . |\overrightarrow{QP}| \cos \angle OQP \Rightarrow \cos \angle OQP = \dfrac{(-7\mathbf{i} + 9\mathbf{j}) . (-4\mathbf{i} + 8\mathbf{j} + 2\mathbf{k})}{|-7\mathbf{i} + 9\mathbf{j}| . |-4\mathbf{i} + 8\mathbf{j} + 2\mathbf{k}|}$

$\Rightarrow \cos \angle OQP = \dfrac{28 + 72}{\sqrt{(49 + 81)} . \sqrt{(16 + 64 + 4)}} = \dfrac{100}{\sqrt{130} . \sqrt{84}} = 0 \cdot 9569$

$\Rightarrow \angle OQP = 16 \cdot 9°$

EXERCISES 12.2

1   A plane passes through the points A, B and C which have position vectors $2\mathbf{i} - \mathbf{j} + \mathbf{k}$, $3\mathbf{i} + 2\mathbf{j} - \mathbf{k}$ and $-\mathbf{i} + 3\mathbf{j} + 2\mathbf{k}$ respectively with respect to a fixed origin O. Find
    a) a vector equation of the plane
    b) a cartesian equation of the plane
    c) a unit vector normal to the plane

2   A plane passes through the point A which has position vector $\mathbf{i} - 2\mathbf{j} + 2\mathbf{k}$ with respect to a fixed origin O. Find the equation of the plane given that the normal to the plane is parallel to the straight line with vector equation $\mathbf{r} = 2\mathbf{i} + \mathbf{j} - \mathbf{k} + \lambda(2\mathbf{i} + 3\mathbf{j} - \mathbf{k})$
    Find the point of intersection of the plane and the line

3   A plane Π has equation $\mathbf{r} . (2\mathbf{j} - 3\mathbf{k}) = 7$. Find
    a) a vector equation of a straight line passing through a point P with position vector
       $\mathbf{i} - 2\mathbf{j} + 5\mathbf{k}$ with respect to a fixed origin O and normal to the plane Π
    b) the distance of P from Π
    c) the cosine of $\angle ONP$ where N is the foot of the perpendicular from P to the plane Π

4   A vector equation of a line l with respect to an origin O is
$\mathbf{r} = (-6\mathbf{i} + 2\mathbf{j}) + \lambda(2\mathbf{i} - \mathbf{j} + 3\mathbf{k})$, where λ is a variable scalar. A plane Π has vector equation
$\mathbf{r} \cdot (-\mathbf{i} + \mathbf{j} + \mathbf{k}) = 8$
a) Show that l lies in the plane Π
b) Find the position vector of N the foot of the perpendicular from O to l
c) Find a vector equation of the plane containing O and l

5   Find a vector equation of the line l in the direction of the vector $3\mathbf{i} + 4\mathbf{k}$ and passing through the point A whose position vector is $\mathbf{i} - 2\mathbf{j} + \mathbf{k}$. Find also the point B where l meets the plane Π with vector equation $\mathbf{r} \cdot (\mathbf{i} - \mathbf{j} + 2\mathbf{k}) = -6$
The point C is at a distance 15 units from B in the direction $3\mathbf{i} + 4\mathbf{k}$. Find the position vector of the point in plane Π which is nearest to the point C

EXAMINATION QUESTIONS

1

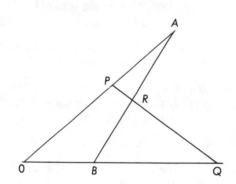

In the above figure, $\overrightarrow{OA} = \mathbf{a}$, $\overrightarrow{OB} = \mathbf{b}$, $OP:PA = 3:2$ and $OQ:QB = 3:2$.

a) Write down $\overrightarrow{AB}$, $\overrightarrow{OP}$, $\overrightarrow{OQ}$ and $\overrightarrow{PQ}$, giving each answer in terms of $\mathbf{a}$, $\mathbf{b}$ or $\mathbf{a}$ and $\mathbf{b}$.
b) Given that $AR = hAB$, where h is a scalar, show that
$\overrightarrow{OR} = (1 - h)\mathbf{a} + h\mathbf{b}$.
c) Given that $PR = kPQ$, where k is a scalar, write down an expression for $\overrightarrow{OR}$ in terms of $\mathbf{a}$, $\mathbf{b}$ and k.
d) Using the two expressions for $\overrightarrow{OR}$ obtained in b) and c), calculate the values of h and k and hence write down the value of the ratio $PR:RQ$.   (L)

2   The line L passes through the point $A(2, -9, 11)$ and is in the direction of the vector $\mathbf{n}$, where
$$\mathbf{n} = \begin{pmatrix} 3 \\ 6 \\ -2 \end{pmatrix}.$$
Write down a vector equation for L and show that L passes through the point $B(14, 15, 3)$.
The plane Π passes through the point $C(4, 7, 13)$ and is at right angles to $\mathbf{n}$. Find an equation for Π.

Verify that the resolved parts of $\overrightarrow{AC}$ and $\overrightarrow{BC}$ in the direction of $\mathbf{n}$ have the same magnitude but opposite signs. Give a geometrical interpretation of this result.   (JMB)

3   With respect to an origin O, the position vectors of the points L, M and N are
$a(4\mathbf{i} + 7\mathbf{j} + 7\mathbf{k})$, $a(\mathbf{i} + 3\mathbf{j} + 2\mathbf{k})$ and $a(2\mathbf{i} + 4\mathbf{j} + 6\mathbf{k})$
respectively, where a is a scalar constant.

a) Find the vectors $\overrightarrow{ML}$ and $\overrightarrow{MN}$.
b) Prove that $\cos\angle LMN = 9/10$.
c) Prove that the area of $\triangle LMN = (3a^2\sqrt{19})/2$.   (L)

4   Show that if **a** and **b** are two non-zero vectors, the vector
    n = (**b**.**b**)**a** − (**a**.**b**)**b**
    is perpendicular to **b**.
    Relative to an origin O the points A and B have position vectors $(c\mathbf{i}+c\mathbf{j}+c\mathbf{k})$ and
    $(c\mathbf{i}-c\mathbf{j}-c\mathbf{k})$ respectively, where c is a constant. Find
    a) a vector **m** which is perpendicular to the line OB and is parallel to the plane OAB,
    b) a vector equation of the line l_1 which passes through A and is parallel to **m**.
    A second line l_2 has equation
    $\mathbf{r} = (c\mathbf{i}-c\mathbf{j}-c\mathbf{k})+t(2\mathbf{i}-\mathbf{j}-\mathbf{k})$,
    where t is a parameter. Show that the lines l_1 and l_2 intersect and find the position
    vector of their point of intersection.   (L)

5   Planes Π_1 and Π_2 have equations given by
    $\Pi_1:\mathbf{r}.(2\mathbf{i}-\mathbf{j}+\mathbf{k}) = 0$,
    $\Pi_2:\mathbf{r}.(\mathbf{i}+5\mathbf{j}+3\mathbf{k}) = 1$
    a) Show that the point $A(2,-2,3)$ lies in Π_2.
    b) Show that Π_1 is perpendicular to Π_2.
    c) Find, in vector form, an equation of the straight line through A which is
       perpendicular to Π_1.
    d) Determine the coordinates of the point where this line meets Π_1.
    e) Find the perpendicular distance of A from Π_1.
    f) Find a vector equation of the plane through A parallel to Π_1.   (L)

6   With respect to a fixed origin O, the points L and M have position vectors
    $6\mathbf{i}+3\mathbf{j}+2\mathbf{k}$ and $2\mathbf{i}+2\mathbf{j}+\mathbf{k}$ respectively.

    a) Form the scalar product $\overrightarrow{OL}.\overrightarrow{OM}$ and hence find the cosine of angle LOM
    b) The point N is on the line LM produced such that angle MON is 90°. Find an
       equation for the line LM in the form $\mathbf{r} = \mathbf{a}+\mathbf{b}t$ and hence calculate the position
       vector of N.   (AEB 1988)

7   The position vectors of the points A, B and C, relative to a fixed origin O, are
    $\mathbf{a} = 6\mathbf{i}+4\mathbf{j}-\mathbf{k}$, $\mathbf{b} = 8\mathbf{i}+5\mathbf{j}-3\mathbf{k}$ and $\mathbf{c} = 2\mathbf{i}+8\mathbf{j}-5\mathbf{k}$ respectively.
    Find
    a) the vector $\overrightarrow{AB}$,
    b) the length of $\overrightarrow{AB}$,
    c) the cosine of angle ABC,
    d) the area of triangle ABC.

    Show that, for all values of the parameter t, the point P with position vector
    $\mathbf{p} = (8+2t)\mathbf{i}+(5+t)\mathbf{j}-(3+2t)\mathbf{k}$ lies on the straight line passing through A and B.
    Determine the value of t for which OP is perpendicular to AB and hence, or
    otherwise, calculate the shortest distance from O to the line AB.   (AEB 1987)

8   The position vectors of the points A, B, C are $\mathbf{a} = 4\mathbf{i}+10\mathbf{j}+6\mathbf{k}$, $\mathbf{b} = 6\mathbf{i}+8\mathbf{j}-2\mathbf{k}$,
    $\mathbf{c} = \mathbf{i}+10\mathbf{j}+3\mathbf{k}$ with respect to a fixed origin O.

    a) Show that angle ACB is a right angle.
    b) Find the area of the triangle ABC and hence, or otherwise, show that the shortest
       distance from C to AB is $3\sqrt{\left(\dfrac{3}{2}\right)}$.
    c) The point D lies on the straight line through A and C. Show that the vector
       $\overrightarrow{AD} = \lambda(\mathbf{i}+\mathbf{k})$ for some scalar λ.
       Given that the lengths AB and AD are equal, determine the possible position
       vectors of D.   (AEB 1989)

9   **i**, **j** and **k** are unit vectors parallel to the x-, y- and z-axes of a cartesian frame of
    reference $Oxyz$, origin O.
    A line L_1, passes through the point $(3, 6, 1)$ and is parallel to the vector $2\mathbf{i}+3\mathbf{j}-\mathbf{k}$.
    A line L_2 passes through the point $(3, -1, 4)$ and is parallel to the vector $\mathbf{i}-2\mathbf{j}+\mathbf{k}$.
    i) Using the form $\mathbf{r} = \mathbf{a}+\mathbf{b}t$, write down the vector equations of the lines L_1 and L_2.
       Show that the lines intersect and find the coordinates of the point of intersection.
    ii) What is the acute angle between the lines?

iii) The point A $(5, 9, 0)$ lies on L_1 whilst the point B $(5, a, b)$ lies on L_2. Find a and b and hence find the point C which lies on the line AB such that $AC:CB = 1:2$.

What is the magnitude of $\overrightarrow{OC}$?

What is the unit vector parallel to $\overrightarrow{OC}$? (NI 1987)

10

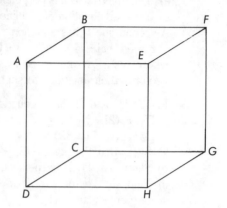

a) The above figure shows a cube. Calculate, to the nearest degree, the angles between
   i) AG and EF,
   ii) AG and BH,
   iii) AG and the plane $ABFE$.

b) The points A, B, C have position vectors
   $$\mathbf{OA} = -2\mathbf{i} + 3\mathbf{j} - 7\mathbf{k}$$
   $$\mathbf{OB} = \mathbf{i} + 7\mathbf{j} + 5\mathbf{k}$$
   $$\mathbf{OC} = 4\mathbf{i} + 3\mathbf{j} + \mathbf{k}$$
   relative to an origin O and a set of mutually perpendicular unit vectors $\mathbf{i}$, $\mathbf{j}$, $\mathbf{k}$. Show that
   $$\mathbf{AB} = 3\mathbf{i} + 4\mathbf{j} + 12\mathbf{k}$$
   and obtain a similar expression for $\mathbf{AC}$. Calculate $|\mathbf{AB}|$, $|\mathbf{AC}|$ and the scalar product $\mathbf{AB}.\mathbf{AC}$ and deduce that
   $$\llcorner BAC = \cos^{-1}\left(\frac{57}{65}\right).$$

   Write down an expression for the position vector of the point D which divides AC in the ratio $\lambda : 1 - \lambda$. By equating the scalar product $\mathbf{AC}.\mathbf{OD}$ to zero, deduce that the position vector of the foot of the perpendicular from O to AC is given by
   $$\frac{52}{25}\mathbf{i} + 3\mathbf{j} - \frac{39}{25}\mathbf{k}.$$
   (WJEC)

11 The points A and B have position vectors $2\mathbf{i} + 3\mathbf{j}$ and $\mathbf{i} + \mathbf{j}$ respectively, relative to the origin O. The point P lies on OA produced and is such that $OP = 2OA$; Q lies on OB produced and is such that $OQ = 3OB$. Find the vector equations of the lines AQ and BP and find also the position vector of the point of intersection of these two lines.

The point Z has position vector $\mathbf{k}$ relative to O. Show that the cosine of the acute angle between ZA and ZB is $\sqrt{(6/7)}$ and find the area of the triangle ZAB, giving your answer in surd form. (C)

12 The points A, B, C, D have position vectors $\mathbf{a}$, $\mathbf{b}$, $\mathbf{c}$, $\mathbf{d}$ given by
   $$\mathbf{a} = \mathbf{i} + 2\mathbf{j} + 3\mathbf{k},$$
   $$\mathbf{b} = \mathbf{i} + 2\mathbf{j} + 2\mathbf{k},$$
   $$\mathbf{c} = 3\mathbf{i} + 2\mathbf{j} + \mathbf{k},$$
   $$\mathbf{d} = 4\mathbf{i} - \mathbf{j} - \mathbf{k},$$
   respectively. The point P lies on AB produced and is such that $AP = 2AB$, and the point Q is the midpoint of AC.
   i) Show that PQ is perpendicular to AQ.
   ii) Find the area of the triangle APQ.
   iii) Find a vector perpendicular to the plane ABC.
   iv) Find the cosine of the acute angle between AD and BD. (C)

ANSWERS TO EXERCISES

EXERCISES 12.1

1 $\mathbf{a}-\mathbf{b}$, $\mathbf{a}-\mathbf{b}$, $-4\mathbf{a}$, $5\mathbf{a}+3\mathbf{b}$, $-4\mathbf{a}-8\mathbf{b}$

2 i) $7\mathbf{i}+\mathbf{j}-8\mathbf{k}$; ii) $13\mathbf{i}-10\mathbf{j}-6\mathbf{k}$;
iii) $-2\mathbf{i}+7\mathbf{j}-2\mathbf{k}$;

iv) $11\mathbf{i}-24\mathbf{j}+9\mathbf{k}$; $\dfrac{1}{\sqrt{13}}(3\mathbf{j}-2\mathbf{k})$

3 $6\mathbf{b}-5\mathbf{a}$

4 $\dfrac{1}{2}(\mathbf{a}+\mathbf{b})$, $\dfrac{2}{7}(\mathbf{a}+\mathbf{b})+\dfrac{3}{7}\mathbf{c}$

5 $\mathbf{r}=(4\mathbf{i}+\mathbf{j}+\mathbf{k})+\lambda(\mathbf{i}-2\mathbf{j}-4\mathbf{k})$, $(3,3,5)$

6 $(-4,3,-4)$

7 $-2\mathbf{i}+5\mathbf{j}+5\mathbf{k}$

EXERCISES 12.2

1 a) $\mathbf{r}=(2\mathbf{i}-\mathbf{j}+\mathbf{k})+\lambda(\mathbf{i}+3\mathbf{j}-2\mathbf{k})$
$\qquad+\mu(-3\mathbf{i}+4\mathbf{j}+\mathbf{k})$;
b) $11x+5y+13z=30$;

c) $\dfrac{1}{3\sqrt{35}}(11\mathbf{i}+5\mathbf{j}+13\mathbf{k})$

2 $[\mathbf{r}-(\mathbf{i}-2\mathbf{j}+2\mathbf{k})].(2\mathbf{i}+3\mathbf{j}-\mathbf{k})=0$;
$(0,-2,0)$

3 a) $\mathbf{r}=(\mathbf{i}-2\mathbf{j}+5\mathbf{k})+\lambda(2\mathbf{j}-3\mathbf{k})$;

b) $2\sqrt{13}$; c) $\dfrac{7}{\sqrt{78}}$

4 b) $-4\mathbf{i}+\mathbf{j}+3\mathbf{k}$;
c) $\mathbf{r}=\lambda(-6\mathbf{i}+2\mathbf{j})+\mu(-4\mathbf{i}+\mathbf{j}+3\mathbf{k})$

5 $\mathbf{r}=(\mathbf{i}-2\mathbf{j}+\mathbf{k})+\lambda(3\mathbf{i}+4\mathbf{k})$; $(-2,-2,-3)$;

$\left(\dfrac{3}{2}\mathbf{i}+\dfrac{7}{2}\mathbf{j}-2\mathbf{k}\right)$

ANSWERS TO EXAMINATION QUESTIONS

1 a) $\mathbf{b}-\mathbf{a}$; $\dfrac{3\mathbf{a}}{5}$, $3\mathbf{b}$, $3\mathbf{b}-\dfrac{3\mathbf{a}}{5}$;

d) $\dfrac{1}{2}$, $\dfrac{1}{6}$; $1:5$.

2 $\mathbf{r}=(2\mathbf{i}-9\mathbf{j}+11\mathbf{k})+\lambda(3\mathbf{i}+6\mathbf{j}-2\mathbf{k})$;
$\mathbf{r}=(4\mathbf{i}-7\mathbf{j}-13\mathbf{k}).(3\mathbf{i}+6\mathbf{j}+2\mathbf{k})=0$

3 a) $a(3,4,5)$, $a(1,1,4)$

4 a) $\mathbf{m}=2\mathbf{i}+\mathbf{j}+\mathbf{k}$;
b) $\mathbf{r}=\mathbf{i}c+\mathbf{j}c+\mathbf{k}c+\lambda(2\mathbf{i}+\mathbf{j}+\mathbf{k})$, $-c\mathbf{i}$

5 c) $\mathbf{r}=2\mathbf{i}-2\mathbf{j}+3\mathbf{k}+\lambda(2\mathbf{i}-\mathbf{j}+\mathbf{k})$;

d) $\left(-1,-\dfrac{1}{2},\dfrac{3}{2}\right)$; e) $\dfrac{3\sqrt{3}}{\sqrt{2}}$;

f) $\mathbf{r}.(2\mathbf{i}-\mathbf{j}+\mathbf{k})=9$

6 a) $\dfrac{20}{21}$; b) $\dfrac{1}{11}(-14,13,2)$

7 a) $2\mathbf{i}+\mathbf{j}-2\mathbf{k}$; b) 3; c) $\dfrac{5}{21}$;

d) $2\sqrt{26}$; $t=-3$; $\sqrt{17}$

8 $10\mathbf{i}+10\mathbf{j}+12\mathbf{k}$, $-2\mathbf{i}+10\mathbf{j}$

9 i) $\mathbf{r}=3\mathbf{i}+6\mathbf{j}+\mathbf{k}+\lambda(2\mathbf{i}+3\mathbf{j}-\mathbf{k})$;
$\mathbf{r}=3\mathbf{i}-\mathbf{j}+4\mathbf{k}+\mu(\mathbf{i}-2\mathbf{j}+\mathbf{k})$; $(1,3,2)$;
ii) $56 \cdot 9°$ iii) $a=-5$, $b=6$;

$C\equiv\left(5,\dfrac{13}{3},2\right)$; $\dfrac{\sqrt{430}}{3}$;

$\dfrac{1}{\sqrt{430}}(15\mathbf{i}+13\mathbf{j}+6\mathbf{k})$

10 a) $55°$, $71°$, $35°$; b) $6\mathbf{i}+8\mathbf{k}$; 13, 10, 114,
$(6\lambda-2)\mathbf{i}+3\mathbf{j}+(8\lambda-7)\mathbf{k}$

11 $\mathbf{r}=2\mathbf{i}+3\mathbf{j}+\lambda\mathbf{i}$;
$\mathbf{r}=(\mathbf{i}+\mathbf{j})+\mu(3\mathbf{i}+5\mathbf{j})$;

$\mathbf{r}=\dfrac{11}{5}\mathbf{i}+3\mathbf{j}$; $\sqrt{\left(\dfrac{6}{7}\right)}$

12 ii) 1; iii) $a\mathbf{j}$; iv) $\dfrac{10}{\sqrt{102}}$

GETTING STARTED

In mathematics we constantly attempt to explain physical happenings by relating a physical quantity x, say, to another physical quantity y. We usually produce sets of ordered pairs of the quantities (x_1, y_1), (x_2, y_2), $(x_3, y_3) \ldots (x_n, y_n)$ to form a relation.

There are various notations for expressing a relation other than listing the set of ordered pairs. If for instance, it is known that each y is the square of the corresponding x value then the relation can be expressed as

$$\{(x, y): y = x^2, \ x, \ y \in \mathbb{R}\}$$

The set of the first elements $\{x_1, x_2, x_3 \ldots x_n\}$ of the pairs of the relation is called the **domain** of the relation. The set of the second elements $\{y_1, y_2, y_3 \ldots y_n\}$ of the pairs of the relation is called the **range** of the relation.

A relation can be represented diagramatically as shown.

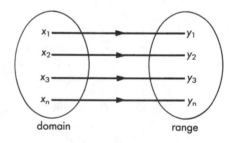

or

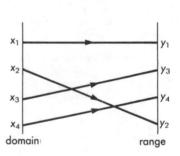

Fig. 13.1

ESSENTIAL PRINCIPLES

FUNCTIONS

A relation in which each element of the domain is associated with one and only one element of the range is called a **function** or **mapping**

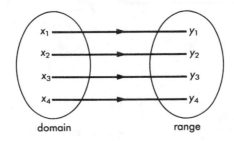

Fig. 13.2 A one–one mapping or function

For example $\{(x, y) : y = x + 1, \; x, \; y \in \mathbb{R}^+\}$ would be

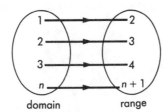

Fig. 13.3

or

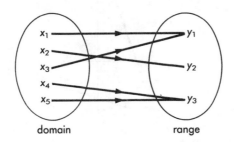

Fig. 13.4 A many–one mapping or function

For example $\{(x, y), \; y = x^2, \; x \in \mathbb{Z}, \; -3 \leqslant x \leqslant 3, \; x \neq 0\}$

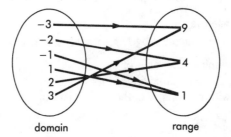

Fig. 13.5 A two–one mapping

Note: a one–many mapping is not a function. Functions are normally denoted by a single letter say f, g, F, G, etc. and consequently a notation frequently used for a function f say is
$$f : x \longmapsto$$
Thus for $y = f(x) = 3x + 2$, we write $f : x \longmapsto 3x + 2$
Often we loosely refer to 'the function $f(x) = 3x + 2$' but strictly speaking $f(x)$ is not the function, but rather the value of the function at x.

INVERSE FUNCTION

If a function f is such that it is one–one and maps an element x_n in the domain to an element y_n in the range then the function which maps the element y_n back to x_n is said to be the **inverse** of the function f and is denoted by f^{-1}. You must note that the function f has to be one–one, (one–many relations are not functions) and that the domain of f is then identical with the range of its inverse function f^{-1} and the range of f is identical with the domain of f^{-1}.

To obtain the inverse function f^{-1}, simply interchange the x and y in the equation of the function and then rearrange the equation to make y the subject.

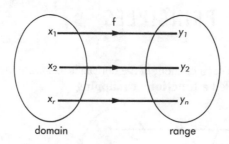
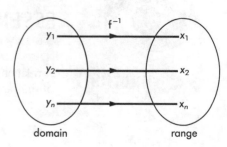

Fig. 13.6

domain   range   domain   range

Worked example

Find the inverse of the function f: $x \mapsto 3x + 2$, $x \in \mathbb{R}$. Sketch the graphs of the functions f and f^{-1}

$f : x \mapsto 3x + 2 \Rightarrow y = 3x + 2$; now interchange x and y

$$\Rightarrow x = 3y + 2 \Rightarrow y = \frac{1}{3}(x - 2) \Rightarrow f^{-1} : x \mapsto \frac{1}{3}(x - 2),\ x \in \mathbb{R}$$

The graphs of $y = f(x)$ and $y = f^{-1}(x)$ in this case are as shown in Fig. 13.7.

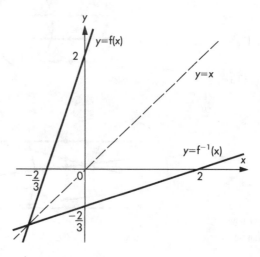

Fig. 13.7

You should note these graphs are reflections (mirror images) of each other in the line $y = x$

Worked example

The function f is defined by $f : x \mapsto x^2 - 6x + 2$, $0 \leqslant x \leqslant 7$

a) Find the minimum value of $f(x)$
b) Sketch the function and state its range
c) State, with reasons, whether f^{-1} exists or not

$y = f(x) = x^2 - 6x + 2$

$$\Rightarrow \frac{dy}{dx} = 2x - 6 \Rightarrow \frac{dy}{dx} = 0 \text{ when } x = 3$$

$$\frac{d^2y}{dx^2} = 2 = \text{positive} \Rightarrow y \text{ has a minimum value of } -7 \text{ when } x = 3$$

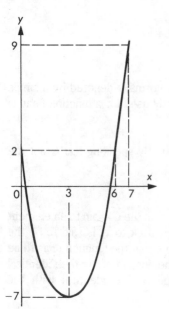

When $x = 0$, $y = 2$, when $x = 7$, $y = 9$, hence the graph of the function is as shown in Fig. 13.8.

The range of f is therefore $[-7, 9]$.

The function is not a one–one mapping for $0 \leqslant x \leqslant 6$, hence f^{-1} does not exist.

Note: 1 The end points of the domain do not necessarily give the end points of the range.

2 The straight line $y = C$, $-7 < C \leqslant 2$, cuts the graph of the function in two distinct points, showing that the function is not a one–one mapping and hence does not have an inverse for $0 \leqslant x \leqslant 6$. This is a good test as to whether a function possesses an inverse. i.e. graph the function and then, if and only if, all straight lines parallel to the x-axis meet the graph of f in at most one point will the function have an inverse.

Fig. 13.8

Worked example

State which of the following relations are functions and determine the inverse function in those cases where one exists

i) $R_1 = \{(x, y), y = x^2, x \in \mathbb{R}, x \neq 0\}$
ii) $R_2 = \{(x, y), y = \pm x^2, x \in \mathbb{R}, x \neq 0\}$
iii) $R_3 = \{(x, y), y = x^2, x \in \mathbb{R}^+\}$

i) $y = x^2$, $x \in \mathbb{R}$, $x \neq 0$

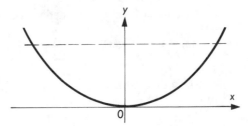

Fig. 13.9

Relation is a two–one mapping $\Rightarrow$ relation is a function, but does not have an inverse.

ii) $y = \pm x^2$, $x \in \mathbb{R}$, $x \neq 0$

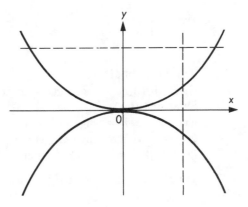

Fig. 13.10

Relation is a two–two mapping $\Rightarrow$ relation is not a function.

iii) $y = x^2$, $x \in \mathbb{R}$, $x \neq 0$

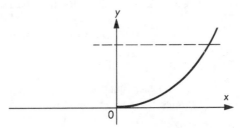

Fig. 13.11

Relation is a one–one mapping $\Rightarrow$ relation is a function which has an inverse.
$y = x^2$, $x \in \mathbb{R}^+ \Rightarrow x = +\sqrt{y} \Rightarrow y = +\sqrt{x}$, $x \in \mathbb{R}^+$
$\Rightarrow f^{-1} = \{(x, y), y = +\sqrt{x}, x \in \mathbb{R}^+\}$

COMPOSITE FUNCTIONS

> Learn the notation for a composite function used by your exam board.

Two functions f and g can be combined to give a **composite function**. Unfortunately, the GCE A-level boards to not accept a uniform notation for a composite function and you must therefore be familiar with that used by your board. The alternative notations are fg = f.g = f∘g ≡ f[g(x)], the f composite of g. For this function to exist the domain of f must be contained in the range of g.

You must note that the order of the composite function is important for in general f[g(x)] ≠ g[f(x)].

The composite function fg is determined by

i) finding g(x), and then
ii) finding f[g(x)]

Worked example

$f = \{(x, y), y = 3x + 1, x \in \mathbb{R}\}$

$g = \{(x, y), y = 2x - 4, x \in \mathbb{R}\}$

Find i) fg, ii) gf

$fg(x) = f(2x - 4) = 3(2x - 4) + 1 = 6x - 11, \Rightarrow fg: x \mapsto 6x - 11$

$gf(x) = g(3x + 1) = 2(3x + 1) - 4 = 6x - 2 \Rightarrow gf: x \mapsto 6x - 2$

Worked example

Given that f and g are functions defined by

$$f(x) = x + \frac{\pi}{2}, \; g(x) = \sin\left(\frac{1}{2}x - \frac{\pi}{4}\right), \; x \in \mathbb{R}$$

show that, when $x = \dfrac{5\pi}{4}$, $f[g(x)] - g[f(x)] = \dfrac{\pi}{2}$

$$f[g(x)] = \sin\left(\frac{1}{2}x - \frac{\pi}{4}\right) + \frac{\pi}{2} \; \Rightarrow \; f\left[g\left(\frac{5\pi}{4}\right)\right] = \sin\left(\frac{3\pi}{8}\right) + \frac{\pi}{2}$$

$$g[f(x)] = \sin\left[\frac{1}{2}\left(x + \frac{\pi}{2}\right) - \frac{\pi}{4}\right] = \sin\frac{1}{2}x \; \Rightarrow \; g\left[f\left(\frac{5\pi}{4}\right)\right] = \sin\left(\frac{5\pi}{8}\right) = \sin\left(\frac{3\pi}{8}\right)$$

$$\text{Hence } f\left[g\left(\frac{5\pi}{4}\right)\right] - g\left[f\left(\frac{5\pi}{4}\right)\right] = \frac{\pi}{2}$$

Worked example

Given that $f(x) = x^3$ and $g(x) = 2^x$, $x \in \mathbb{R}$, state the range of a) f, b) g, c) fg

Show that $(fg)^{-1} = g^{-1}f^{-1}$

$f(x) = x^3, \; x \in \mathbb{R} \Rightarrow$ range of f is $\mathbb{R}$

$g(x) = 2^x, \; x \in \mathbb{R} \Rightarrow$ range of g is $\mathbb{R}^+$

$fg(x) = f(2^x) = (2^x)^3 = 2^{3x} \Rightarrow$ range of fg is $\mathbb{R}^+$

$y = 2^{3x} \Rightarrow$ for the inverse $(fg)^{-1}$, $x = 2^{3y}$ or $y = \dfrac{1}{3}\log_2 x$

i.e. $(fg)^{-1}(x) = \dfrac{1}{3}\log_2 x$

$f(x) = x^3 \Rightarrow y = x^3 \Rightarrow$ for the inverse f^{-1}, $x = y^3$ or $y = x^{\frac{1}{3}}$

$g(x) = 2^x \Rightarrow y = 2^x \Rightarrow$ for the inverse g^{-1}, $x = 2^y$ or $y = \log_2 x$

$$\Rightarrow g^{-1}f^{-1}(x) = g^{-1}\left(x^{\frac{1}{3}}\right) = \log_2 x^{\frac{1}{3}} = \frac{1}{3}\log_2 x = (fg)^{-1}(x)$$

$$\Rightarrow (fg)^{-1} = g^{-1}f^{-1}$$

Composite functions can be extended to include for instance fgh or (fg)h for any f, g, h for which the composites are defined.

Worked example

Given $f: x \mapsto (2 + x)^3, \; x \in \mathbb{R}$

$g: x \mapsto \sin x, \; x \in \mathbb{R}$

$h: x \mapsto \sqrt{x}, \; x \in \mathbb{R}^+$

Find a) fgh(x),   b) (fg)h(x)

$gh(x) = g(\sqrt{x}) = \sin \sqrt{x}$

$\Rightarrow fgh(x) = f(\sin \sqrt{x}) = (2 + \sin \sqrt{x})^3$

$h(x) = \sqrt{x}, \; fg(x) = (2 + \sin x)^3$

$\Rightarrow (fg)h(x) = (fg)(\sqrt{x}) = (2 + \sin \sqrt{x})^3$

Note: in this case $(fg)h(x) \equiv fgh(x)$

This result is not only true for the above functions, but is true for any f, g, h for which the composites are defined.

THE EXPONENTIAL AND LOGARITHMIC FUNCTIONS

A function of the form $f: x \mapsto a^x$, $x \in \mathbb{R}$ and 'a' a positive constant is called an **exponential function**.

The graphs of the function when $a = 2, 3, 4$ are shown in Fig. 13.12.

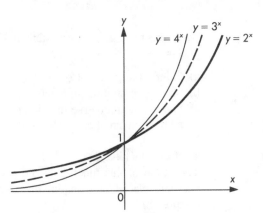

Fig. 13.12

You should note that whilst all of the curves pass through the point $(0, 1)$ they have a different gradient at this point, and that the higher the value of 'a' the greater is the value of this gradient. By plotting the curves accurately you can show that the gradient of the tangent to the curve $y = 2^x$ at $(0, 1)$ is approximately 0.7 whilst that for the curve $y = 3^x$ is approximately 1.1. There is therefore a value of 'a' between $a = 2$ and $a = 3$ such that the gradient of the tangent to the curve $y = a^x$ at $(0, 1)$ is 1. We use the letter e to denote this value of 'a' and call the special function e^x *the* exponential function.

It can be shown that e is the sum of the infinite series $1 + \dfrac{1}{1!} + \dfrac{1}{2!} + \dfrac{1}{3!} + \ldots + \dfrac{1}{n!} + \ldots$ which is approximately 2.71828.

> **Remember the logarithmic function is the inverse.**

The function $f: x \mapsto e^x$, $x \in \mathbb{R}$ is a one–one mapping and so possesses an *inverse*. Its inverse is the **logarithmic function**

$y = e^x \Rightarrow$ for the inverse $x = e^y$
$\Rightarrow \log_e x = \log_e e^y \Rightarrow y \log_e e = y \Rightarrow y = \log_e x = \ln x$

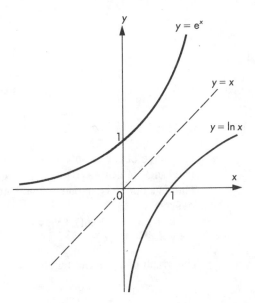

Fig. 13.13

The graphs of the exponential function and the logarithmic function are as shown in Fig. 13.13, each being the reflection of the other in the line $y = x$. You should note that no part of the curve with equation $y = \ln x$ lies in the 2nd or 3rd quadrant, i.e. $\ln x$ does not exist for $x < 0$.

EXAMINATION TYPE QUESTIONS COMPLETE WITH MARKING SCHEME

Functions f and g, each with domain $D = \{x : x \in \mathbb{R}, \ x > -1\}$, are defined by $f(x) = \ln(1+x)$, $g(x) = 1 + x^2$.

a) Sketch the graph of each function and state their ranges
b) Show that an inverse function f^{-1} exists and sketch the curve with equation $y = f^{-1}(x)$
c) Show that an inverse function $g^{-1}(x)$ does not exist, but that with a suitable restriction on the domain of g, which should be stated, an inverse will exist   (*12 marks*)

$y = f(x) = \ln(1+x)$
The graph of this function is the same as that of $y = \ln x$ but with the curve displaced one unit in the negative direction of the x-axis (see Chapter 8)

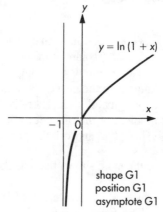

Fig. 13.14

$y = g(x) = 1 + x^2$
The graph of this function is part of a parabola (see Chapter 8)

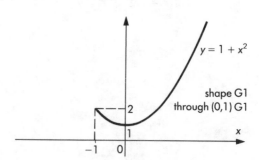

Fig. 13.15

The range of $f : x \mapsto \ln(1+x)$, $x > -1$ is $(-\infty, +\infty)$.   B1
The range of $g : x \mapsto 1 + x^2$, $x > -1$ is $(1, \infty)$.   B1
The function f is a one–one mapping and therefore f possesses an inverse.
$y = \ln(1+x) \Rightarrow x = \ln(1+y) \Rightarrow e^x = 1 + y$
    $\Rightarrow y = e^x - 1, \ x \in \mathbb{R}$ B1

The graph of f^{-1} is the mirror image of the graph of f in the line $y = x$

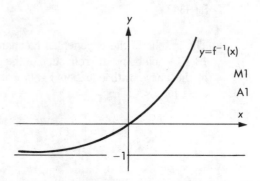

Fig. 13.16

The function g is not a one–one mapping and therefore g does not possess an inverse.   B1

A restriction of the domain of g to $x>0$, [or $-1<x<0$] would enable an inverse of g to exist.   B1

EXAMINATION QUESTIONS

1   The functions f and g are defined by

$$f:x \mapsto \frac{1}{2}x+3, \ x\in\mathbb{R}$$

$$g:x \mapsto \frac{1}{x}, \ x\in\mathbb{R} \text{ and } x \neq 0.$$

Write down, in a similar form,
a) the composite function fg,
b) the inverse function f^{-1} (L)

2   Two functions f and g are defined by

$$f:x \mapsto \frac{25}{3x-2}, \ x\in\mathbb{R}, \ 1<x\leqslant9,$$
$$g:x \mapsto x^2, \ x\in\mathbb{R}, \ 1<x\leqslant3.$$

Find
a) the range of f,
b) the inverse function f^{-1}, stating its domain,
c) the composite function fg, stating its domain,

d) the solutions of the equation $fg(x) = \dfrac{2}{x-1}$ (L)

3   The functions f and g are defined by
$$f:x \mapsto x-2, \ x>2,$$
$$g:x \mapsto e^x, \ x>0.$$
Given that the function h is defined by
$$h = gf, \ x>2,$$
state
a) the range of h,
b) the domain and range of h^{-1}
Sketch the curves with equations
c) $y = h(x),$
d) $y = h^{-1}(x).$ (L)

4   The functions f and g are defined by
$$f:x \mapsto x^2-3, \ x\in\mathbb{R},$$
$$g:x \mapsto 2x+5, \ x\in\mathbb{R}.$$
Find in a similar form the composite function $f\circ g$.
Sketch on separate axes the graphs of f and $f\circ g$.
Hence, or otherwise, show that the range of f corresponding to the domain $-4\leqslant x\leqslant4$ is $-3\leqslant f(x)\leqslant13$, and find the range of $f\circ g$ corresponding to this domain   (AEB 1986)

5   A quadratic function f, defined by $f:x \mapsto x^2+bx+c, \ x\in\mathbb{R}$, has the following properties:
$$f(3) = -17,$$
$$f'(3) = 0.$$
Find the value of b and the value of c and state the range of f.
Sketch the graph of f.
A second quadratic function g has the same rule as f and has domain all real $x\geqslant3$. Find

the inverse function g^{-1} of g in the form $g^{-1}:x \mapsto g^{-1}(x)$, stating clearly $g^{-1}(x)$ in terms of x *and* stating the domain of this inverse function.
Sketch in the same diagram the graphs of g and g^{-1} and by observing the symmetry of your graphs, or otherwise, calculate the value of x for which
$$g(x) = g^{-1}(x)$$
(AEB 1988)

6 The figure shows a sketch of the part of the graph of $y = f(x)$ for $0 \leqslant x \leqslant 2a$.

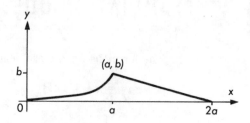

The line $x = 2a$ is a line of symmetry of the graph.
Sketch on separate axes the graphs of
a) $y = f(x)$ for $0 \leqslant x \leqslant 4a$,
b) $y = -f(2x)$ for $0 \leqslant x \leqslant a$,
c) $y = 3f(\frac{1}{2}x)$ for $0 \leqslant x \leqslant 2a$
d) $y = f(x - a)$ for $2a \leqslant x \leqslant 4a$
(AEB 1987)

7 The functions f and g are defined by
$$f(x) = x + 4, \ x > -4$$
$$g(x) = \ln x, \ x > 0.$$
Denoting by h the composite function gf, write down h(x) and state the domain and range of h. Find $h^{-1}(x)$.
Sketch the graphs of h(x) and $h^{-1}(x)$ on one diagram, labelling each graph clearly and indicating on the graphs the coordinates of any intersections with the axes.
(JMB)

8 The function F is defined for real x by
$$F(x) = 3e^x - 4.$$
Express F as a composite function of f, g and h, which are defined for real x by
$$f(x) = e^x, \ g(x) = x - 4, \ h(x) = 3x.$$
Sketch the graph of $y = F(x)$, indicating the coordinates of the points of intersection with the axes. State the equation of the asymptote.
The inverse function of F is F^{-1}. Find $F^{-1}(x)$ and state the domain of F^{-1}. (JMB)

9

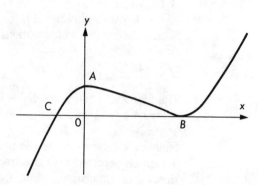

The curve shown in the diagram has equation $y = f(x)$. There is a maximum at the point $A(0, 1)$, a minimum at the point $B(4, 0)$ and the curve cuts the x-axis at the point $C(-1, 0)$. Sketch on separate diagrams, the graphs of

i) $y = f(x + 2)$,
ii) $y = -3f(x)$,
iii) $y = f(2x)$,

labelling each graph clearly and showing the coordinates of the points corresponding to A, B and C.
(C)

10 a) The functions f and g are defined by
$$f: x \mapsto e^x, \ x \in \mathbb{R}$$
and
$$g: x \mapsto x + 2, \ x \in \mathbb{R}.$$
  i) Express in a similar form the composite functions $f \circ g$ and $g \circ f$.
  ii) Show that $g \circ f(0) - f \circ g(-2) = 2$.
  iii) Sketch on the same axes the graphs of the functions f and $f \circ g$ indicating clearly which is which.
  Specify a simple transformation which maps the graph of f onto the graph of $f \circ g$.

  b) The function h is defined by $h: x \mapsto 2 + e^x, \ x \in \mathbb{R}$.
  Express in a similar form h^{-1}, the inverse function of h.
  State the domain of h^{-1}.
  Sketch h^{-1}.   (AEB 1989)

11 The functions, f, g and h are defined by
$$f: x \mapsto \ln x, \ (x \in \mathbb{R}^+),$$
$$g: x \mapsto \frac{1}{x}, \ (x \in \mathbb{R}^+),$$
$$h: x \mapsto x^2, \ (x \in \mathbb{R}).$$
  i) Give definitions of each of the functions fg and f^{-1}, and state, in each of the following cases, a relationship between the graphs of
   a) f and fg,
   b) f and f^{-1}.

  ii)

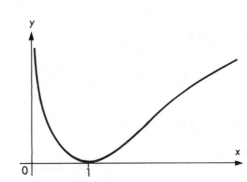

  The diagram shows a sketch of the graph of $y = hf(x)$. State how the sketch shows that hf is not one–one, and prove that, if α and β, where $0 < \alpha < \beta$, are such that $hf(\alpha) = hf(\beta)$, then $\alpha = g(\beta)$.
  iii) The function ϕ is defined by
$$\phi: x \mapsto hf(x), \qquad (0 < x \le 1).$$
  Sketch the graph of ϕ^{-1}, and give an explicit expression in terms of x for $\phi^{-1}(x)$.
                                                                        (C)

12 The functions f, g, h are defined for $x \in \mathbb{R}$ by
$$f(x) = x^2,$$
$$g(x) = \sqrt{x}, \ (x \ge 0)$$
$$h(x) = x + 2.$$
  Sketch on separate, clearly labelled diagrams the graphs of
   i) $y = f(x)$,
   ii) $y = g(x)$,
   iii) $y = gf(x)$,
   iv) $y = fh(x)$,
   v) $y = hf(x)$.
  Find $(gh)^{-1}(y)$, given that $y \ge 0$.   (C)

ANSWERS TO EXAMINATION QUESTIONS

1 a) $\text{fg}: x \mapsto \dfrac{1}{2x} + 3$;   b) $\text{f}^{-1}: x \mapsto 2x - 6$

2 a) $1 \leqslant f < 25$;

   b) $\text{f}^{-1}: x \mapsto \dfrac{25 + 2x}{3x}$, $1 \leqslant x < 25$;

   c) $\text{fg}: x \mapsto \dfrac{25}{3x^2 - 2}$;   d) $\dfrac{7}{6}$, 3

3 a) $(1, \infty)$;   b) $(1, \infty)$; $(2, \infty)$

4 $\text{f} \circ \text{g}: x \mapsto (2x + 5)^2 - 3$; $-3 \leqslant \text{f} \circ \text{g}(x) \leqslant 166$

5 $-6, -8$; $\geqslant -17$; $\text{g}^{-1}: x \mapsto 3 + (x + 17)^{\frac{1}{2}}$;
   $x \geqslant -17$; $x = 8$

7 $\text{h}: x \mapsto \ln(x + 4)$, $x > -4$, $(-\infty, \infty)$;
   $\text{h}^{-1}: x \mapsto \text{e}^x - 4$

8 $\text{f}(x) = \text{ghf}(x)$; $y = -4$;
   $\text{f}^{-1}: x \mapsto \ln\left(\dfrac{x + 4}{3}\right)$, $x > -4$

10 a) $\text{f} \circ \text{g}: x \mapsto \text{e}^{x+2}$, $\text{g} \circ \text{f}: x \mapsto \text{e}^x + 2$;
   b) $\text{h}^{-1}: x \mapsto \ln(x - 2)$; $x > 2$.

11 i) $-\ln x$, e^x;   iii) $\phi^{-1}: x \mapsto \text{e}^{-x^{\frac{1}{2}}}$, $x > 0$

12 $y^2 - 2$.

CHAPTER

LINEAR KINEMATICS

DISPLACEMENT–TIME SKETCHES

SPEED–TIME SKETCHES

UNIFORM ACCELERATION IN A STRAIGHT LINE

VERTICAL MOTION UNDER GRAVITY

NON-UNIFORM ACCELERATION IN A STRAIGHT LINE

SIMPLE HARMONIC MOTION

GETTING STARTED

Linear kinematics is concerned with the motion of a particle or body without reference to the forces acting or consideration of the laws under which motion takes place.

Remember that in mechanics we are trying to solve a practical problem by setting up what is called a mathematical model. As practical situations are usually very complicated we make assumptions about our models and terms such as 'frictionless horizontal plane', 'ignoring air resistance' and 'inextensible string' are often used. All of these are impossible to achieve in practice! These assumptions are made so that the equations we obtain from the models are within our capability to solve and they do give reasonable approximations to the real thing.

In nearly every case you will find it helpful to draw a *sketch diagram* which condenses the data given and assists you to write down meaningful equations.

Further, you will often need to introduce some symbols of your own to represent unknown quantities. You should never have more unknown quantities than equations in examination questions. If you have, then either something has been missed or something has been introduced unnecessarily. So check back and read the question again.

ESSENTIAL PRINCIPLES

DISPLACEMENT – TIME SKETCHES

Suppose data is collected, measured from a particular instant (say $t = 0$) where t is the time in seconds, for a particle P moving in a straight line. At $t = 1, 2, 3, 4, \ldots$ the displacement of P from O, the starting point, is noted in metres. A sketch *curve* can then be drawn and it may look like Fig. 14.1.

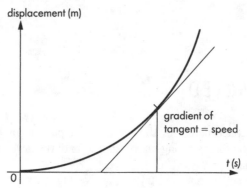

Fig. 14.1

The gradient of the tangent at a point on the curve is the speed of P in ms^{-1} at that instant.

SPEED–TIME SKETCHES

In a similar way, the speed of P can be recorded at different times t and from these data a **speed–time** sketch (often called a **velocity–time** sketch by examiners) can be drawn (see Fig. 14.2).

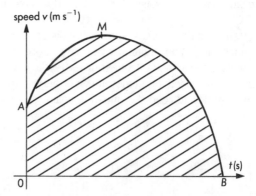

Fig. 14.2

The value of v at A is the initial speed (in ms^{-1}) and the value of t at B is the time (in s) when $v = 0$. The gradient of the tangent at a point on the curve gives the numerical value of the **acceleration**. Note that this is zero at the top M, positive between A and M and negative (indicating retardation) between M and B.

The numerical value of the area of the shaded region gives the distance (in metres) covered by P in moving from the point A (where $t = 0$) to B (where the speed is zero).

Worked example

The speed of a car at 2-second intervals is recorded in the following table. Estimate the distance covered in metres

time (s)	0	2	4	6	8	10
speed (ms^{-1}	8	13	17	19·5	21·5	23

A sketch of speed against time is drawn, and the data inserted as shown in Fig. 14.3.

The distance is estimated by using the trapezium rule to find the area under the curve

$$\text{Distance} \approx 2\left[\frac{8+13}{2} + \frac{13+17}{2} + \frac{17+19\cdot5}{2} + \frac{19\cdot5+21\cdot5}{2} + \frac{21\cdot5+23}{2}\right]\text{m}$$

$$= [8 + 23 + 2(13 + 17 + 19\cdot5 + 21\cdot5)]\,\text{m}$$

$$\approx 173\,\text{m}$$

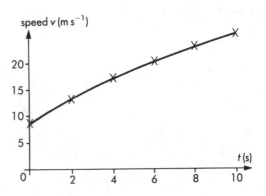

Fig. 14.3

UNIFORM ACCELERATION IN A STRAIGHT LINE

General formulae for motion in a line under constant acceleration

$$s = \left(\frac{u+v}{2}\right)t$$

$$v = u + at$$
$$v^2 = u^2 + 2as$$

❝ Some useful formulae. ❞

$$s = ut + \frac{1}{2}at^2$$

$$s = vt - \frac{1}{2}at^2$$

These formulae are used when a particle P moves in a straight line with acceleration of constant magnitude a, covering a distance s, in time t, with initial speed u and final speed v

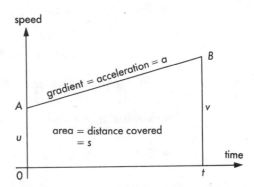

Fig. 14.4

With a speed–time sketch, the results are easily derived.

Gradient of $AB = a = \dfrac{v-u}{t} \Rightarrow v - u = at$ \hfill (1)

Distance covered $= s = \dfrac{1}{2}(v+u)t \Rightarrow v + u = \dfrac{2s}{t}$ \hfill (2)

Add (1) and (2) to get $s = vt - \dfrac{1}{2}at^2$

Subtract (1) from (2) to get $s = ut + \dfrac{1}{2}at^2$

Multiply (1) and (2) to get $v^2 - u^2 = 2as$

Many exam questions on uniform acceleration are best answered using a speed–time sketch, but many students prefer to use the formulae.
The following worked example illustrates both methods.

Worked example

A car is uniformly accelerated from rest at a set of traffic lights A and attains a speed of $10\,\mathrm{m\,s^{-1}}$ in $5\,\mathrm{s}$. The car then moves at constant speed $10\,\mathrm{m\,s^{-1}}$ for $6\,\mathrm{s}$ before it is uniformly retarded at $5\,\mathrm{m\,s^{-2}}$ and is brought to rest at a second set of traffic lights B

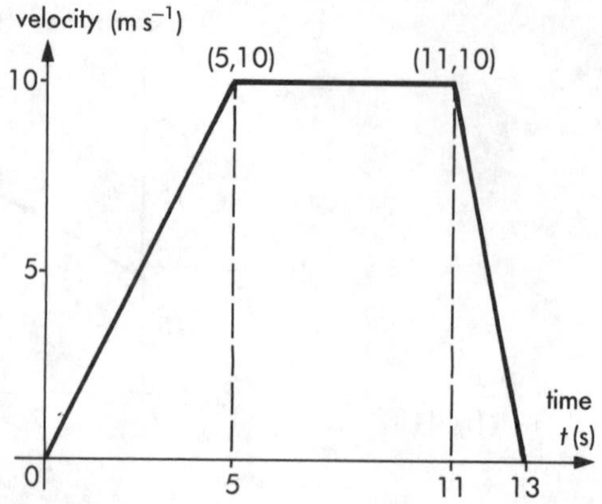

Fig. 14.5

We first draw a $v-t$ sketch

- **Acceleration stage** is from 0 to $(5, 10)$
- **Constant speed stage** is from $(5, 10)$ to $(11, 10)$
- **Retardation stage** is from $(11, 10)$ to $(13, 0)$ because it will take 2s under constant retardation $5\mathrm{ms}^{-2}$ to reduce speed from $10\mathrm{ms}^{-1}$ to zero.
- **Total distance** = area under graph from $t = 0$ to $t = 13$

$$= \frac{1}{2}(13 + 6) \times 10\mathrm{m}$$

$$= 95\mathrm{m}$$

Alternatively we could say by using the formulae:

- **Acceleration stage** $\quad s_1 = \left(\frac{u+v}{2}\right)t$ where $u = 0,\ v = 10,\ t = 5$

$$s_1 = \left(\frac{0+10}{2}\right)5\mathrm{m} = 25\mathrm{m}$$

> The formulae give precise results.

- **Constant speed stage** $s_2 = ut$, where $u = 10,\ t = 6$
  $$s_2 = 10(6)\mathrm{m} = 60\mathrm{m}$$
- **Retardation stage** $\quad v^2 = u^2 + 2as$ where $v = 0,\ u = 10,\ a = -5$
  $$0 = 100 - 2 \times 5 \times s_3 \Rightarrow s_3 = 10\mathrm{m}$$
- **Total distance** $= s_1 + s_2 + s_3 = (25 + 60 + 10)\mathrm{m} = 95\mathrm{m}$

VERTICAL MOTION UNDER GRAVITY

The formulae apply also when a particle is projected vertically upwards or allowed to fall freely under gravity when it is assumed that air resistance is ignored.

Worked example

A particle P is projected vertically upwards with initial speed $14\mathrm{ms}^{-1}$ from a point A. It comes to instantaneous rest at the point B. Calculate the distance AB and the time taken for P to reach B from A. Find also the time, in seconds, for which P is more than $7\mathrm{m}$ above the ground

In examples such as this the magnitude of g, the acceleration due to gravity, will be given. Here, we take the value as $9{\cdot}8\mathrm{ms}^{-2}$.
$u = 14,\ v = 0,\ a = -9{\cdot}8$
Using $v^2 = u^2 + 2as$ we obtain
$$0 = 196 - 2 \times 9{\cdot}8 \times AB$$
$$AB = 10\mathrm{m}$$
Using $v = u + at$
$$0 = 14 - 9{\cdot}8 \times t$$

Time from A to $B = \dfrac{14}{9{\cdot}8} = \dfrac{10}{7}\mathrm{s} = 1\dfrac{3}{7}\mathrm{s}$

At B, P is $10\mathrm{m}$ above the ground, and P will fall freely from rest at B. Time taken to fall $3\mathrm{m}$ (i.e. it is more than $7\mathrm{m}$ above ground) is, using

$$s = ut + \frac{1}{2}at^2 \Rightarrow 3 = \frac{1}{2} \times 9{\cdot}8 \times t^2$$

$$t = \sqrt{\frac{6}{9 \cdot 8}} = 0 \cdot 78 \, \text{s}$$

Time required is twice this $= 1 \cdot 56 \, \text{s}$ because it goes up and down.

EXERCISES 14.1

1 A car starts from rest at a point O and moves with constant acceleration of magnitude $3 \, \text{ms}^{-2}$ until it is moving with speed $12 \, \text{ms}^{-1}$ at the instant it passes through a point A. The car then moves with constant acceleration of magnitude $2 \, \text{ms}^{-2}$ for $5 \, \text{s}$, when it passes through a point B. Calculate
   a) the speed of the car at B
   b) the distance between O and B

2 At the instant when a car, travelling at constant speed $18 \, \text{ms}^{-1}$, passes a check-point, a motor cyclist starts from rest and moves with uniform acceleration of magnitude $3 \, \text{ms}^{-2}$ from the check-point. Calculate
   a) the time taken to overtake the car
   b) the distance covered by the motor cyclist before the car is overtaken

3 A lift accelerates from rest at a constant rate $a \, \text{ms}^{-2}$ to reach a speed of $v \, \text{ms}^{-1}$. It ascends with constant speed $v \, \text{ms}^{-1}$ for a certain time and is then uniformly retarded to rest at a constant rate $a \, \text{ms}^{-2}$. Altogether, the lift rises $H \, \text{m}$ in time $T \, \text{s}$. Sketch a speed–time graph and hence, or otherwise, show that v is the smaller root of the equation $v^2 - vTa + aH = 0$

4 A ball is thrown vertically upwards from ground level with initial speed $35 \, \text{ms}^{-1}$ and moves freely under gravity ($|g| = 9 \cdot 8 \, \text{ms}^{-2}$). Find the speed and direction of motion of the ball after a) $3 \, \text{s}$, b) $5 \, \text{s}$. Find also the *total* distance travelled by the ball between these instants

5 A car uniformly accelerates from rest at $0 \cdot 8 \, \text{ms}^{-2}$ for $10 \, \text{s}$. The car then uniformly retards through a distance of $15 \, \text{m}$ and comes to rest. Calculate
   a) the greatest speed of the car
   b) the total time during which the car is moving

6 The top of a cliff is $140 \, \text{m}$ above the beach. A pebble is dropped from rest at the top of the cliff and moves freely under gravity ($|g| = 9 \cdot 8 \, \text{ms}^{-2}$). Calculate
   a) the speed with which the pebble hits the beach
   b) the time of fall. (Give answers to 1 decimal place)

7 A particle P is projected from a point O vertically upwards with initial speed $49 \, \text{ms}^{-1}$. P moves freely under gravity ($|g| = 9 \cdot 8 \, \text{ms}^{-2}$) and comes to instantaneous rest at the point Y. Calculate
   a) the distance OY
   b) the time for which P is more than $60 \, \text{m}$ above O

NON-UNIFORM ACCELERATION IN A STRAIGHT LINE

For a particle P moving in a straight line, it is convenient to measure both time t and displacement x from a fixed point O, say, in the line of motion. That is, we choose O to be the initial position of P, thus implying that $t \geqslant 0$ and that $x = 0$ when $t = 0$.

Further, the point O partitions the line of motion, because for all points in the line on one side of 0, $x > 0$ and for all points on the other side $x < 0$.

The **velocity** of P at time t is defined to be v, where $v = \dfrac{dx}{dt}$

If $v > 0$, P is moving in the direction for which x is increasing.

The **acceleration** of P at time t is defined to be a, where $a = \dfrac{dv}{dt} = \dfrac{d^2x}{dt^2}$

If $a > 0$, the acceleration of P is acting in the direction for which x is increasing.

❝ Learn this expression for non-uniform acceleration. ❞

Since $a = \dfrac{dv}{dt} = \dfrac{dv}{dx}\dfrac{dx}{dt}$ and $\dfrac{dx}{dt} = v$, we have $\qquad a = v\dfrac{dv}{dx}$

This is an important alternative form in which the acceleration may be expressed, and it is used frequently and should be memorised. In examination questions the acceleration of a particle may be given in various ways. From this information a differential equation is formed and then solved to find the information required. The following examples illustrate the methods used.

Worked example

A particle moves in a straight line so that its acceleration at time t seconds is $(2t-6)\,\mathrm{m\,s^{-2}}$. Given that the initial velocity of the particle is $8\,\mathrm{m\,s^{-1}}$, find
a) the values of t at the points A and B when the particle is instantaneously at rest
b) the distance between A and B

a) Let the speed and displacement of the particle at time ts be $v\,\mathrm{m\,s^{-1}}$ and xm

The differential equation describing the motion is $\dfrac{dv}{dt} = 2t-6$

Integrating with respect to t
$v = t^2 - 6t + C$, where C is a constant.
We are given that $v = 8$ when $t = 0$, and therefore $C = 8$
$\Rightarrow v = t^2 - 6t + 8$
The particle is at rest when $v = 0$, $\Rightarrow t^2 - 6t + 8 = 0$
Since $t^2 - 6t + 8 = (t-2)(t-4)$, the required values of t are 2 and 4

b) Since $v = \dfrac{dx}{dt}$, we now have the second differential equation relating x and t

$$\frac{dx}{dt} = t^2 - 6t + 8$$

Integrating with respect to t
$x = t^3/3 - 3t^2 + 8t + D$, where D is another constant
Since $x = 0$ when $t = 0$ we have $D = 0$, and we take O as the initial position of the particle.
The relation between x and t is $x = t^3/3 - 3t^2 + 8t$
If $t = 2$ when the particle is at A, $OA = (8/3 - 12 + 16)$ m $= 20/3$m
Also $t = 4$, when the particle is at B, $OB = (64/3 - 48 + 32)$ m $= 16/3$m
Hence $AB = OA - OB = 4/3$m; i.e. the distance between A and B is $4/3$m

Worked example

A particle starts from the point O with initial speed u and moves in a straight line. At any instant the acceleration of the particle is c/v, where v is the speed of the particle and c is a constant. The particle passes through the point A with speed $2u$. Prove that the particle moves from O to A in time $(3u^2)/(2c)$ and that the distance OA is $(7u^3)/(3c)$
The acceleration of the particle is related to the time t that it has been in motion from O by

the differential equation $\dfrac{dv}{dt} = \dfrac{c}{v}$ which on separating the variables gives

$$t = \frac{1}{c}\int v\,dv$$

Since the speed changes from u at O to $2u$ at A, the time required to move from O to A is
$$\frac{1}{c}\int_u^{2u} v\,dv = \frac{1}{2c}\Big[v^2\Big]_u^{2u} = \frac{3u^2}{2c}$$

The acceleration is also given by the differential equation $v\dfrac{dv}{dx} = \dfrac{c}{v}$

which relates distance and speed. In this case, separating the variables gives

$$x = \frac{1}{c}\int v^2\,dv$$

Using the limits u and $2u \Rightarrow OA = \dfrac{1}{c}\int_u^{2u} v^2\,dv$

$$= \frac{1}{3c}\Big[v^3\Big]_u^{2u} = \frac{7u^3}{3c}$$

EXERCISES 14.2

1   The speed, $v\,\mathrm{m\,s^{-1}}$, of a particle P at time $t\,\mathrm{s}$, where $t>0$, is given by $v=t^2-2t+4$. Given that P moves in a straight line, calculate
a) the value of t when the acceleration is zero
b) the distance covered by P between the instants when $t=1$ and $t=4$

2   A particle P moves along the positive x-axis with acceleration of magnitude c^2x, directed towards O, where c is a positive constant and x is the distance of P from O. Given that P is projected from O with speed u at time $t=0$ in the direction of increasing x, show that, at time t and before P first comes to instantaneous rest,
$$cx=u\sin ct$$

3   A particle P, moving in a straight line, is subjected to a retardation of magnitude $(v^2+2s)\,\mathrm{m\,s^{-2}}$, where $v\,\mathrm{m\,s^{-1}}$ is the speed of P at time $t\,\mathrm{s}$. Given that $v=12$ when $t=0$,

show that P moves through a distance $\ln\left(\dfrac{13}{5}\right)$ m before coming to rest. Find also the

time taken for P to cover the distance, giving your answer to the nearest tenth of a second

4   A particle P moves in a straight horizontal line and is subject to a retardation whose magnitude is proportional to v^3, where $v\,\mathrm{m\,s^{-1}}$ is the speed of P at time $t\,\mathrm{s}$, where $t\geqslant 0$. The initial speed of P is $10\,\mathrm{m\,s^{-1}}$ and the magnitude of the initial retardation is $2\,\mathrm{m\,s^{-2}}$.

Show that $v^2=\dfrac{500}{2t+5}$

Find the distance covered by P in the first 10 seconds of its motion, giving your answer to the nearest metre

5   A car is moving along a straight horizontal road and the engine is switched off at the instant it passes a point O. At time $t\,\mathrm{s}$ after passing O, the speed $v\,\mathrm{m\,s^{-1}}$ of the car is given by

$\dfrac{1}{v}=\mathrm{A}+\mathrm{B}t$, where A and B are constants.

Show that the retardation of the car under these conditions is proportional to the square of its speed. Given that the speed is $30\,\mathrm{m\,s^{-1}}$ at O and the retardation is $2\,\mathrm{m\,s^{-2}}$, find A and B. Hence find the speed of the car, $10\,\mathrm{s}$ after it passes through O

SIMPLE HARMONIC MOTION

A particle P moves along the x-axis in such a way that its acceleration is always directed towards O and is of magnitude $w^2|x|$, where $OP=x$, and w is a constant. When this situation occurs, P is said to move along the x-axis with **simple harmonic motion** (SHM).

We have, $\dfrac{\mathrm{d}^2x}{\mathrm{d}t^2}=-w^2x$

that is $v\dfrac{\mathrm{d}v}{\mathrm{d}x}=-w^2x$

Integrating with respect to x gives

$$\frac{1}{2}v^2=-\frac{1}{2}w^2x^2+C$$

Let $v=0$ when $x=a$ so that
$$v^2=w^2(a^2-x^2)$$

$$\Rightarrow \frac{\mathrm{d}x}{\mathrm{d}t}=\pm w(a^2-x^2)^{\frac{1}{2}}\text{ and integrating again}$$

$$\pm wt+C=\int\frac{\mathrm{d}x}{(a^2-x^2)^{\frac{1}{2}}}=-\arccos\left(\frac{x}{a}\right)$$

Let $x=a$ when $t=0$ to give $C=0$ and we can take the solution under these conditions as
$$x=a\cos wt$$

From our analysis we see that P moves along the x-axis between the extreme points $(a, 0)$ and $(-a, 0)$ completing an oscillation in time $\dfrac{2\pi}{w}$: a is called the **amplitude** of the motion.

The maximum speed of P occurs as it passes through O and is wa.
The maximum acceleration occurs at the extreme points and is of magnitude w^2a.

SUMMARY OF USEFUL RESULTS

(for $x = a$, $v = 0$ at $t = 0$)

$$\frac{d^2x}{dt^2} = v\frac{dv}{dx} = -w^2x$$

$$v^2 = w^2(a^2 - x^2)$$

$$x = a \cos wt$$

$$\text{Period} = \frac{2\pi}{w}$$

Worked example

A particle P moves along the x-axis with simple harmonic motion. The period is 2·4 s and the distance between the two points A and B at which P is instantaneously at rest is 0·9 m. Calculate   a) the greatest speed,   b) the greatest acceleration of P in this motion
Since period is 2·4 s, we can take

$$\frac{2\pi}{w} = 2\cdot4 \Rightarrow w = \frac{\pi}{1\cdot2}$$

a) As the amplitude is 0·45 m

$$\text{Greatest speed} = wa = \frac{\pi}{1\cdot2} \times 0\cdot45\,\text{m s}^{-1} \approx 1\cdot18\,\text{m s}^{-1}$$

b) Magnitude of greatest acceleration $= w^2a = \left(\dfrac{\pi}{1\cdot2}\right)^2 \times 0\cdot45\,\text{m s}^{-2}$

$$\approx 3\cdot08\,\text{m s}^{-2}$$

SIMPLE HARMONIC MOTION AS THE PROJECTION OF UNIFORM MOTION IN A CIRCLE

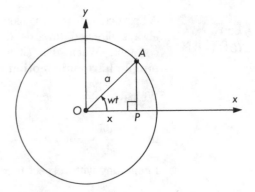

Fig. 14.6

In Fig. 14.6 the particle A moves around the circle $x^2 + y^2 = a^2$ with constant angular speed w, so that at time t

$$\angle AOx = wt$$

The point P on Ox is the foot of the perpendicular from A to Ox.
Take $OP = x$, then $x = a \cos wt$

$$\frac{d^2x}{dt^2} = -aw^2 \cos wt = -w^2x$$

From the definition of SHM, P moves along Ox with SHM of period $\dfrac{2\pi}{w}$ and amplitude a.

This representation of SHM as the projection onto a fixed diameter of uniform motion in a circle is very useful in finding the time between various positions of a particle moving with SHM. The circle, centre O, radius a, is often called **the reference circle.**

Worked example

A particle P moves in a straight line with SHM of period T, amplitude a and centre O. Find the time for P to move directly from the point H, distance $a/3$ from O, to the point K, distance $4a/5$ from O, given that H and K are on opposite sides of O (see Fig. 14.7)

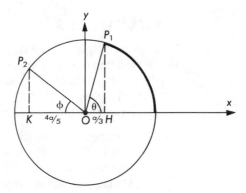

Fig. 14.7

Consider the motion as the projection of the motion of a particle moving round the circle $x^2 + y^2 = a^2$ with uniform angular speed w

The period $T = 2\pi/w$ and so $w = 2\pi/T$

The time from P_1 to P_2, and therefore from H to K, is $(\angle P_1OP_2)/w$

$$\Rightarrow \text{required time} = (\pi - \theta - \phi)/w = T\left[\pi - \arccos\frac{1}{3} - \arccos\frac{4}{5}\right]/(2\pi)$$

$$\approx 0{\cdot}20\,T$$

Worked example

On a certain day, low water for a harbour occurs at 11·30 and high water occurs at 17·50 the corresponding depths being 4 m and 11 m. Assuming that the tidal rise and fall of the water level is simple harmonic, find the earliest time, to the nearest minute, that a ship drawing 8 m can enter the harbour

The period of the motion is $2 \times 6\frac{1}{3}$ h $= 12\frac{2}{3}$ h, remembering that the period is the *total* time for the water to rise from its lowest level to its highest level and then fall again to its lowest level.

The amplitude of the motion is $\frac{1}{2} \times 7$ m $= 3\frac{1}{2}$ m

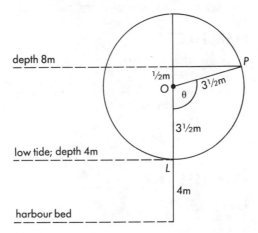

Fig. 14.8

In Fig. 14.8 we consider the motion of the water level in the harbour as the projection of the motion of a particle moving with constant angular speed in a circle of radius $3\frac{1}{2}$ m and completing each revolution in time $12\frac{2}{3}$ h. The points L and P correspond to the depths of water at low tide and when the ship can first enter the harbour

$$\angle LOP = \theta = \pi/2 + \arcsin(1/7)$$

Time taken by water level to rise from low tide until depth is 8 m
= time taken for the particle moving in the circle to go from L to P

$$= \frac{\theta}{2\pi} \times (\text{period})$$

$$= \left[\frac{\pi/2 + \arcsin{(1/7)}}{2\pi} \right] \times (12\tfrac{2}{3})\,\text{h}$$

$\approx 3\,\text{h}\ 27\,\text{min}$ (nearest minute)

Time for earliest entry of ship is 14 57

EXERCISES 14.3

1 A particle P is moving in a straight line with SHM, centre O, between extreme positions A and B, where $AB = 6\,\text{m}$. The greatest speed of P is $8\,\text{ms}^{-1}$. Calculate
   a) the period of an oscillation
   b) the speed of P at M, the mid-point of OA
   c) the least time required by P to move from A to M

2 A particle P is moving in a straight line with SHM, centre O between extreme positions A and B. When $OP = 4\,\text{m}$, the speed of P is $4\,\text{ms}^{-1}$ and when $OP = 2\,\text{m}$, the speed of P is $8\,\text{ms}^{-1}$.
   Calculate OA and the time required for P to move from A to B

3 A particle P moves in a straight line with simple harmonic motion, centre O, between extreme positions D and E. When P is at F, where $OF = 2\,\text{m}$, the speed of P is $12\,\text{ms}^{-1}$ and the magnitude of the acceleration of P is $162\,\text{ms}^{-2}$. Calculate
   a) the number of oscillations made by P each minute
   b) the distance OD
   c) the time taken by P to move directly from O to F
   d) the greatest speed of P

4 At a certain harbour entrance the depth of water at high tide is $30\,\text{m}$ and the depth at low tide is $8\,\text{m}$. The time between high tide and low tide is $6\,\text{h}$. Assuming that the tide rises and falls with simple harmonic motion find
   a) the rate in ms^{-1} at which the water level is moving when the depth of water is $20\,\text{m}$
   b) the time, to the nearest minute, for the water level to rise from depth $12\,\text{m}$ to depth $24\,\text{m}$ in the same tide

NB: Examination questions on these topics can be found in Chapter 19.

ANSWERS TO EXERCISES

EXERCISES 14.1

1 a) $22\,\text{ms}^{-1}$; b) $109\,\text{m}$
2 a) $12\,\text{s}$; b) $216\,\text{m}$
4 a) $5\cdot6\,\text{ms}^{-1}\uparrow$; b) $14\,\text{ms}^{-1}\downarrow$; $11\cdot6\,\text{m}$
5 a) $8\,\text{ms}^{-1}$; b) $13\cdot75\,\text{s}$
6 a) $52.4\,\text{m s}^{-1}$; b) $5.3\,\text{s}$
7 a) $122.5\,\text{m}$; b) $7.14\,\text{s}$

EXERCISES 14.2

1 a) 1; b) $18\,\text{m}$
3 $0\cdot2\,\text{s}$
4 $62\,\text{m}$

5 $A = \dfrac{1}{30}, B = \dfrac{1}{450}$; $18\,\text{ms}^{-1}$

EXERCISES 14.3

1 a) $\dfrac{3\pi}{4}\,\text{s}$; b) $4\sqrt{3}\,\text{ms}^{-1}$; c) $\dfrac{\pi}{8}\,\text{s}$

2 $4\sqrt{5}\,\text{m}$; $2\pi\,\text{s}$
3 a) 86; b) $2\cdot4\,\text{m}$; c) $0\cdot11\,\text{s}$;
   d) $21\cdot6\,\text{ms}^{-1}$
4 a) $9\cdot14 \times 10^{-3}\,\text{ms}^{-1}$; b) $133\,\text{min}$

MOTION IN TWO DIMENSIONS

GETTING STARTED

PROJECTILES – PARABOLIC MOTION UNDER GRAVITY

When air resistance is neglected, a particle moves freely under gravity when thrown or projected. The particle has acceleration of magnitude $g\,\mathrm{m\,s^{-2}}$ vertically downwards and is usually called a **projectile**.

When solving problems we consider the horizontal and the vertical parts of the motion separately. A particle projected with speed v at an angle of elevation θ has an unchanged horizontal speed component $v\cos\theta$ and initial vertical speed component $v\sin\theta$ as shown in Fig. 15.1.

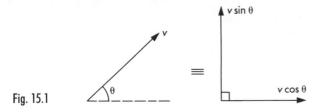

Fig. 15.1

ESSENTIAL PRINCIPLES

Worked example

Suppose that a particle is projected from a point O and, that after time t, it is at P, a distance x_1 horizontally and a distance y_1 vertically from O as shown in Fig. 15.2

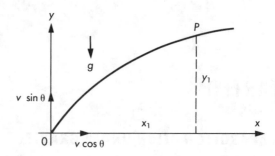

Fig. 15.2

Horizontally there is no acceleration
$$\Rightarrow x = (v\cos\theta)t \qquad\qquad (1)$$
Vertically the acceleration is $-g$

$$\Rightarrow y = (v\sin\theta)t - \frac{1}{2}gt^2 \qquad\qquad (2)$$

From (1) $t = \dfrac{x}{v\cos\theta}$ and putting this in (2)

gives $y = (v\sin\theta)\left(\dfrac{x}{v\cos\theta}\right) - \dfrac{1}{2}g\left(\dfrac{x}{v\cos\theta}\right)^2$

$$\Rightarrow y = x\tan\theta - \frac{1}{2}\frac{gx^2}{v^2\cos^2\theta}$$

From pure maths, we know that an equation of the form $y = ax - bx^2$ is a parabola. The path (or trajectory) of a particle moving under gravity is a parabola, provided that air resistance is neglected.

In (2) we see that $y = 0$ gives $t = 0$ (at start) and $t = \dfrac{2v\sin\theta}{g}$

(when the particle is level with O again, at A, say)

Time of flight $= \dfrac{2v\sin\theta}{g}$

In (1) with $x = OA$ and $t = \dfrac{2v\sin\theta}{g}$

we have $OA =$ horizontal range $= (v\cos\theta)\left(\dfrac{2v\sin\theta}{g}\right) = \dfrac{2v^2\sin\theta\cos\theta}{g}$

Horizontal range $= \dfrac{v^2\sin 2\theta}{g}$ (since $\sin 2\theta \equiv 2\sin\theta\cos\theta$)

For a given speed of projection v, the greatest horizontal range is achieved when $\dfrac{v^2}{g}\sin 2\theta$ is greatest.

That is when $\sin 2\theta = 1$ and $\theta = 45°$

Suppose that $y = H$ when the vertical speed component is zero.
Using $v^2 = u^2 + 2as$ with $v = 0$, $u = v\sin\theta$, $a = -g$ and $s = H$
$$0 = v^2\sin^2\theta - 2gH$$

Greatest height $= H = \dfrac{v^2\sin^2\theta}{2g}$

SOLVING QUESTIONS ON PROJECTILES

All of the preceding work must be learnt thoroughly. Experience shows that examination questions on projectiles are best answered by working out each stage rather than resorting to formulae.

Worked example

A particle P is projected from a point O with speed $160\,\mathrm{m\,s^{-1}}$ at an angle of elevation $\arctan\left(\frac{4}{3}\right)$ and moves freely under gravity. Calculate the time of flight, the horizontal range and the greatest height above the level of O achieved by P in its flight (take $g = 9.8\,\mathrm{m\,s^{-2}}$)

Start with a clear diagram containing all the information (Fig. 15.3)

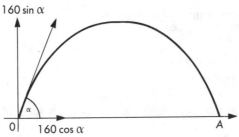

> Again, a clear diagram can help you make progress.

Fig. 15.3

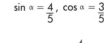
$\sin\alpha = \frac{4}{5}$, $\cos\alpha = \frac{3}{5}$

Fig. 15.4

$\tan\alpha = \frac{4}{3} \Rightarrow$

Initial vertical component of velocity $= 160 \times \frac{4}{5}\,\mathrm{m\,s^{-1}}$

$= 128\,\mathrm{m\,s^{-1}}$

Using $s = ut + \frac{1}{2}at^2$ with $s = 0$, $u = 128$, $a = -9\cdot8$

$\Rightarrow 0 = 128\,t - 4\cdot9\,t^2$

This gives *time of flight* $= \frac{128}{4\cdot9} \approx 26\cdot12\,\mathrm{s}$

Using $s = ut$ with $s = OA$, $u = 160 \times \frac{3}{5} = 96$

$\Rightarrow OA = 96 \times 26\cdot12\,\mathrm{m}$

Horizontal range $\approx 2508\,\mathrm{m}$

Using $v^2 = u^2 + 2as$ with $v = 0$, $u = 128$

and $a = -9\cdot8$ we have, where H is greatest height

$0 = 128^2 - 2 \times 9\cdot8 \times H$

Greatest height $= \frac{128^2}{2 \times 9\cdot8}\,\mathrm{m} \approx 836\,\mathrm{m}$

Working from first principles is even more important when the data is changed slightly as in the next worked example.

Worked example

A particle P is projected with speed $70\,\mathrm{m\,s^{-1}}$ from the top of a cliff at a height $40\,\mathrm{m}$ above sea-level. P strikes the sea at a horizontal distance of $200\,\mathrm{m}$ from the point of projection.

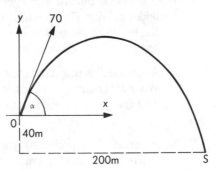

Fig. 15.5

Taking $g = 9 \cdot 8 \, \text{ms}^{-2}$, find the two possible angles of projection

Considering the horizontal motion from O to S
$$200 = (70 \cos \alpha)t \qquad (1)$$

Using $s = ut + \dfrac{1}{2}at^2$ with $s = -40$

$u = 70 \sin \alpha$ and $g \, (= a) = -9 \cdot 8$ vertically we have
$$-40 = (70 \sin \alpha)t - 4 \cdot 9t^2 \qquad (2)$$
Eliminating t from (1) and (2) gives
$$-40 = 70 \sin \alpha \left(\frac{200}{70 \cos \alpha} \right) - 4 \cdot 9 \left(\frac{200}{70 \cos \alpha} \right)^2$$

Using $\dfrac{\sin \alpha}{\cos \alpha} = \tan \alpha$ and $\dfrac{1}{\cos^2 \alpha} = \sec^2 \alpha = 1 + \tan^2 \alpha$

we get
$$-40 = 200 \tan \alpha - 40(1 + \tan^2 \alpha)$$
which reduces further to
$$\tan \alpha(\tan \alpha - 5) = 0$$
P is either projected horizontally ($\alpha = 0$) or at an angle of elevation $\arctan 5 \approx 78 \cdot 7°$.

Worked example

A particle P is projected with speed $20 \, \text{ms}^{-1}$ at an angle of elevation $30°$ from a point O and moves freely under gravity, ($g = 9 \cdot 8 \, \text{ms}^{-2}$). Find the speed and direction of motion of P after 2s

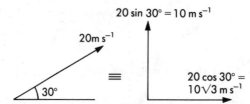

Fig. 15.6

After 2s, vertical speed is found by using $v = u + at$ where $u = 10$, $t = 2$, $a = -9 \cdot 8$
$v = 10 - 2 \times 9 \cdot 8 = -9 \cdot 6$ (− means downwards);
horizontal speed unchanged $= 10 \, \sqrt{3} \, \text{ms}^{-1}$

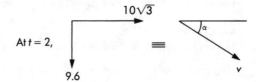

Fig. 15.7

$v \cos \alpha = 10 \, \sqrt{3}$
$v \sin \alpha = 9 \cdot 6$
$v^2 (\cos^2 \alpha + \sin^2 \alpha) = v^2 = 300 + 92 \cdot 16$
$\Rightarrow$ speed after 2s $\approx 19 \cdot 8 \, \text{ms}^{-1}$

Direction $= \alpha = \arctan \left(\dfrac{9 \cdot 6}{10 \, \sqrt{3}} \right) \approx 29°$ to the horizontal as shown

EXERCISES 15.1

1 A pebble is thrown with speed $15 \, \text{ms}^{-1}$ at an angle of elevation $40°$ and moves freely under gravity. Find the time taken for the pebble to travel 10m horizontally. Find also the height of the pebble above the level of the point from which it was thrown when it has travelled a distance of 10m horizontally. (Take $g = 9 \cdot 8 \, \text{ms}^{-2}$)

2 A particle P is projected at an angle of elevation α from a point O on a horizontal plane. When P is moving upwards at an angle β to the horizontal, P passes through a point A. The line OA is inclined at an angle θ to the horizontal. Show that
$$2 \tan \theta = \tan \alpha + \tan \beta$$

3 A particle P is projected from a point O at the top of a cliff, 52m above the sea. P moves freely under gravity and strikes the sea at a point S. The velocity of projection of

P is $25\,\mathrm{m\,s^{-1}}$ at an angle of elevation $\arctan\left(\dfrac{7}{24}\right)$. Calculate:

a) the time taken for P to reach S from O.
b) the horizontal distance between O and S,
c) the speed of P when it strikes the sea. (Take $g = 10\,\mathrm{m\,s^{-2}}$).

4   A particle P is projected from a point O with speed u and angle of elevation θ and moves freely under gravity. At time t after its projection, its horizontal and vertical displacements from O are a and h respectively. At this instant P has horizontal and vertical speed components v and w respectively. Find expressions in terms of u, θ, t and g for a, h, v and w. Deduce that the direction of motion of P is perpendicular to OP when t satisfies the equation
$$g^2t^2 - 3ugt\sin\theta + 2u^2 = 0$$

5   A particle P is projected from a fixed point O with speed v and elevation α. P passes through points at horizontal distances of $6a$ and $12a$ when at heights of $2a$ and $3a$ above the level of O respectively. Show that

a) $\tan\alpha = \dfrac{5}{12}$

b) $v = \dfrac{13}{2}\sqrt{(ag)}$

Find, in terms of a, the greatest height above the level of O, achieved by P

6   A gun projects shells at speed $500\,\mathrm{m\,s^{-1}}$. Assuming that shells move freely under gravity ($g = 10\,\mathrm{m\,s^{-2}}$) find
a) the greatest horizontal range of a shell
b) the possible angles of projection when half the maximum horizontal range is achieved.

UNIFORM CIRCULAR MOTION

> 66 Here the speed does not vary. 99

In uniform circular motion, a particle P describes a circle, centre O, with uniform speed. The line OP rotates with constant angular speed, measured in radians per second (rad s^{-1}). Taking $OP = a$, the information in Fig. 15.8 shows the position of P at time ts, where t is measured from when P was at A, and w rad s^{-1} is the constant angular speed.

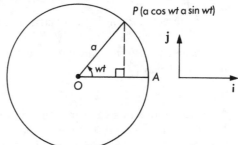

Fig. 15.8

Taking unit vectors $\mathbf{i}$ along $\overrightarrow{OA}$ and $\mathbf{j}$ perpendicular to $\overrightarrow{OA}$, the position vector, referred to O, at time t of P is

$$\overrightarrow{OP} = (a\cos wt)\mathbf{i} + (a\sin wt)\mathbf{j}$$

The velocity $\mathbf{v}$ of P at this instant is obtained by differentiating with respect to t

$$\Rightarrow \mathbf{v} = (-aw\sin wt)\mathbf{i} + (aw\cos wt)\mathbf{j}$$
$$|\mathbf{v}| = (a^2w^2\sin^2 wt + a^2w^2\cos^2 wt)^{\frac{1}{2}} = aw$$

Speed of $P = aw = |\mathbf{v}|$
Since $\mathbf{v}.\mathbf{r}$ $[(-aw\sin wt)\mathbf{i} + (aw\cos wt)\mathbf{j}]\,.\,[(a\cos wt)\mathbf{i} + (a\sin wt)\mathbf{j}] = 0$

the velocity of P is at right angles to OP and is therefore directed along the tangent to the circle at P.

The acceleration of P is obtained by differentiating $\mathbf{v}$ with respect to time.

$$\Rightarrow \frac{d\mathbf{v}}{dt} = (-aw^2\cos wt)\mathbf{i} + (-aw^2\sin wt)\mathbf{j}$$

$$= -aw^2\overrightarrow{OP} = aw^2\overrightarrow{PO}$$

The magnitude of the acceleration is $aw^2 = \dfrac{v^2}{a}$ and the acceleration is directed along $\overrightarrow{PO}$, that is towards the centre of the circle.

Worked example

A particle is moving in a circular path with constant speed $30\,\mathrm{m\,s^{-1}}$. The radius of the circle is $70\,\mathrm{m}$. Find the magnitude of the acceleration and the time taken for the particle to complete one revolution.

$$\text{Acceleration} = \frac{v^2}{a} \ \text{(from above)}$$

$$= \frac{30^2}{70}\,\mathrm{m\,s^{-2}} \approx 12{\cdot}86\,\mathrm{m\,s^{-2}}$$

Circumference of circle $= 2\pi \times 70\,\mathrm{m}$

$$\text{Time to complete this distance} = \frac{2\pi \times 70}{30}\,\mathrm{s}$$

$$= 14{\cdot}67\,\mathrm{s}$$

NON-UNIFORM CIRCULAR MOTION

> Here the speed does vary.

> Note these amendments to the formulae.

A few GCE A-level syllabuses require students to extend their work on uniform circular motion to consider cases where the speed v varies. The motion of a particle moving in a vertical circle under gravity is a typical case.

The formulae derived for uniform circular motion require amendment in these cases to the following:

At a particular instant for speed v we have:

- *Velocity of particle* is of magnitude v directed along the tangent to the circle.
- *Acceleration of particle* has components:

$\dfrac{dv}{dt}$ directed along the tangent to the circle

$\dfrac{v^2}{a}$ directed towards the centre O of the circle

At this instant also, $v = aw$, but it must be appreciated that w is now the angular speed at this instant only.

Worked example

A particle moves in a circular path, centre O and radius $2\,\mathrm{m}$, starting with angular speed $10\,\mathrm{rad\,s^{-1}}$ at time $t = 0$. At time $t\,\mathrm{s}$, the angular speed w of the particle is $(10 - 2t)\,\mathrm{rad\,s^{-1}}$. Calculate, at the instant when $t = 3$

a) the linear speed of the particle
b) the resultant acceleration of the particle

At the instant when $t = 3$
a) Angular speed $= 4\,\mathrm{rad\,s^{-1}}$
So linear speed $= aw = 2 \times 4\,\mathrm{m\,s^{-1}} = 8\,\mathrm{m\,s^{-1}}$

b) Acceleration component along tangent $= \dfrac{dv}{dt}$

$$= a\frac{dw}{dt}$$

$$= 2 \times (-2)\,\mathrm{m\,s^{-2}}$$

$$= -4\,\mathrm{m\,s^{-2}}$$

Acceleration component directed towards centre

$$= \frac{v^2}{a} = \frac{8^2}{2} = 32\,\mathrm{m\,s^{-2}}$$

Resultant acceleration is of magnitude $\sqrt{4^2 + 32^2}\,\mathrm{m\,s^{-2}}$

$$\approx 32{\cdot}25\,\mathrm{m\,s^{-2}}$$

At this instant the direction of the acceleration is $\arctan\left(\dfrac{4}{32}\right)$ with the radius of the circle;

i.e. $7 \cdot 1°$ approximately.

EXERCISES 15.2

1 Find the speed and magnitude of the acceleration of a particle P moving in a circle, centre O, radius $30\,\text{cm}$ with constant angular speed $6\ \text{rad s}^{-1}$

2 Taking the earth to be a sphere of radius $6400\,\text{km}$, calculate the magnitude of the acceleration, in m s^{-2} to 2 significant figures, of a point on the equator due to the earth's rotation

3 A particle is moving in a horizontal circular path with constant speed $0 \cdot 9\,\text{m s}^{-1}$ and the particle has acceleration $3\,\text{m s}^{-2}$ directed towards the centre of the circle. Calculate the rate at which the particle is rotating in rad s^{-1}. Hence, find the radius of the circle in m

4 A particle moving at constant speed rotates once around a circle of radius $2\,\text{m}$ in $4\,\text{s}$. Calculate the magnitude of the velocity and the acceleration of the particle

5 A particle is moving in a circle of radius $5\,\text{m}$ and its speed at time t seconds is $t^2\,\text{m s}^{-1}$. Find the magnitude of the acceleration of the particle at the instant when $t = 2$

6 Referred to a fixed origin O, the position vector of a particle P is $\mathbf{r}\,\text{m}$ at time $t\,\text{s}$, where
$$\mathbf{r} = (\cos 2t)\,\mathbf{i} + (\sin 2t)\,\mathbf{j}$$
Calculate the magnitudes of the velocity and the acceleration of P at the instant when $t = \dfrac{\pi}{6}$

Show that P is moving with constant speed in a circle and calculate the time taken for P to complete one revolution of this circle

RELATIVE MOTION

Consider two particles P and Q moving along the same line. At time t, we take the displacements of P and Q from some fixed origin O in the line of motion to be x_P and x_Q respectively and we assume that x_P and x_Q are functions of t. The displacement of Q relative to P is $x_Q - x_P$, and we *define* the velocity of Q relative to P to be

$$\frac{\mathrm{d}x_Q}{\mathrm{d}t} - \frac{\mathrm{d}x_P}{\mathrm{d}t},$$

the time derivative of the displacement. That is, the velocity of Q relative to P is the velocity Q appears to have when observed from P. The notion of one-dimensional relative motion just described can be generalised to motion in two or three dimensions by using vectors as follows. Let $\mathbf{r}_P$ and $\mathbf{r}_Q$ be the position vectors of two particles P and Q at time t and with respect to a fixed origin O.

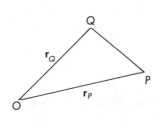

Fig. 15.9

In Fig. 15.9, we have $\overrightarrow{PQ} = \mathbf{r}_Q - \mathbf{r}_P$.

The vector $\mathbf{r}_Q - \mathbf{r}_P$ is the position vector of Q relative to P at time t. We define the velocity of Q relative to P to be

$$\frac{\mathrm{d}\mathbf{r}_Q}{\mathrm{d}t} - \frac{\mathrm{d}\mathbf{r}_P}{\mathrm{d}t},$$

the time derivative of the displacement vector $\overrightarrow{PQ}$.

This can be written $(\mathbf{V}_Q - \mathbf{V}_P)$, where $\mathbf{V}_Q$ and $\mathbf{V}_P$ are the velocities of Q and P respectively.

That is, the velocity of Q relative to P, often written $_Q\mathbf{V}_P$, is the velocity Q appears to have when observed from P. (Often, $\dfrac{\mathrm{d}\mathbf{r}}{\mathrm{d}t}$ is written as $\dot{\mathbf{r}}$ and $\dfrac{\mathrm{d}^2\mathbf{r}}{\mathrm{d}t^2}$ as $\ddot{\mathbf{r}}$)

$$_Q\mathbf{V}_P = \mathbf{V}_Q - \mathbf{V}_P = \dot{\mathbf{r}}_Q - \dot{\mathbf{r}}_P = -_P\mathbf{V}_Q$$

Similarly, the acceleration $_Q\mathbf{a}_P$ of Q relative to P is defined as

$$_Q\mathbf{a}_P = \ddot{\mathbf{r}}_Q - \ddot{\mathbf{r}}_P$$

Worked example

The velocities of particles A and B are $(12\mathbf{i}-3\mathbf{j})\,\mathrm{m\,s^{-1}}$ and $(4\mathbf{i}-9\mathbf{j})\,\mathrm{m\,s^{-1}}$ respectively. Find the velocity of A relative to B

The velocity of a particle C relative to A is $(\mathbf{i}+2\mathbf{j})\,\mathrm{m\,s^{-1}}$. Find the velocity of C

We have $_A\mathbf{V}_B = \mathbf{V}_A - \mathbf{V}_B = 12\mathbf{i}-3\mathbf{j}-(4\mathbf{i}-9\mathbf{j})\,\mathrm{m\,s^{-1}} = (8\mathbf{i}+6\mathbf{j})\,\mathrm{m\,s^{-1}}$

The velocity of A relative to B is $(8\mathbf{i}+6\mathbf{j})\,\mathrm{m\,s^{-1}}$

We have $\mathbf{V}_C = \mathbf{V}_A + _C\mathbf{V}_A = (12\mathbf{i}-3\mathbf{j})+(\mathbf{i}+2\mathbf{j})\,\mathrm{m\,s^{-1}} = (13\mathbf{i}-\mathbf{j})\,\mathrm{m\,s^{-1}}$

The velocity of C is $(13\mathbf{i}-\mathbf{j})\,\mathrm{m\,s^{-1}}$

Worked examples

The wind has velocity of magnitude $U\,\mathrm{m\,s^{-1}}$ *from* $\theta°$ west of north.

a) To a cyclist travelling due east with speed $10\,\mathrm{m\,s^{-1}}$, the wind appears to blow from due north. Show this information in a sketch

b) When the cyclist increases speed to $15\,\mathrm{m\,s^{-1}}$ while travelling due east, the wind appears to blow *from* $30°$ to the east of north. Show this information in a second sketch.

c) Use your sketches to find values for U and θ

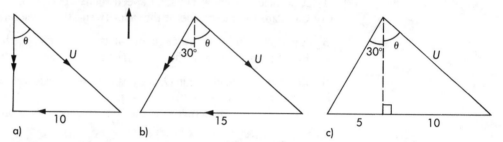

Fig. 15.10   a)   b)   c)

The sketch describing situation a) is shown in Fig. 15.10(a), where the double-arrow vector shows the apparent velocity of the wind to the cyclist. The sketch describing situation b) is shown in Fig. 15.10(b) and again the double-arrow vector shows the apparent velocity of the wind to the cyclist. We now combine these two diagrams to give Fig. 15.10(c) and

$U \sin\theta = 10$

$U \cos\theta = 5\tan 60° = 5\sqrt{3}$

From these equations we have, $U = 5\sqrt{7}$ and $\theta = \tan^{-1}(2/\sqrt{3})$

Once the analysis of the question has been done, it is also possible to obtain answers for U and θ by drawing to scale because the question asks you to *find* values which implies that a graphical approach is acceptable. If you do this, you should obtain an answer for U in the range $13{\cdot}0 - 13{\cdot}4$ and an answer for θ in the range $40°$–$42°$.

Worked example

At time 15 00, relative to a fixed tracking station S, the position vectors of two ships A and B are $(18\mathbf{i}+6\mathbf{j})\,\mathrm{km}$ and $(4\mathbf{i}+10\mathbf{j})\,\mathrm{km}$ respectively. A and B are moving with constant velocities $(\mathbf{i}+4\mathbf{j})\,\mathrm{km\,h^{-1}}$ and $(3\mathbf{i}+2\mathbf{j})\,\mathrm{km\,h^{-1}}$ respectively. Write down the displacement of A relative to B at time $t\,\mathrm{h}$ after 15 00 and hence, or otherwise, find the shortest distance between the ships and the time at which this occurs.

Relative to S, the position vectors of A and B at time t, $\mathbf{r}_A$ and $\mathbf{r}_B$, are

$\mathbf{r}_A = (18\mathbf{i}+6\mathbf{j})+t(\mathbf{i}+4\mathbf{j})$

$\mathbf{r}_B = (4\mathbf{i}+10\mathbf{j})+t(3\mathbf{i}+2\mathbf{j})$

Displacement of A relative to B is

$\mathbf{r}_A - \mathbf{r}_B = (14\mathbf{i}-4\mathbf{j})+t(-2\mathbf{i}+2\mathbf{j})$

Fig. 15.11 shows the path of A relative to B and the shortest distance between A and B *occurs* when:

$[14\mathbf{i}-4\mathbf{j}+t(-2\mathbf{i}+2\mathbf{j})].(-2\mathbf{i}+2\mathbf{j}) = 0$

$-2(14-2t)+2(2t-4) = 0$

$8t = 36 \Rightarrow t = 4\tfrac{1}{2}$

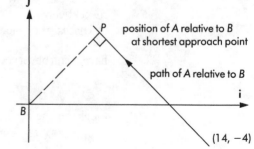

Fig. 15.11

position of A relative to B at shortest approach point

path of A relative to B

When $t = 4\frac{1}{2}$, distance between A and B is least

$$= |[(14\mathbf{i} - 4\mathbf{j}) + \frac{9}{2}(-2\mathbf{i} + 2\mathbf{j})| = |5\mathbf{i} + 5\mathbf{j}|$$

$$|5\mathbf{i} + 5\mathbf{j}| = (5^2 + 5^2)^{\frac{1}{2}} = \sqrt{50} = 5\sqrt{2}$$

The shortest distance between A and B is $5\sqrt{2}$ km and occurs at 19 30

SUPPLEMENTARY NOTES ON METHODS OF SOLUTION

Questions concerned with relative motion requiring closest approach (or collision) can be solved by a variety of methods. However, *all these* methods require full understanding of the diagram given in this solution. The method given here uses the scalar product to determine the value of t when $\overrightarrow{BP}$ and the direction of $\mathbf{r}_A - \mathbf{r}_B$ are perpendicular.

Some students prefer to write
$$|\mathbf{r}_A - \mathbf{r}_B|^2 = (14 - 2t)^2 + (2t - 4)^2$$
$$= 8t^2 - 72t + 212$$

The value of t for which $|\mathbf{r}_A - \mathbf{r}_B|$ is a minimum can then be found by completing the square or by differentiation. All approaches lead to $t = 4\frac{1}{2}$. Graphical solutions are only permissible when the examiner tells you to *find* the results required. If you are told *to calculate*, graphical attempts will obviously attract heavy penalties.

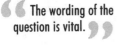 **The wording of the question is vital.**

EXERCISES 15.3

1   The velocities of two particles P and Q are given, in ms^{-1}, by $\mathbf{V}_P$ and $\mathbf{V}_Q$ where
$$\mathbf{V}_P = 5\mathbf{i} - 7\mathbf{j} \text{ and } \mathbf{V}_Q = -3\mathbf{i} + 8\mathbf{j}$$
Find the magnitude of the velocity of P relative to Q

2   A cyclist moving with velocity $(3\mathbf{i} + 4\mathbf{j})\,\mathrm{ms}^{-1}$ notes that the wind appears to blow from the direction $\mathbf{i}$. The cyclist doubles his speed and maintains the same direction. The wind now appears to blow from the direction $\mathbf{i} + \mathbf{j}$. Prove that the true velocity of the wind is $(2\mathbf{i} + 4\mathbf{j})\,\mathrm{ms}^{-1}$

3   At noon, aircraft A is 20 km due north of a fixed point O and is moving due east with steady speed 300 kmh^{-1}. At this instant aircraft B leaves O with constant speed 400 kmh^{-1} flying due north. Find the magnitude and the direction of the velocity of A relative to B. Hence determine the least distance between A and B and the time at which this occurs

4   The position vectors, $\mathbf{r}_A$ and $\mathbf{r}_B$, of two particles A and B at time t are
$$\mathbf{r}_A = (a\cos wt)\mathbf{i} + (a\sin wt)\mathbf{j}$$
$$\mathbf{r}_B = a(1 - w^2 t^2)\mathbf{i} - (aw^2 t^2)\mathbf{j}$$
where a and w are constants.

Calculate, at time $t = \dfrac{\pi}{4w}$, the magnitude of

i) the velocity of A relative to B
ii) the acceleration of A relative to B

5   At noon, ship B is 10 km due west of ship A. Both ships are moving with constant velocity, the velocity of A being 9 kmh^{-1} due north. Given that the ships would eventually collide at 14 20, find the velocity of B

6   A river flows at 4 ms^{-1} from west to east between parallel banks which are at a distance 500 m apart. A boat can move at 7 ms^{-1} in still water. Find the direction in which the boat should be steered in order to cross the river from the south bank to the north bank by the shortest possible route. Find the time taken and the distance covered by the boat for this crossing

7   Relative to a fixed origin O, at time ts, the position vectors of the points P and Q are $\mathbf{r}_1$ m and $\mathbf{r}_2$ m respectively, where
$$\mathbf{r}_1 = (3\sin wt)\mathbf{i} - (2\cos wt)\mathbf{j}$$
$$\mathbf{r}_2 = (2\cos wt)\mathbf{i} + (5\cos wt)\mathbf{j}$$

and w is a constant. Find the velocity and speed of P relative to Q when $t = \dfrac{\pi}{4w}$

NB: Examination questions on these topics can be found in Chapter 19.

ANSWERS TO EXERCISES

EXERCISES 15.1

1 $0.87\,$s, $4.68\,$m
3 a) 4s; b) 96m; c) $40.8\,\text{ms}^{-1}$
4 $v = u\cos\theta$, $w = u\sin\theta - gt$, $a = ut\cos\theta$,
   $h = ut\sin\theta - \tfrac{1}{2}gt^2$

5 $\dfrac{25a}{8}$

6 a) 25000m; b) 15° or 75°

EXERCISES 15.2

1 $1.8\,\text{ms}^{-1}$, $10.8\,\text{ms}^{-2}$
2 $3.4 \times 10^{-2}\,\text{ms}^{-2}$

3 $\dfrac{10}{3}\text{rad s}^{-1}$, $0.27\,$m

4 $\pi\,\text{ms}^{-1}$, $\tfrac{1}{2}\pi^2\,\text{ms}^{-2}$

5 $\dfrac{4}{5}\sqrt{41}\,\text{ms}^{-2}$

6 $2\,\text{ms}^{-1}$, $4\,\text{ms}^{-2}$; $\pi\,$s

EXERCISES 15.3

1 $17\,\text{ms}^{-1}$
3 $500\,\text{kmh}^{-1}$, $143.1°$; 12km, 12 02
4 i) $aw[1 + \tfrac{1}{2}\pi^2]^{\frac{1}{2}}$; ii) $3aw^2$
5 $9.75\,\text{kmh}^{-1}$ in the direction 022·6°
6 330·3°, 87s, 578m

7 $\dfrac{w}{\sqrt{2}}(5\mathbf{i} - 3\mathbf{j})\,\text{ms}^{-1}$, $w\sqrt{17}\,\text{ms}^{-1}$

BASIC KINETICS

GETTING STARTED

We analyse the effect of the application of forces on particles. This work is based on three laws, first stated by Newton in the seventeenth century.

NEWTON'S LAWS OF MOTION

1 Every body remains at rest or in uniform motion in a straight line unless it is made to change that state by external forces.
2 The rate of change of linear momentum of a body is proportional to the force being applied to the body and acts in the same direction as this force.
3 The force exerted by body A on body B is equal in magnitude and opposite in direction to the force exerted by body B on body A.

The standard units of length, mass and time are the metre (m), the kilogram (kg) and the second (s) respectively. The units of other quantities are derived from these.

We define the momentum of a particle to be the product of its mass and its velocity. This definition, where mass is a scalar quantity and velocity is a vector quantity, means that momentum is a vector, whose direction is the same as that in which the particle is moving.

Newton's second law for a particle, of mass m, moving with velocity $\mathbf{v}$, under the action of a force $\mathbf{F}$ is

$$\mathbf{F} \propto \frac{\mathrm{d}}{\mathrm{d}t}(m\mathbf{v}) = m\frac{\mathrm{d}\mathbf{v}}{\mathrm{d}t} \text{ assuming that } m \text{ does not change.}$$

$$\Rightarrow \mathbf{F} = km\frac{\mathrm{d}\mathbf{v}}{\mathrm{d}t} \text{ where } k \text{ is a constant.}$$

The units used to measure the magnitude of a force are chosen to make the constant unity.

$$\mathbf{F} = m \times \text{acceleration} = m\mathbf{a}$$

where m is measured in kg, acceleration $\mathbf{a}$ in m s^{-2} and then the force $\mathbf{F}$ is measured in newtons (N).

When you are working through exercises it is essential to use these units consistently.

ESSENTIAL PRINCIPLES

THE NATURE OF FORCES

There are two distinct types of forces. The first type are external to the body under consideration, the most important of these being **gravitational forces**. The gravitational force with which the earth attracts a body is called *the weight of the body*. When we are considering distances which are small compared with the radius of the earth, we use a constant vector $\mathbf{g}$ for the acceleration due to gravity and we take a value such as $9\cdot8\,\mathrm{ms}^{-2}$ or $10\,\mathrm{ms}^{-2}$ for the magnitude of $\mathbf{g}$ in numerical questions. It should be stressed however that the gravitational force of attraction is *not constant*, but is dependent on the bodies themselves and their surroundings as described by Newton's law of universal gravitation. If you need to use this law, the question will tell you how to do this.

The second type of force is related to the body or bodies under consideration. The most common are

a) forces exerted by one body on another body in contact with it, according to Newton's third law,
b) frictional forces,
c) the tension in a light string or the thrust in a light rod connected to the body,
d) specific forces applied to the body and described in a particular question.

DIAGRAMS

> Check to make sure your sketch includes all the information.

In nearly every question your solution will depend on a clear diagram (or diagrams) which display all the forces acting on a body. A clear freehand sketch is sufficient unless you are drawing to scale for a graphical solution. After completing your sketch *always* read through a question again to make sure that you have included all the data and that you have not made any unjustified assumptions.

Worked example

A force of magnitude $2\,\mathrm{N}$ acts on a particle P of mass $0\cdot4\,\mathrm{kg}$. Given that P moves through $2\,\mathrm{m}$ from rest in time $T\,\mathrm{s}$ in a straight line, find the magnitude of the acceleration of P and the value of T

$$\boxed{0\cdot4} \rightarrow 2\,\mathrm{N}$$

Using $F = ma$ we have $\quad a = \dfrac{F}{m} = \dfrac{2}{0\cdot4}\,\mathrm{ms}^{-2}$

$$\text{acceleration} = 5\,\mathrm{ms}^{-2}$$

Using $s = ut + \dfrac{1}{2}at^2$ with $s = 2$, $t = T$, $a = 5$, $u = 0$

we have $2 = \dfrac{1}{2} \times 5 \times T^2 \Rightarrow T^2 = 0\cdot8$

$$T = 0\cdot89$$

Worked example

A car is moving along a straight level road with constant speed. The engine of the car is producing a tractive force of magnitude $65\,\mathrm{N}$. Discuss the motion of the car

Since the car is moving *with constant speed in a straight line*, we have, by Newton's first law, a total net force of zero acting on the car. We know also that the car has zero acceleration because it is moving with constant speed in a straight line. The car's engine is producing a constant force of magnitude $65\,\mathrm{N}$ and this means that a second force equal in magnitude and opposite in direction to this must be acting on the car. This second force would be made up of various resistances like air resistance, friction etc. acting on the car, and opposing its motion.

Worked example

At time $t\,\mathrm{s}$, the velocity of a particle P, of mass $0\cdot3\,\mathrm{kg}$, is $(3t\mathbf{i} + \mathrm{e}^t\mathbf{j} + t^2\mathbf{k})\,\mathrm{ms}^{-1}$
Find the magnitude of the force acting on P when $t = 2$

$$\mathbf{v} = 3t\mathbf{i} + \mathrm{e}^t\mathbf{j} + t^2\mathbf{k}$$

$$\text{Acceleration} = \frac{\mathrm{d}\mathbf{v}}{\mathrm{d}t} = 3\mathbf{i} + \mathrm{e}^t\mathbf{j} + 2t\mathbf{k} \equiv \mathbf{a}$$

Using $\mathbf{F} = m\mathbf{a}$
$$\mathbf{F} = 0\cdot3\,(3\mathbf{i} + e^t\mathbf{j} + 2t\mathbf{k})$$
When $t = 2$, $\mathbf{F} = 0\cdot3\,(3\mathbf{i} + e^2\mathbf{j} + 4\mathbf{k})$
$$\Rightarrow |\mathbf{F}| = \sqrt{(0.9^2 + 0.09e^4 + 1\cdot2^2)}\ \text{N} \approx 2.68\text{N}$$

CONNECTED PARTICLES

" Avoid taking short cuts. "

Many examination questions are set which require the application of Newton's laws to particles connected by light, inextensible strings passing over smooth, light pulleys. Under these ideal conditions, it may be assumed that the strings do not stretch and the magnitude of the tension throughout a string is constant.

The following two worked examples show the methods used. Always apply Newton's second law to each member of the system, writing down an equation for each. Those who try to take short cuts often make serious fundamental errors.

Worked example

Two particles A and B of mass $0\cdot1$kg and $0\cdot2$kg respectively are connected by a light inextensible string which passes over a small, smooth, fixed pulley C. The particles are released from rest with the hanging parts of the string taut and vertical. Taking $g = 9\cdot8\,\text{ms}^{-2}$, calculate the acceleration of either particle and the tension in the string

Using the notation in Fig. 16.1, it is clear that on release from rest A will move vertically upwards and B will move vertically downwards. The tension in the string is TN and $a\,\text{ms}^{-2}$ is the acceleration of either particle.

For A, $T - 0\cdot1g = 0\cdot1a$ (Newton's second law)
For B, $0\cdot2g - T = 0\cdot2a$ (Newton's second law)

Adding, $0\cdot3a = 0\cdot1g \Rightarrow a = \dfrac{1}{3}g \approx 3\cdot27$

and $T = 0\cdot1g + 0\cdot1\left(\dfrac{1}{3}g\right) \approx 1\cdot31$

The tension in the string is $1\cdot31$N and the acceleration of either particle is $3\cdot27\,\text{ms}^{-2}$.

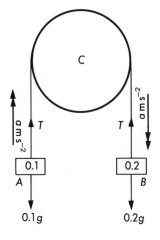

Fig. 16.1

Worked example

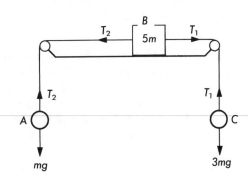

Fig. 16.2

The particles A, B, C of mass m, $5m$, $3m$ respectively are connected by two light inextensible strings as shown in Fig. 16.2. The strings pass over small, smooth pulleys, fixed at opposite ends of a smooth, horizontal table. The particles are released from rest. Calculate the acceleration of B and the tension in each string in terms of m and g.

Take a as the acceleration of B and T_1, T_2 as the tensions in BC and AB respectively. Using Newton's second law for each particle in turn, we have

A $T_2 - mg = ma$
B $T_1 - T_2 = 5ma$
C $3mg - T_1 = 3ma$

Adding we have $9ma = 2mg$

which gives $a = \dfrac{2}{9}g$

In the equation of motion for A

$$T_2 = mg + \frac{2}{9}mg = \frac{11}{9}mg$$

In the equation of motion for C

$$3mg - T_1 = 3m\left(\frac{2}{9}g\right)$$

$$\Rightarrow T_1 = \frac{7}{3}mg$$

EXERCISE 16.1

1   A particle P, of mass 0·05kg, moves from rest in a straight line under the action of a constant force of magnitude 0.4N. Calculate
   a) the magnitude of the acceleration of P
   b) the speed of P after 2s
   c) the speed of P after it has moved through 5m from rest

2   A car, of mass 800kg, is retarding at $4\,\mathrm{ms^{-2}}$ under the action of a constant force. Given that the car is moving in a straight horizontal line, calculate:
   a) the magnitude of the retarding force
   b) the time taken and the distance covered by the car in coming to rest if the force starts to act when the car is moving at $36\,\mathrm{ms^{-1}}$

3   A car and trailer, of mass 700kg and 200kg respectively, are connected by a light horizontal tow-bar. The engine of the car is producing a constant tractive force of magnitude 600N and resistances of magnitudes 300N for the car and 100N for the trailer oppose the motion of the car and trailer. Calculate the magnitude of the acceleration of the car and the tension in the tow-bar when the car moves along a straight horizontal road

4   A student of mass 80kg is standing in a lift. Find the magnitude of the force exerted by the lift on the student:
   a) when the lift is accelerating vertically upwards at $2\,\mathrm{ms^{-2}}$
   b) when the lift is accelerating vertically downwards at $3\,\mathrm{ms^{-2}}$
   (Take $g = 9·8\,\mathrm{ms^{-2}}$)

5   A light inextensible string has particles of mass 0·3kg and 0·1kg tied at its ends. The string passes over a smooth light fixed pulley. The particles are released from rest with the string taut and the hanging parts vertical. Calculate:
   a) the acceleration of either particle
   b) the tension in the string
   c) the force exerted by the particles on the pulley during motion
   Assuming the motion is not interrupted, calculate the distance moved through by either particle in the first half second of motion
   (Take $g = 10\,\mathrm{ms^{-2}}$)

6   Particle A is on a smooth plane inclined at 30° to the horizontal. A is connected to a particle B by a light inextensible string which passes over a fixed smooth pulley P placed at the top of the inclined plane, as shown in Fig. 16.3. The particles are released from rest and A moves along a line of greatest slope of the inclined plane. Given that A has mass 0·5kg and B has mass 0·2kg find the magnitude and direction of the acceleration of A. Find also the tension in the string
   (Take $g = 10\,\mathrm{ms^{-2}}$)

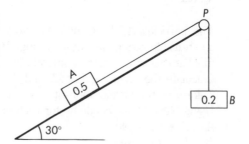

Fig. 16.3

WORK AND ENERGY

A particle of mass m is moving in a straight line under the action of a constant force of magnitude F. The particle covers a distance s in time t, the initial speed is u, the final speed v and the acceleration has magnitude a.

We have $F = ma$
   and $v^2 = u^2 + 2as$, since the motion is uniformly accelerated.

By combining these two equations and eliminating a we obtain

$$Fs = \frac{1}{2}mv^2 - \frac{1}{2}mu^2 \qquad (1)$$

The product force × distance, where the force is constant, is called the **work** done by the force over the space interval covered. The unit of work is the newton-metre (Nm) and the unit is called the **joule**, denoted by J. The expression $\frac{1}{2}mv^2$ is defined as the **kinetic energy** (KE) of a particle of mass m moving with speed v.

Equation (1) illustrates a particular case of a more general result which may be given as:

Work done by a force = change in KE produced by the force.

For example, suppose that a particle, of mass m, is moving along the x-axis under the action of a force of magnitude P, where P depends only on x, the distance of the particle from the origin.

The equation of motion is

$$P = mv\frac{dv}{dx} \qquad (2)$$

$$\left(\text{remember acceleration can be written as } v\frac{dv}{dx}\right)$$

Give that $v = v_1$, when $x = x_1$ and $v = v_2$ when $x = x_2$, equation (2) may be integrated with respect to x to give

$$\int_{x_1}^{x_2} P\,dx = \int_{x_1}^{x_2} mv\frac{dv}{dx}\,dx$$
$$= \int_{v_1}^{v_2} mv\,dv$$
$$= \left[\frac{1}{2}mv^2\right]_{v_1}^{v_2}$$
$$= \frac{1}{2}mv_2^2 - \frac{1}{2}mv_1^2$$

We define $\int_{x_1}^{x_2} P\,dx$ to be the work done by the force P over the space interval $x = x_1$ to

This is the general result.

$x = x_2$, and, we then have

Work done by a force = change in KE produced by the force

GRAVITATIONAL POTENTIAL ENERGY

We define the **gravitational potential energy** of a particle situated at a point A to be the work which would be done by its weight if it were to move from A to a point B, placed at some fixed level. For example, if the point B is situated at height h above the level of A, the work required *against* gravity to move a particle, of mass m, from A to B is mgh and we should say that the potential energy of the particle at A with respect to the level of B is $-mgh$. Potential energy has no absolute value, but the difference between the potential energies of a particle at different levels can be determined absolutely.

THE PRINCIPLE OF CONSERVATION OF MECHANICAL ENERGY

For bodies in motion under the action of a **conservative system of forces**, the sum of the kinetic and the potential energies of the bodies is constant. If the work done by a force on a particle which is moved from a point A to a point B is independent of the path joining A and B, the force is said to be conservative. In particular, the uniform gravitational field is a conservative field of force.

We include two simple examples to illustrate the main difference between a **conservative** system and a **non-conservative** system of forces.

Worked example

Example 1   Consider again the two points A and B, where B is at vertical height h above A, and suppose the particle of mass m is moved from A to B and then from B to A. The total work done in this complete operation $= -mgh + mgh = 0$, thus agreeing with our original contention that the work done = force × distance, and in this case, the distance is zero.

Worked example

Example 2   Consider now the points A and B at a distance h apart on a rough horizontal table. The particle of mass m is moved along the table from A to B and then from B to A. Let us suppose that the frictional force opposing the movement of the particle on the surface of the table is proportional to the weight of the particle in magnitude and equal to kmg, where k is a numerical constant. As the particle is moved from A to B the frictional force acts in the direction $\overrightarrow{BA}$ and the work required against friction to move the particle directly from A to B is $kmgh$. As the particle returns to A from B a further amount of work equal to $kmgh$ is required to achieve this because the frictional force acts in the direction $\overrightarrow{AB}$. In all, the total work required is $2kmgh$ and none of this is 'recoverable' as in Example 1 because it is transferred into other forms of energy, e.g. heat because of the friction. Frictional forces are non-conservative as this example illustrates.

Often a particle is subjected to both conservative and non-conservative forces simultaneously and in such cases only part of the work is recoverable. We may however, extend the principle of conservation of mechanical energy when frictional forces are involved and the following statement is often called the work–energy principle.

$$\text{Increase in mechanical energy} = \text{work done by external forces} - \text{work done against friction}$$

> **Look carefully at the signs.**

We may need to adjust the signs in this equation when work is done against external forces rather than by them and there may be an overall decrease in the total initial mechanical energy in these cases.

Worked example

The total mass of a toboggan and rider is 90 kg. The toboggan and rider slide down a line of greatest slope of a plane inclined at arcsin (1/4) to the horizontal. Over a distance of 60 m the speed of the toboggan increases from $5\,\mathrm{ms}^{-1}$ to $12\,\mathrm{ms}^{-1}$. Taking $g = 10\,\mathrm{ms}^{-2}$, calculate the magnitude of the frictional force, assumed constant, which is opposing the motion of the toboggan.

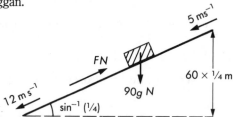

Fig. 16.4

The information given in the question is shown in Fig. 16.4.
Increase in kinetic energy is $(\frac{1}{2} \times 90 \times 12^2 - \frac{1}{2} \times 90 \times 5^2)\,\mathrm{J}$
Work done by gravitational forces is $90 \times 10 \times 60 \times 1/4\,\mathrm{J}$
Work done against the frictional force F is $(60F)\,\mathrm{J}$
Increase in KE = work done by gravitational forces − work done against friction, which gives $F = 135{\cdot}75$
The magnitude of the frictional force is $135{\cdot}75\,\mathrm{N}$

EXERCISES 16.2

(Take $g = 10\,\mathrm{ms}^{-2}$)

1   Find, in J, the kinetic energy of a body of mass 3 kg moving with speed $8\,\mathrm{ms}^{-1}$

2   A particle of mass 0·3 kg is slowly raised through a vertical height of 4 m. Find the work done in achieving this

3   A pump is raising 20 kg of water per second from a reservoir and delivering the water with speed $6\,\mathrm{ms}^{-1}$. Given that the height through which the water is raised is 20 m, find the work done, in J, in each second

4   One end of an inextensible string, of length 1·5 m, is tied at a fixed point O. A particle, tied at the other end of the string A, moves in vertical circles under gravity about centre O with $OA = 1.5\,\mathrm{m}$. Given that the greatest speed of the particle is $14\,\mathrm{ms}^{-1}$, find the least speed. Also find the speed of the particle at an instant when it is moving vertically

5

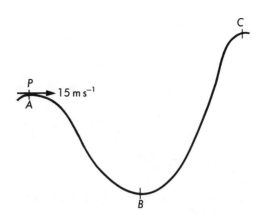

Fig. 16.5

A smooth ring P is threaded on the smooth wire ABC shown in Fig. 16.5, where A, B and C are in a vertical plane. The level of B is 5m below the level of A and the level of C is 3m above the level of A. The ring passes through A with speed 15ms^{-1}. Calculate the speed of the ring at the instant when
a) it passes through B
b) it passes through C

6 A cyclist who, with his machine, has mass 90kg freewheels down a hill inclined at $35°$ to the horizontal. The only force acting on the cyclist in the line of motion is the gravitational component of his weight. Find
a) the distance travelled from rest before the cyclist is moving at 20ms^{-1}
b) the potential energy lost in this distance by the cyclist

7 A bullet, of mass 5g, passes directly through a board of thickness 2cm and the speed is reduced from 400ms^{-1} to 200ms^{-1}. Assuming the resistive force exerted by the board on the bullet to be uniform, find the magnitude of the force in N to 2 significant figures

8 A child, of mass 50kg, sits on a light mat and slides from rest down a plane inclined at $\arcsin\left(\dfrac{12}{25}\right)$ to the horizontal through a distance of 20m in 4s. Find the speed of the child at the end of this descent. Use the work–energy equation to find the magnitude of the constant frictional resistance during the child's descent

---

FORCES AND CIRCULAR MOTION

We know that when a particle is moving in a circle, centre O and radius a, with constant speed v, the acceleration is of magnitude $\dfrac{v^2}{a}$ and is directed towards O.

Now we consider situations where circular motion occurs and we discuss the forces involved.

Worked example
A particle P of mass m is attached to a fixed point O by means of a light inextensible string of length l. P moves in a horizontal circle at constant speed v. The centre of the circle is C, where C is at distance h vertically below O. Calculate the tension in the string and show that P completes one circular orbit in time $2\pi\sqrt{\dfrac{h}{g}}$

P has two forces acting on it; these are its weight mg and the tension in the string T. As P is moving in a *horizontal* circle, the vertical component of T is equal to mg.
That is $T \cos \angle POC = mg$

$$\Rightarrow T\left(\frac{h}{l}\right) = mg \Rightarrow T = \frac{mgl}{h}$$

Applying Newton's second law horizontally we have

$$T\cos\angle OPC = m\frac{v^2}{PC}$$

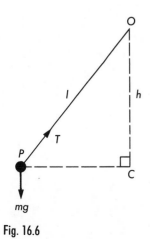

Fig. 16.6

In $\triangle POC$, $\cos \angle OPC = \dfrac{PC}{l}$

and $PC^2 = l^2 - h^2$

$\Rightarrow T\left(\dfrac{PC^2}{l}\right) = mv^2$ and since $T = \dfrac{mgl}{h}$

$\Rightarrow \dfrac{mgl}{h}\left(\dfrac{PC^2}{l}\right) = mv^2 \Rightarrow v^2 = \dfrac{g(PC)^2}{h}$

$\Rightarrow v = PC\sqrt{\dfrac{g}{h}}$

The circumference of the circular path of P is $2\pi PC$ and P will move around the circumference once in time

$$\dfrac{2\pi PC}{PC\sqrt{\dfrac{g}{h}}} = 2\pi\sqrt{\dfrac{h}{g}}$$

as required.

Worked example

A particle, of mass $0\cdot1$ kg, is moving on the inside surface of a fixed hemispherical bowl in a horizontal circle. The radius of the inner surface of the bowl is 2 m. Given that the constant speed of the particle is $3\,\text{ms}^{-1}$, find the radius of the circle in which the particle is moving and the magnitude of the force exerted by the bowl on the particle (Take $g = 10\,\text{ms}^{-2}$)

O is the centre of the rim of the hemisphere, C the centre of the circle in which the particle P is moving, and the radius is r m. The magnitude of the force exerted by the bowl on P is R N and this force acts along PO.
$\angle COP = \theta$

Resolving vertically $R\cos\theta = 0\cdot1g = 1$ (1)

Horizontally applying Newton's second law

$R\sin\theta = \dfrac{0\cdot1\times3^2}{r} = \dfrac{0\cdot9}{r}$ (2)

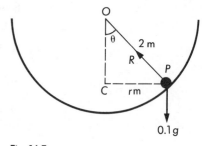

Fig. 16.7

Dividing (2) by (1) gives

$\tan\theta = \dfrac{0\cdot9}{r}$ (3)

In $\triangle OCP$, $\tan\theta = \dfrac{r}{OC} = \dfrac{r}{\sqrt{4-r^2}}$ (4)

From (3) and (4) $= \dfrac{0\cdot9}{r} = \dfrac{r}{\sqrt{4-r^2}}$

$81(4-r^2) = 100r^4$
$100r^4 + 81r^2 - 324 = 0$
$(25r^2 - 36)(4r^2 + 9) = 0$
$(5r-6)(5r+6)(4r^2+9) = 0$

The only value of r applicable is $\dfrac{6}{5}$

The radius of the circle is $1\cdot2$ m

$\tan\theta = \dfrac{0\cdot9}{r} = \dfrac{0\cdot9}{1\cdot2} = \dfrac{3}{4}$

So $\cos\theta = \dfrac{4}{5}$

and $R = \dfrac{1}{\cos\theta} = \dfrac{5}{4} = 1\cdot25$ [using (1)]

The force exerted by P on the bowl has magnitude $1\cdot25$ N

EXERCISES 16.3

1 A light inextensible string of length $5a$ has one end O tied at a fixed point. A particle P, of mass m, is tied at the other end of the string. P moves in a horizontal circle whose centre is $3a$ vertically below O with constant speed u. Find:
a) the tension in the string
b) an expression for u in terms of g and a

2 A car is moving horizontally in a circle of radius r with constant speed v. Given that the road on which the car is moving is banked at θ to the horizontal, show that there is no tendency for the car to side-slip when $v^2 = gr\tan\theta$

3 A particle is attached to a fixed point O by means of a light inextensible string. The particle describes a horizontal circle with uniform speed once every $2\,\mathrm{s}$. Given that the string is inclined at $30°$ to the horizontal, find the length of the string, giving your answer to the nearest cm (Take $g = 10\,\mathrm{m\,s}^{-2}$)

4 A particle P is attached to a fixed point O by means of a light inextensible string, of length a. The particle P is hanging at rest when it is given a blow which gives P an initial speed u, where $2u^2 = 7ga$. Find the angle made by the string with the upward vertical at the instant when the tension in the string vanishes

FORCES IN LIGHT ELASTIC STRINGS AND SPRINGS

Hooke's law states that if a light string is stretched in such a way that it assumes its *natural length* on release, then the **tension** is proportional to the extension. Similarly, if a spring is compressed, it is in thrust, and the **thrust** is proportional to the compression.

An elastic string, of natural length l, is stretched to a length $l+x$. In this case, the tension T in the string is given by

$$T = \lambda\left(\frac{x}{l}\right),$$

where λ is a constant for the string, called the **modulus of elasticity**.

Suppose now that the string is stretched by a further small distance δx. The work done is approximately $T\delta x = \left(\dfrac{\lambda x}{l}\right)\delta x$.

Therefore, the work done on the string in stretching it from its natural length l to length $l+c$ is

$$\int_0^c \left(\frac{\lambda x}{l}\right)\,\mathrm{d}x = \left[\frac{1}{2}\lambda\frac{x^2}{l}\right]_0^c = \frac{1}{2}\lambda\frac{c^2}{l}$$

We say that the potential energy stored in an elastic string (or spring) of natural length l and modulus of elasticity λ, when it is extended (or contracted for a spring) by a length c is $\dfrac{1}{2}\lambda\dfrac{c^2}{l}$

Worked example

A light elastic string, of natural length l and modulus of elasticity $4mg$, has one end tied at a fixed point O. A particle P, of mass m, is tied at the other end of the string.
a) Find the distance OP when P hangs at rest vertically below O in equilibrium
b) P is held at O and released to fall vertically. Find the greatest speed of P during its fall and the greatest distance between O and P.

Let T be the tension in the string

Fig. 16.8

a) In equilibrium, as shown in Fig. 16.8,
   $$T = mg \qquad (1)$$
   and by Hooke's law
   $$T = \frac{4mg(OP - l)}{l} \qquad (2)$$
   Eliminating T from (1) and (2) we have
   $$\frac{4mg(OP - l)}{l} = mg \Rightarrow OP = \frac{5l}{4}$$

b) During the fall, P will have greatest speed at the instant it passes through the equilibrium position found in a). By then it will have lost gravitational potential energy $\dfrac{5mgl}{4}$, gained kinetic energy $\dfrac{1}{2}mv^2$ and gained elastic potential energy $\dfrac{1}{2l}(4mg)\left(\dfrac{l}{4}\right)^2$

Energy is conserved, so we have

$$\frac{5mgl}{4} = \frac{1}{2}mv^2 + \frac{1}{8}mgl$$

$$\Rightarrow v^2 = \frac{9}{4}gl \Rightarrow v = \frac{3}{2}\sqrt{gl}$$

The greatest speed of P is $\dfrac{3}{2}\sqrt{gl}$

When P reaches its lowest position, say y below O, it will have lost gravitational potential energy and gained elastic potential energy. Energy is conserved so we have

$$mgy = \frac{1}{2} \times 4mg \frac{(y-l)^2}{l}$$

$$\Rightarrow yl = 2y^2 - 4yl + 2l^2$$
$$2y^2 - 5yl + 2l^2 = 0$$
$$(2y - l)(y - 2l) = 0 \Rightarrow y = 2l$$

We reject $y = \dfrac{l}{2}$ because this is clearly inapplicable and conclude the greatest distance between O and P is $2l$

Worked example

Consider the same model as in the last worked example, where P hangs at rest at E, distance $\dfrac{5l}{4}$ below O.

Suppose now that P is pulled down a further distance $\dfrac{l}{4}$ and released from rest. Show that P moves with simple harmonic motion and find the period. Find also the greatest speed of P.

At time t after release, we suppose that P is a distance x from E, as shown in Fig. 16.9. The tension T in the string by Hooke's law is

$$T = \frac{4mg(x + \frac{1}{4}l)}{l}$$

Fig. 16.9

Using Newton's second law at this instant

$$m\frac{d^2x}{dt^2} = mg - T = mg - \frac{4mg(x + \frac{1}{4}l)}{l}$$

$$\frac{d^2x}{dt^2} = -\left(\frac{4g}{l}\right)x$$

Therefore P moves with simple harmonic motion, period $2\pi\sqrt{\dfrac{l}{4g}} = \pi\sqrt{\dfrac{l}{g}}$

Writing $\dfrac{d^2x}{dt^2}$ as $v\dfrac{dv}{dx}$, where v is the speed of P at time t, we have

$$v\frac{dv}{dx} = -\left(\frac{4g}{l}\right)x$$

Integrating $\dfrac{1}{2}v^2 = -\dfrac{2g}{l}x^2 + C$, where C is a constant

At $t = 0$, $x = \dfrac{l}{4}$ and $v = 0 \Rightarrow C = \dfrac{2g}{l}\left(\dfrac{l}{4}\right)^2 = \dfrac{1}{8}gl$

This gives $v^2 = \dfrac{4g}{l}\left(\dfrac{l^2}{16} - x^2\right)$

Greatest speed of P occurs at $x = 0$ and this speed is $\frac{1}{2}\sqrt{gl}$

EXERCISES 16.4

(Take $g = 10\,\mathrm{m\,s^{-2}}$ in numerical questions)

1   a) A light elastic string of natural length $1\,\mathrm{m}$ and modulus of elasticity $4\,\mathrm{N}$ is attached to two points A and B at the same level and distance $1\cdot5\,\mathrm{m}$ apart. Calculate the tension in the string and the energy stored in the string

    b) The string described in a) has one end tied at a fixed point O and a particle of mass $0\cdot4\,\mathrm{kg}$ hangs in equilibrium vertically below O, tied to the other end. Find the extension in the string and the tension in the string

2   A light elastic string, of natural length $1\cdot2\,\mathrm{m}$, has one end attached to a fixed point A. A particle, of mass $0\cdot2\,\mathrm{kg}$, is tied to the other end B of the string. When the particle hangs in equilibrium, $AB = 1\cdot5\,\mathrm{m}$. Calculate the modulus of elasticity of the string.

    B is now held at A and the particle falls vertically. Calculate the speed of the particle when $AB = 1\cdot5\,\mathrm{m}$

3   A particle P of mass m is attached to one end of a light spring, the other end being attached at a fixed point A. When P hangs at rest below A, the extension in the spring is a. P is pulled down a further distance $\dfrac{1}{10}a$ and released from rest. Show that P

moves with simple harmonic motion of period $2\pi\sqrt{\dfrac{a}{g}}$

Find also:

    a) the speed of P when it is $\dfrac{1}{20}a$ above the lowest point of its motion

    b) the tension in the spring at the lowest point of its motion

4   One end of a light elastic string of natural length $0\cdot7\,\mathrm{m}$ is attached to a fixed point O on a smooth horizontal plane. A particle P, of mass $0\cdot3\,\mathrm{kg}$, is tied to the other end and released from rest at a point on the plane at distance $1\,\mathrm{m}$ from O. The tension in the string at the instant when P is released from rest is $9\,\mathrm{N}$. Calculate the speed of P when its distance from O is $0\cdot7\,\mathrm{m}$ and find the time, to 2 significant figures, for P to reach O from the instant it was released from rest

5   The ends of a light elastic string of natural length $1\cdot2\,\mathrm{m}$, are tied to points A and B respectively where A and B are fixed points at a horizontal distance $1\cdot2\,\mathrm{m}$ apart. A particle of mass $0\cdot4\,\mathrm{kg}$ is tied at the mid-point M of the string and the particle hangs in equilibrium at a depth $0\cdot8\,\mathrm{m}$ below the level of AB. Calculate
    a) the tension in either half of the string
    b) the modulus of elasticity of the string

POWER

Power is defined as a rate of doing work.
The standard unit of power is taken as a rate of working of $1\,\mathrm{J\,s^{-1}}$ and this unit is called the **watt** (W). As the watt is a fairly small unit, the **kilowatt** (kW) and the **megawatt** (MW) are often used in practice.

Consider a body of mass m moving in a straight line under the action of a constant force of magnitude F. Suppose that the velocity of the body is v at time t. The work done by this force in moving the body from displacement x to displacement $x + \delta x$, measured from some fixed point in the line of motion of the body, is $F\delta x$.

The average rate of working over this displacement is $F\,\dfrac{\delta x}{\delta t}$

The power at time t is $\displaystyle\lim_{\delta t \to 0}\left(F\,\dfrac{\delta x}{\delta t}\right) = F\dfrac{\mathrm{d}x}{\mathrm{d}t} = Fv$

$$F = ma$$
$$= 0$$
$$1 \frac{S \times d}{t}$$

Worked example

A cyclist is moving along a level road at constant speed $12\,\text{ms}^{-1}$ against resistive forces which are of total magnitude $15\,\text{N}$. Find the rate at which the cyclist is working

In order to maintain constant velocity, the cyclist must produce a tractive force of magnitude $15\,\text{N}$ (equal in magnitude and opposite in direction to the resistive forces). This tractive force is moved through a distance $12\,\text{m}$ each second.

Rate of working of the cyclist $= 15 \times 12\,\text{W} = 180\,\text{W}$

Worked example

An engine is required to pump water from a reservoir to a position which is $80\,\text{m}$ vertically higher than the level of water in the reservoir. Given that a mass of $300\,\text{kg}$ of water is raised each second and discharged with speed $20\,\text{ms}^{-1}$, find the power, in kW, required by the engine to achieve this. Take $g = 10\,\text{ms}^{-2}$.

Work required per second to raise water $= 300 \times 10 \times 80\,\text{J} = 240\,000\,\text{J}$
Work required per second to produce speed of discharge $= \frac{1}{2} \times 300 \times 20^2\,\text{J} = 60\,000\,\text{J}$
Total work per second required $= 300\,000\,\text{J}$
Power required by engine $= 300\,\text{kW}$

EFFICIENCY

❝ **Know how to deal with percentage figures.** ❞

Supplementary Note: the water has to be raised and given a speed of discharge, that is, the work required is determined by finding the increases in the kinetic and potential energies of the water being raised each second. In practice, the amount of work required by the engine of the pump would have to be appreciably greater than $300\,\text{kW}$, and often, the efficiency, that is the fraction of the total work which is effective, is given as a percentage. If, for example, we were told that the efficiency is 60%, the total amount of work required by the engine of the pump would be $300 \times 100/60\,\text{kW} = 500\,\text{kW}$.

Worked example

A lorry of mass 8 tonnes moves at constant speed $12\,\text{ms}^{-1}$ up a road inclined at θ to the horizontal, where $\sin\theta = 1/20$. The non-gravitational resistances are of magnitude $2500\,\text{N}$. Taking the acceleration due to gravity to be $10\,\text{ms}^{-2}$, calculate the rate, in kW, at which the engine of the lorry is working

$$RF = 0$$
$$\therefore T_F = F = 2500 + \frac{1}{20} 8000 \times 10$$
$$= 6500\,N$$
$$P = 78\,kW$$

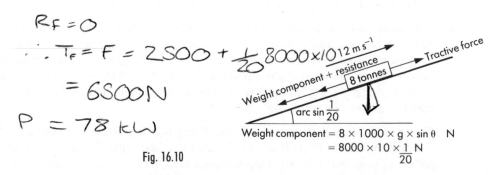

Weight component $= 8 \times 1000 \times g \times \sin\theta$ N
$= 8000 \times 10 \times \frac{1}{20}$ N

Fig. 16.10

Let us suppose that the engine of the lorry is working at $H\,\text{kW}$
The tractive force at $12\,\text{ms}^{-1}$ is $(1000H/12)\,\text{N}$
The forces opposing the motion of the lorry are the resistances and the component of the weight of the lorry along the line of motion.

We have, therefore, $\quad 1000H/12 = 2500 + 8000 \times 10 \times 1/20$
$\qquad\qquad\qquad$ that is $\quad H = 78$
The engine of the lorry is working at $78\,\text{kW}$

Worked example

A car, of mass $800\,\text{kg}$, moves down a road inclined at an angle θ to the horizontal at constant speed $30\,\text{ms}^{-1}$. At this speed the resistive forces are of magnitude $1200\,\text{N}$ and the engine of the car is working at $30\,\text{kW}$. Taking the acceleration due to gravity to be of magnitude $10\,\text{ms}^{-2}$, calculate $\sin\theta$. On another occasion the car is moving along a horizontal straight road with the engine working at a constant rate $H\,\text{kW}$. At a particular instant the speed of the car is $15\,\text{ms}^{-1}$ and the acceleration is of magnitude $0\cdot5\,\text{ms}^{-2}$. Given that the resistive forces are of magnitude $300\,\text{N}$ at this speed, calculate the value of H

We consider the forces acting on the car in the line of motion as shown in Fig. 16.11.

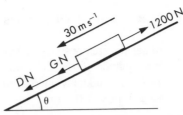

Fig. 16.11

These forces are the driving force of magnitude DN, the resistive forces of magnitude 1200N and the weight component of the car of magnitude GN.
Since Power = Force × Speed, $30D = 30 \times 1000 \Rightarrow D = 1000$
Also $G = 800g \sin \theta = 8000 \sin \theta$
Since the car is moving with constant speed, by Newton's first law
$$D - 1200 + G = 0$$
because the net force acting on the car in the line of motion is zero.
We have therefore $1000 - 1200 + 8000 \sin \theta = 0$
$$\Rightarrow \sin \theta = 1/40$$

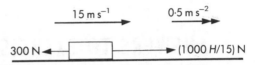

Fig. 16.12

In Fig 16.12, we show the forces acting on the car in the line of motion when it is moving along a level road with *instantaneous* speed 15ms^{-1} and *instantaneous* acceleration of magnitude 0·5ms^{-2}, with the engine of the car working at HkW. At this *instant*, the driving force is of magnitude $(1000H/15)$N, the resistive forces are of magnitude 300N and the force producing acceleration is of magnitude $800 \times 0·5$N (from Newton's second law). The instantaneous difference between the magnitudes of the driving force and the resistive forces is equal to the magnitude of the force making the car accelerate,
and so $1000H/15 - 300 = 400$
$$\Rightarrow H = 10·5$$

> **Look for key words in exam questions.**

Supplementary notes: several points are worth making from this last example, particularly you should note the use of the word *instantaneous* in the second part because the situation is continuously changing and we are discussing the motion of the car *at a particular instant*. If the car were to continue moving along the level road with its engine working at the constant rate of HkW, the car would eventually attain maximum speed for this rate of working of the engine. When this maximum speed is attained, the motion is steady and since the acceleration is then zero, we have

$$1000H = Rv$$

where RN is the magnitude of the resistive forces and $v$ms^{-1} is the maximum speed. *When the engine of a car is working at a constant rate, there is a finite greatest speed attainable*. Also, you will find in many examination questions that you are told to take the resistive forces as constant, although in practice the magnitude of these forces depends on the speed of the car and other factors.

EXERCISES 16.5

(Take $g = 10$ms^{-2})

1   A car is moving along a level road with constant speed 15ms^{-1} in a straight line against constant resistances of magnitude 246N. Calculate, in kW, the rate at which the engine of the car is working

2   A lorry, of mass 2500kg, climbs a straight hill inclined at arcsin $\left(\dfrac{1}{20}\right)$ to the horizontal at constant speed 15ms^{-1}. The non-gravitational resistances opposing the motion of the lorry are constant and total 2000N. Find, in kW, the rate at which the engine of the lorry is working

3   A pump which is only 50% efficient is delivering petrol at 5kgs^{-1} with speed 2ms^{-1}. This petrol is being raised through a vertical height of 8m. Calculate, in W, the rate at which the pump is working

4   A car, of mass 1600 kg, moves up a straight road inclined at $\arcsin\left(\frac{1}{25}\right)$ to the horizontal at a steady speed of $15\,\mathrm{ms}^{-1}$. The non-gravitational resistances are constant and of magnitude 400 N. Calculate, in kW, the power developed by the car's engine. If the power is suddenly increased by 10 kW, find the magnitude of the instantaneous acceleration of the car

5   A train is moving along a straight level track at a steady speed of $10\,\mathrm{ms}^{-1}$. The power exerted by the engine is 600 kW. Calculate, in N, the tractive force exerted by the engine. The power is suddenly increased to 800 kW. Given that the resistances have the same magnitude at all speeds, find the maximum speed of the train for this increased rate of working of its engine

6   A pump which works at an effective rate of 20 kW raises water from a depth of 25 m and delivers it with speed $15\,\mathrm{ms}^{-1}$. Calculate, in kg, the mass of water raised per minute

NB: Examination questions on these topics can be found in Chapter 19.

ANSWERS TO EXERCISES

EXERCISES 16.1

1   a) $8\,\mathrm{ms}^{-2}$; b) $16\,\mathrm{ms}^{-1}$; c) $8\cdot94\,\mathrm{ms}^{-1}$
2   a) 3200 N; b) 9 s, 162 m
3   $\frac{2}{9}\,\mathrm{ms}^{-2}$, $144\frac{4}{9}\,\mathrm{N}$
4   a) 944 N; b) 544 N
5   a) $5\,\mathrm{ms}^{-2}$; b) 1·5 N; c) 3 N; $\frac{5}{8}\,\mathrm{m}$
6   $\frac{5}{7}\,\mathrm{ms}^{-2}$ down the slope, $2\frac{1}{7}\,\mathrm{N}$

EXERCISES 16.2

1   96 J
2   12 J
3   4360 J
4   $11\cdot7\,\mathrm{ms}^{-1}$, $12\cdot9\,\mathrm{ms}^{-1}$
5   a) $18\cdot03\,\mathrm{ms}^{-1}$; b) $12\cdot8\,\mathrm{ms}^{-1}$
6   a) 34·9 m; b) 18 kJ
7   15000 N
8   $10\,\mathrm{ms}^{-1}$, 115 N

EXERCISES 16.3

1   a) $\frac{5}{3}mg$; b) $4\sqrt{\frac{ga}{3}}$
3   203 cm
4   60°

EXERCISES 16.4

1   a) 2 N, 0·5 J; b) 1 m, 4 N
2   8 N, $3\sqrt{3}\,\mathrm{ms}^{-1}$
3   a) $\frac{1}{20}\sqrt{3ga}$; b) $\frac{11}{10}mg$
4   $3\,\mathrm{ms}^{-1}$, 0·26 s
5   a) 2·5 N; b) 3·75 N

EXERCISES 16.5

1   3·69 kW
2   48·75 kW
3   820 W
4   15·6 kW, $0\cdot42\,\mathrm{ms}^{-2}$
5   60000 N, $13\frac{1}{3}\,\mathrm{ms}^{-1}$
6   3310 kg

GETTING STARTED

THE IMPULSE OF A FORCE

The **impulse** of a force is a vector quantity, let us call it $\mathbf{J}$, and if the velocity of a particle P changes from $\mathbf{v}_1$ to $\mathbf{v}_2$ in a time interval t, then we define $\mathbf{J} = m\mathbf{v}_2 - m\mathbf{v}_1$, where m is the mass of P. That is, the impulse of a force acting on a particle over an interval of time is the change in momentum produced by the force.

If the force $\mathbf{F}$ is constant, then we have $\mathbf{v}_2 = \mathbf{v}_1 + \mathbf{a}t$

where $\mathbf{a}$ is the acceleration of the particle

Also $\mathbf{F} = m\mathbf{a}$ (Newton's second law)

Combining these two equations by eliminating $\mathbf{a}$ gives

$$\mathbf{F}t = m(\mathbf{v}_2 - \mathbf{v}_1) = \mathbf{J}$$

If the force $\mathbf{F}$ is variable, we write

$$\mathbf{J} = \int_0^t \mathbf{F}\,\mathrm{d}t$$

$$= \int_0^t m\frac{\mathrm{d}\mathbf{v}}{\mathrm{d}t}\,\mathrm{d}t \left(\text{Newton's second law, } \mathbf{F} = m\frac{\mathrm{d}\mathbf{v}}{\mathrm{d}t}\right)$$

$$= m\left[\mathbf{v}\right]_{\mathbf{v}_1}^{\mathbf{v}_2} = m(\mathbf{v}_2 - \mathbf{v}_1)$$

The units of the impulse of a force are Newton-seconds, written Ns.

ESSENTIAL PRINCIPLES

Worked example

A particle P is moving with constant velocity $(2\mathbf{i} - 3\mathbf{j})\,\mathrm{ms}^{-1}$. The particle receives an impulse after which P moves with constant velocity $(6\mathbf{i} - 6\mathbf{j})\,\mathrm{ms}^{-1}$. Given that the mass of P is $0.3\,\mathrm{kg}$, calculate the magnitude and the direction of the impulse

Impulse = change in momentum

$$= [0.3(6\mathbf{i} - 6\mathbf{j}) - 0.3(2\mathbf{i} - 3\mathbf{j})]\,\mathrm{Ns}$$
$$= (1.2\mathbf{i} - 0.9\mathbf{j})\,\mathrm{Ns}$$

Magnitude of impulse $= \sqrt{[(1.2)^2 + (0.9)^2]}\,\mathrm{Ns}$
$$= 1.5\,\mathrm{Ns}$$

The direction of the impulse is directed along the vector $4\mathbf{i} - 3\mathbf{j}$

Worked example

A particle P, of mass $0.2\,\mathrm{kg}$, falls freely from rest at a height of $10\,\mathrm{m}$ above a horizontal floor. P bounces back from the floor with speed $4\,\mathrm{ms}^{-1}$. Calculate the impulse of the force exerted by P on the floor. (Take $g = 9.8\,\mathrm{ms}^{-2}$.)

Using $v^2 = u^2 + 2as$ with $u = 0$, $a = 9.8$ and $s = 10$ we have
$v^2 = 2 \times 9.8 \times 10 = 196 \Rightarrow v = 14$

P strikes the floor with speed $14\,\mathrm{ms}^{-1}$ and rebounds with speed $4\,\mathrm{ms}^{-1}$

Impulse = change in momentum
$$= 0.2[14 - (-4)]\,\mathrm{Ns}$$
$$= 3.6\,\mathrm{Ns}$$

Worked example

Water emerges horizontally from a pipe whose cross-sectional area is $50\,\mathrm{cm}^2$ and is delivered at a constant rate of 200 litres$\,\mathrm{s}^{-1}$. All of the water strikes at right angles a vertical wall from which it does not rebound. Calculate, in N, the thrust exerted on the wall by the water (1 litre $= 10^3\,\mathrm{cm}^3$; 1 litre of water has mass $1\,\mathrm{kg}$)

Speed of water $= 200 \times \dfrac{1000}{50}\,\mathrm{cms}^{-1}$
$$= 40\,\mathrm{ms}^{-1}$$

Mass of water emerging each second $= 200\,\mathrm{kg}$
Momentum of water each second $= 200 \times 40\,\mathrm{Ns}$
$$= 8000\,\mathrm{Ns}$$

Suppose the thrust on the wall due to the water is $F\,\mathrm{N}$
Using $Ft = m(v_2 - v_1)$
we have for $t = 1$,
$F \times 1 = 8000 \Rightarrow F = 8000$
The normal thrust exerted on the wall by the water is of magnitude $8000\,\mathrm{N}$

EXERCISES 17.1

1   A ball, of mass $0.2\,\mathrm{kg}$, is moving at $12\,\mathrm{ms}^{-1}$ horizontally when it strikes a vertical wall at right angles. The ball rebounds with speed $11\,\mathrm{ms}^{-1}$ horizontally. Calculate the magnitude of the impulse exerted by the ball on the wall. Given that the ball and wall were in contact for $0.1\,\mathrm{s}$, calculate the magnitude of the force exerted by the ball on the wall

2   A particle P, of mass $0.2\,\mathrm{kg}$, moving with constant velocity $(-3\mathbf{i} - 6\mathbf{j})\,\mathrm{ms}^{-1}$ receives an impulse and, as a result, moves with constant velocity $(2\mathbf{i} + 6\mathbf{j})\,\mathrm{ms}^{-1}$. Calculate the magnitude of the impulse and the angle the direction of the impulse makes with $\mathbf{j}$

3   A golf ball, of mass $0.1\,\mathrm{kg}$, is standing at rest on a horizontal plane. The ball receives a blow from a golf club and starts to move with speed $50\,\mathrm{ms}^{-1}$ at $30°$ to the horizontal. Calculate the magnitude of the impulse received by the ball. The golf club and ball were in contact for $0.05\,\mathrm{s}$. Calculate the magnitude of the force exerted by the club on the ball

4   Find the magnitude of the force which would be required to give a ball of mass $0.15\,\mathrm{kg}$ a speed of $14\,\mathrm{ms}^{-1}$ if the force acted for $0.1\,\mathrm{s}$

5   A jet of water emerges from a pipe and strikes a vertical wall horizontally and at right angles. Given that 6 kg of water is delivered per second and that, on impact, the water is moving with speed $8\,\text{ms}^{-1}$, estimate the magnitude of the force exerted by the water on the wall when no water rebounds from the wall

6   The particle P, of mass $0\cdot3\,\text{kg}$, is moving parallel to the vector $3\mathbf{i}+4\mathbf{j}$ with speed $10\,\text{ms}^{-1}$ when it receives a blow. After this blow, P moves parallel to the vector $\mathbf{i}$ with speed $8\,\text{ms}^{-1}$. Calculate the magnitude of the impulse

CONSERVATION OF MOMENTUM

Consider two particles, A and B of mass m_A and m_B, moving initially with velocity $\mathbf{u}_A$ and $\mathbf{u}_B$. The particle A is acted on by an external force $\mathbf{F}_A$ and the particle B is acted on by an external force $\mathbf{F}_B$. In addition, there is a force of interaction between A and B, $\mathbf{T}$ on A and $-\mathbf{T}$ on B, according to Newton's third law. The final velocities of A and B are $\mathbf{v}_A$ and $\mathbf{v}_B$.

Using the impulse–momentum equation on each particle in turn we have for this system

$A:\ \int(\mathbf{F}_A+\mathbf{T})\mathrm{d}t = m_A\mathbf{v}_A - m_A\mathbf{u}_A$
$B:\ \int(\mathbf{F}_B-\mathbf{T})\mathrm{d}t = m_B\mathbf{v}_B - m_B\mathbf{u}_B$

By adding these equations we have

$$\int\mathbf{F}_A\mathrm{d}t + \int\mathbf{F}_B\mathrm{d}t = (m_A\mathbf{v}_A + m_B\mathbf{v}_B) - (m_A\mathbf{u}_A + m_B\mathbf{u}_B)$$

In words, this result shows that *the sum of the impulses of the external forces on a pair of particles is equal to their total change in momentum.* When no external forces are acting, we see that the total momentum remains unchanged. This important result is called **the principle of conservation of momentum** and it can clearly be extended to systems containing three or more particles.

PRINCIPLE OF CONSERVATION OF MOMENTUM

> Note this "principle".

The total momentum in a given direction of all the bodies in a system is not changed by any interaction between them.

Worked example

Two particles, of mass $3m$ and $2m$, move in opposite directions in a straight line with speed $4u$ and $2u$ respectively. The particles collide and coalesce to form a single particle P of mass $5m$. Calculate
a) the speed of P in terms of u
b) the loss in kinetic energy due to the collision

Figure 17.1 shows the situations just before and just after the collision in the sense shown

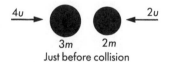

4u   2u
3m   2m
Just before collision

5m
Just after collision

Fig. 17.1

a) By the conservation of momentum principle
$12mu - 4mu = 5mv$

$$v = \frac{8u}{5}$$

The speed of P after the collision is $\dfrac{8u}{5}$

b) The loss in kinetic energy due to the collision is
  KE before $-$ KE after

$$= \frac{1}{2}\times 3m \times 16u^2 + \frac{1}{2}\times 2m \times 4u^2 - \frac{1}{2}\times 5m \times \frac{64u^2}{25}$$

$$= \frac{108}{5}mu^2$$

Worked example

A pile, of mass 1 tonne, is driven into the ground by blows from a body, of mass 100 kg, which falls freely from rest through a vertical distance of 5 m. After each blow, the pile and body move through 0.2 m vertically before coming to rest. Calculate:
a) the speed of the pile and body immediately after a blow
b) the resistance force exerted by the ground on the pile and body, assuming that the magnitude of this force is constant

Fig. 17.2

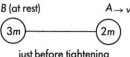

Fig. 17.3

a) Using $v^2 = u^2 + 2as$ with $u = 0$, $a = 10$, $s = 5$, we have $v^2 = 2 \times 10 \times 5$
   Speed of falling body on impact $= 10\,\mathrm{m\,s}^{-1}$

b) Using the principle of conservation of momentum

$$10 \times 100 = 1100v \Rightarrow v = \frac{10}{11}$$

Speed of pile and body just after impact $= \frac{10}{11}\,\mathrm{m\,s}^{-1}$

The kinetic energy of the pile and body just after impact is $\frac{1}{2} \times 1100 \times \left(\frac{10}{11}\right)^2$ J

This kinetic energy is used up in a distance of $0.2\,\mathrm{m}$, so we have the following forces (Fig. 17.3) acting on the combined body over this distance, $R\,\mathrm{N}$ being the resistive force exerted by the ground. From the work-energy principle we have

$$0.2(R - 1100\,g) = \frac{1}{2} \times 1100 \times \left(\frac{10}{11}\right)^2$$

$\Rightarrow R \approx 13\,300$ (3 significant figures)
Resistive force by ground is $13\,300\,\mathrm{N}$

Worked example

Two particles A and B of masses $2m$ and $3m$ respectively, are connected by a light inextensible string and placed side by side on a smooth horizontal plane. A is given a horizontal velocity of magnitude v directly away from B. Calculate:
a) the impulse of the tension in the string at the instant the string tightens
b) the loss in kinetic energy due to the tightening of the string

The diagrams in Fig. 17.4 show what is happening at the stages just before the string tightens, when the string tightens and just after the string tightens, I is the impulse, V the final speed.

Fig. 17.4

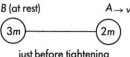 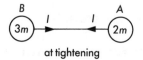 

just before tightening   at tightening   just after tightening

a) For A, $-I = 2mV - 2mv$
   For B, $I = 3mV$

Solving, we have $I = \frac{6mv}{5}$, $V = \frac{2v}{5}$

b) Loss in kinetic energy $= \frac{1}{2} \times 2m \times v^2 - \frac{1}{2} \times 5m \left(\frac{2v}{5}\right)^2$

$$= \frac{3}{5}mv^2$$

EXERCISES 17.2

1 Truck A of mass $800\,\mathrm{kg}$ is moving with speed $10\,\mathrm{m\,s}^{-1}$ along smooth horizontal rails towards a similar truck B, of mass $600\,\mathrm{kg}$, which is at rest on the same rails, but free to move. After the collision between the trucks, B moves with speed $8\,\mathrm{m\,s}^{-1}$. Calculate:
   a) the speed of A after the collision
   b) the impulse of the force exerted by A on B due to the collision
   c) the loss in kinetic energy due to the collision

2 A block, of mass $8\,\mathrm{kg}$, stands at rest on a smooth horizontal plane and it is free to move. A pellet, of mass $0.04\,\mathrm{kg}$, is fired horizontally into the block with speed $300\,\mathrm{m\,s}^{-1}$. Find the speed with which the block and pellet start to move and the loss in kinetic energy due to the impact, giving your answers to 3 significant figures

3 Particle A, of mass $3m$, moving with speed $7u$ collides with particle B, of mass $5m$, which is at rest. After the collision A and B move in the same direction with speeds $2v$

and $3v$ respectively. Express v in terms of u and hence find the loss in kinetic energy due to the collision in terms of m and u

4   Two particles, A and B, of mass $2m$ and $3m$ respectively, are connected by a light inelastic string of length l. The particles lie at rest on a smooth horizontal plane at distance l apart. The particle B is given an impulse in the direction shown in Fig. 17.5 and starts to move with speed u. Find the initial speed of A and the impulse of the tension in the string

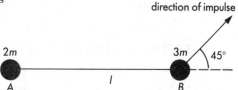

Fig. 17.5

5   Particles A and B, of mass $0{\cdot}1\,\mathrm{kg}$ and $0{\cdot}3\,\mathrm{kg}$ respectively, are connected by a light inextensible string of length $5\,\mathrm{m}$ (Fig. 17.6). Initially A is at rest on a horizontal floor and B is held at height $2\,\mathrm{m}$ above the floor, and the string passes over a small smooth fixed pulley. The length of string in contact with the small pulley is negligible. The particle B is released from rest and after falling freely for $1\,\mathrm{m}$, it jerks A into motion when the string tightens. Calculate, taking $g = 10\,\mathrm{ms}^{-2}$:
a) the initial speed of A
b) the total time taken by B to reach the floor, giving answers to 2 decimal places

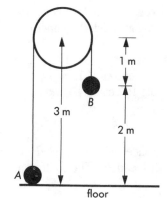

Fig. 17.6

6   Particle A moving with speed $3\,\mathrm{ms}^{-1}$ in a straight line collides with particle B which is moving with speed $1\,\mathrm{ms}^{-1}$ in the same line and in the same direction. On collision, the particles coalesce, forming a single particle C. Given that A is of mass $0{\cdot}2\,\mathrm{kg}$ and B is of mass $0{\cdot}3\,\mathrm{kg}$, find:
a) the speed of C after the collision
b) the loss in kinetic energy due to the collision

NEWTON'S EXPERIMENTAL LAW AND DIRECT COLLISIONS

When two bodies collide directly, the relative speed at which the bodies separate after their collision is in a constant ratio to the relative speed of closure before the collision. This constant ratio for two particular bodies is called their **coefficient of restitution** and is usually denoted by e, where $0 \leqslant e \leqslant 1$.

Suppose we have two particles A and B moving in the same straight line with constant speeds u_A and u_B, and that they collide. After the collision A and B move in the same line, as shown in Fig. 17.7, with speeds v_A and v_B.

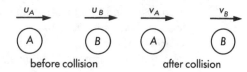

Fig. 17.7   before collision   after collision

By the law stated above

$$\text{Coefficient of restitution} = e = \frac{v_B - v_A}{u_A - u_B}$$

If we also know the masses of A and B, suppose they are m_A and m_B, then by the conservation of momentum principle, we have

$$m_A u_A + m_B u_B = m_A v_A + m_B v_B$$

In questions it is usual to have m_A, m_B, u_A, u_B and e given and by using both Newton's experimental law and the conservation of momentum principle, v_A and v_B can be found.

When solving problems, draw a clear diagram for each stage; that is, one for before the collision and another for after the collision. Write down clear equations and state the principle being applied. *Marks are awarded for these vital stages.* Then, make sure each step in your solution is clear. Often you will be asked to find the loss in kinetic energy due to the collision. In the case just presented this loss is:

"" Show each step in your solution. ""

$$\frac{1}{2}m_A u_A{}^2 + \frac{1}{2}m_B u_B{}^2 - \frac{1}{2}m_A v_A{}^2 - \frac{1}{2}m_B v_B{}^2$$

and you would be expected to substitute in the values of v_A and v_B found earlier before giving the final answer.

Also you may be asked to find the magnitude of the impulse exerted by A on B in the collision and this is *the change in the momentum of B alone*.

$$\text{Impulse exerted by } A \text{ on } B = m_B(v_B - u_B)$$

Note: the *loss in kinetic energy* is found from the *whole system* under consideration but the *impulse* is found by consideration of the change in momentum of *one member of the pair*.

EXTENSION TO OBLIQUE COLLISIONS

Read your syllabus carefully to see if oblique collisions are needed. Most syllabuses require only the study of direct collisions, but in the case where two bodies collide obliquely, the extension is simple to incorporate, and we include it for reference.

When two smooth spheres collide obliquely, Newton's experimental law *holds for component speeds along the line of centres only* (that is along the line in which the impulse is imparted from one body to the other).

Also there is no change in the component speed of either body at right angles to this line of centres because no interaction exists except in the line of centres. Momentum is conserved in all directions. Using these principles, it is possible to set up solutions for these problems too.

Worked example

A particle A, of mass $6m$, moving at speed $4u$ in a straight line, collides with a particle B, of mass $4m$, moving in the same line and in the same direction with speed $2u$. The coefficient of restitution between A and B is $\dfrac{1}{4}$. Calculate:

a) the speeds of A and B after the collision
b) the impulse transmitted from A to B in the collision
c) the kinetic energy lost in the collision

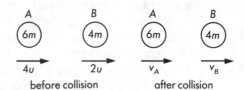

Fig. 17.8   before collision   after collision

Using notation in the diagrams (Fig. 17.8) we have,

a) Conservation of momentum gives
$$24mu + 8mu = 6mv_A + 4mv_B$$
$$16u = 3v_A + 2v_B \qquad (1)$$
Newton's experimental law gives
$$\frac{v_B - v_A}{4u - 2u} = \frac{1}{4}$$
$$2v_B - 2v_A = u \qquad (2)$$
Solving (1) and (2) for v_A and v_B
$$v_A = 3u, \; v_B = \frac{7}{2}u$$

which are the speeds of A and B after the collision.

b) Impulse transmitted from A to B
= change in momentum of A
$= 6m \times 4u - 6m \times v_A$
$= 24mu - 18mu = 6mu$

c) Loss in kinetic energy due to collision
$$= \frac{1}{2} \times 6m \times (4u)^2 + \frac{1}{2} \times 4m \times (2u)^2 - \frac{1}{2} \times 6m \times (3u)^2 - \frac{1}{2} \times 4m \times \left(\frac{7u}{2}\right)^2$$
$$= mu^2 [48 + 8 - 27 - 24{\cdot}5]$$
$$= \frac{9}{2}mu^2$$

Note: in solutions of this type, keep your processing under good control at each stage. For example, tidy up equations 1 and 2 before solving them and write out the full statement in c) for loss in kinetic energy before simplifying. *Also* take care to use brackets to help you. Those who try to jump stages usually make unforced and unnecessary errors. Questions on collisions are both popular and good mark earners, but you need to practise very thoroughly in order to master them.

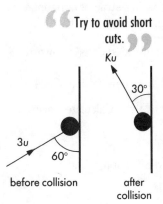

"Try to avoid short cuts."

Fig. 17.9

3u

60°

before collision

Ku

30°

after collision

Worked example

A particle P, of mass m, hits a smooth vertical wall and with speed $3u$ at the instant when it is moving upwards at 60° to the vertical and so that the direction of motion of P is in a plane normal to the wall. Given that P rebounds with speed Ku at 30° to the vertical, as shown in Fig. 17.9, calculate:

a) the value of K
b) the coefficient of restitution between P and the wall
c) the magnitude of the impulse exerted by the wall on P

a) Because the wall is smooth, there is no interaction between P and the wall parallel to the wall on impact, so the momentum in this direction is unchanged.
$$3mu\cos 60° = mKu\cos 30°$$
$$\Rightarrow K = \sqrt{3}$$

b) At right-angles to the wall we have by Newton's experimental law
$$e\,3u\sin 60° = Ku\sin 30°$$
$$\Rightarrow e = \frac{K\sin 30°}{3\sin 60°} = \frac{1}{3}$$

c) Impulse = change in momentum, using
$$= m[(3u\sin 60°)\mathbf{i} + (3u\cos 60°)\mathbf{j}] - m[-(Ku\sin 30°)\mathbf{i} + (Ku\cos 30°)\mathbf{j}]$$
$$= mu[(2\sqrt{3})\mathbf{i}]$$
Magnitude of impulse $= 2\sqrt{3}mu$

Note: many candidates will appreciate that the impulse must be acting at right-angles to the wall and that we only need to find the change in momentum in this direction. Others will prefer to set it out, as in the solution given.

EXERCISES 17.3

1   A particle A, of mass $10m$, is moving with speed $8u$ in a straight line when it collides with a particle B, of mass $8m$, moving in the same line and in the same direction with speed $5u$. Given that the coefficient of restitution between A and B is $\frac{1}{2}$, calculate:

   a) the speeds of A and B after their collision
   b) the loss in kinetic energy due to the collision

2   The particles A and B, of mass $2m$ and $3m$ respectively, are moving towards each other in the same line with speeds $6u$ and $8u$ respectively. The coefficient of restitution between A and B is $\frac{11}{14}$.
   Calculate:
   a) their speeds and directions of motion after the collision
   b) the magnitude of the impulse exerted by B on A due to the collision
   c) the loss in kinetic energy due to the collision

3   Three beads, A, B, C of masses $3m$, $2m$, $2m$, are threaded on a smooth fixed straight horizontal wire of indefinite length, and are at rest with B between A and C. In any collision between the beads, the coefficient of restitution is $0\cdot25$. Initially A is given a speed of $128u$ towards B and they collide. Show that, in all, three collisions occur between the beads and that after these collisions the final speeds of A, B and C are $50u$, $57u$ and $60u$ respectively.

4   Two particles A and B of mass m and $2m$ are moving in a straight line and they collide. At the instant before the collision, the speed of A relative to B is u. Given that the coefficient of restitution between A and B is $\frac{1}{2}$ show that the loss in kinetic energy due to the collision is $\frac{1}{4}mu^2$.

5 Three particles A, B, C, of equal mass, lie on a smooth table in a straight horizontal line with B between A and C. The coefficient of restitution between any two of these particles is $\frac{1}{3}$. The particle A is made to move with constant speed $6\,\text{m}\,\text{s}^{-1}$ towards B. Show that, after the collisions between A and B and between B and C, A, B, C are moving with speeds $2\,\text{m}\,\text{s}^{-1}$, $\frac{4}{3}\,\text{m}\,\text{s}^{-1}$, $\frac{8}{3}\,\text{m}\,\text{s}^{-1}$ respectively. Calculate also the percentage loss in kinetic energy due to these two collisions

NB: Examination questions on these topics can be found in Chapter 19.

ANSWERS TO EXERCISES

EXERCISES 17.1

1 4.6Ns, 46N
2 2·6Ns, 22·6°
3 5Ns, 100N
4 21N
5 48N
6 2·47Ns

EXERCISES 17.2

1 a) $4\,\text{ms}^{-1}$; b) 4800Ns; c) 14·4kJ
2 1·49ms⁻¹, 1790J

3 $v = u$; $45mu^2$
4 $\dfrac{u}{\sqrt{2}}$, $mu\sqrt{2}$
5 a) $3\cdot35\text{ms}^{-1}$; b) 0·94s
6 a) $1\cdot8\text{ms}^{-1}$; b) 0·24J

EXERCISES 17.3

1 a) $6u$, $\dfrac{15u}{2}$; b) $15mu^2$

2 a) $9u$, $2u$ both in opposite directions to initial directions; b) $30mu$; c) $45mu^2$
5 35·8%

PARTICLES AND RIGID BODIES IN EQUILIBRIUM

GETTING STARTED

A body which is at rest is in **equilibrium**. We study systems of forces acting in a plane on a body which remains at rest. To start we consider coplanar forces acting on a particle.

A particle P is in equilibrium if and only if the vector sum (called the **resultant**) of the forces acting on it is the null vector.

Worked example

Forces $(3\mathbf{i} + 4\mathbf{j})\,\text{N}$, $(a\mathbf{i} + 7\mathbf{j})\,\text{N}$ and $(-6\mathbf{i} + b\mathbf{j})\,\text{N}$ act on a particle P which is in equilibrium. Find the values of a and b,

For equilibrium
$$3\mathbf{i} + 4\mathbf{j} + a\mathbf{i} + 7\mathbf{j} - 6\mathbf{i} + b\mathbf{j} = \mathbf{0}$$
$$(3 + a - 6)\mathbf{i} + (4 + 7 + b)\mathbf{j} = \mathbf{0}$$
The $\mathbf{i}$ components must be zero
$$3 + a - 6 = 0 \Rightarrow a = 3$$
The $\mathbf{j}$ component must also be zero
$$4 + 7 + b = 0 \Rightarrow b = -11$$

Worked example

Forces $(5\mathbf{i} - 9\mathbf{j})\,\text{N}$ and $(-2\mathbf{i} + 5\mathbf{j})\,\text{N}$ act on a particle P and a third force $\mathbf{F}$ is applied to P which produces equilibrium. Calculate the magnitude of $\mathbf{F}$.

$$\mathbf{F} + 5\mathbf{i} - 9\mathbf{j} - 2\mathbf{i} + 5\mathbf{j} = \mathbf{0}$$
$$\Rightarrow \mathbf{F} = -3\mathbf{i} + 4\mathbf{j}$$
$$|\mathbf{F}| = \sqrt{(3^2 + 4^2)} = 5$$

The third force $\mathbf{F}$ has magnitude $5\,\text{N}$

ESSENTIAL PRINCIPLES

Three forces acting at a point are in equilibrium if and only if they can be represented by the sides of a triangle taken in order.

Any number of forces acting at a point are in equilibrium if and only if they can be represented by the sides of a polygon taken in order.

RESOLUTIONS OF A FORCE

As in earlier work, a force P acting at an angle α to the horizontal is equivalent to forces $P\cos\alpha$ horizontally and $P\sin\alpha$ vertically, as shown in Fig. 18.1.

Fig. 18.1

SOLUTION OF PROBLEMS

Using the above results, the resultant of a system of coplanar forces can be found by:

i) drawing a force polygon in which *the resultant force* is represented by *the line required to complete the polygon*, but remembering that the resultant will act in the opposite sense to all the other forces

ii) using the sum of the components of all the forces in two directions, usually perpendicular, say ΣF_x and ΣF_y, then the resultant has magnitude

$$\sqrt{[(\Sigma F_x)^2 + (\Sigma F_y)^2]}$$

and the direction is at

$$\arctan\left[\frac{\Sigma F_y}{\Sigma F_x}\right] \text{ with the } x\text{-direction}$$

Worked example

A particle P, of mass $0\cdot4\,\text{kg}$, is in equilibrium, suspended by two light inextensible strings which are inclined at $30°$ and $45°$ to the horizontal. Find, giving your answer to $0\cdot1\,\text{N}$, the tension in each string

There are several methods of solving this problem and we consider three solutions to show some of these.

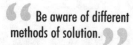
Be aware of different methods of solution.

a) A graphical solution using the triangle of forces.

Taking $g = 10\,\text{ms}^{-2}$, the weight of P is $4\,\text{N}$

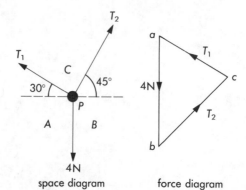

Fig. 18.2   space diagram   force diagram

The space diagram (Fig. 18.2) shows the forces acting on P.

The regions into which the plane is partitioned by the lines of action of the weight of P, T_1 and T_2 are labelled A, B and C in the space diagram. Then, since the directions of T_1 and T_2 are known, we can draw to scale the force diagram abc, where ab represents the weight of P, T_1 and T_2 represent the tensions in the strings.

We obtain by measuring $T_1 = 2\cdot9\,\text{N}$, $T_2 = 3\cdot6\,\text{N}$

BOW'S NOTATION IN RELATING SPACE AND FORCE DIAGRAMS

> Bow's notation has advantages.

The notation used of labelling the spaces in the space diagram with capital letters, followed by the respective small letters for vertices in the force polygon, provides an easy check on the directions of the forces and it also ensures that the forces are taken in order. This notation is called **Bow's notation**.

b) A sketch of the force diagram in a) is sufficient when a solution is given by calculation using the sine rule

$$\frac{T_1}{\sin 45°} = \frac{T_2}{\sin 60°} = \frac{4}{\sin 75°}$$

$$T_1 = \frac{4 \sin 45°}{\sin 75°} \approx 2·9\,\text{N}$$

$$T_2 = \frac{4 \sin 60°}{\sin 75°} \approx 3·6\,\text{N}$$

c) Since the forces are in equilibrium many students solve this type of problem by 'balancing' the resolved parts of forces in two directions (often horizontal and vertical) and solving the resulting equations.

Horizontally $\leftrightarrow$ $T_1 \cos 30° = T_2 \cos 45°$ (1)

Vertically $\updownarrow$ $T_1 \sin 30° + T_2 \sin 45° = 4$ (2)

Solving (1) and (2) simultaneously then gives
$T_1 \approx 2·9\,\text{N}$, $T_2 \approx 3·6\,\text{N}$, as before

FRICTION

Frictional forces are not considered when the surfaces of two bodies in contact are described as *smooth*. When the surfaces of two rough bodies are in contact, frictional forces need to be considered and the following laws apply.

> Important laws of frictional forces.

1   The magnitude of the limiting frictional force is proportional to the magnitude of the normal contact force exerted by one body on the other and it acts at right-angles to the common normal in the opposite direction to that in which the body is moving or tending to move.

2   The coefficient of friction between the surfaces of two bodies in contact is the ratio of the magnitude of the limiting frictional force and the magnitude of the normal contact force. The value of the coefficient of friction depends only on the roughness of the contact surfaces and *not* on the areas of these surfaces.

3   When no movement takes place, the magnitude of the frictional force is just sufficient to prevent relative motion between the surfaces in contact and this magnitude will only take the limiting value when motion is on the point of occurring. Note that when motion occurs, the magnitude of the frictional force takes the limiting value and is independent of the speed at which one body is sliding over the other.

The coefficient of friction is often denoted by μ. For a limiting frictional force of magnitude F and corresponding normal contact force of magnitude N, we write $F = \mu N$. In all other cases where we do not know that equilibrium is limiting, the frictional force of magnitude F and the corresponding normal contact force of magnitude N obey the relation

$$F < \mu N$$

and the magnitude of F is just sufficient to prevent movement.

> Note this 'good practice'.

In working exercises, it is very important to follow good practice by writing $F < \mu N$ unless you are told equilibrium is limiting when you may write $F = \mu N$.

The following diagrams (Fig. 18.3) show two equivalent systems of forces acting at a point O where friction is limiting, the coefficient of friction being μ.

Fig. 18.3

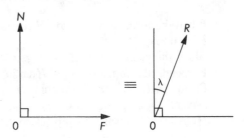

Since friction is limiting $F = \mu N$, so $\dfrac{F}{R} = \mu$

Resolving $\leftrightarrow F = R\sin\lambda$
$\quad\quad\quad\updownarrow N = R\cos\lambda$

Dividing $\dfrac{F}{N} = \dfrac{\sin\lambda}{\cos\lambda} = \tan\lambda = \mu$

and $F^2 + N^2 = R^2(\sin^2\lambda + \cos^2\lambda) = R^2$

The angle λ is called **the angle of friction** between the surfaces in contact. The magnitude of the resultant force (or total force) exerted by one body on the other in contact is R.

Worked example

An ice hockey puck moves across the ice in a straight line, moving with initial speed $12\,\text{ms}^{-1}$ and final speed $8\,\text{ms}^{-1}$, while covering a distance of $20\,\text{m}$. Calculate the coefficient of friction between the puck and the ice

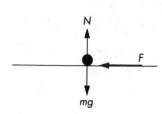

Fig. 18.4

The diagram (Fig. 18.4) shows the forces acting on the puck.
$\updownarrow N = mg$, where m kg is the mass of the puck, N newtons the magnitude of the normal contact force and F newtons the magnitude of the frictional force.
Since the puck is moving, $F = \mu N$
Also $F = ma$, where $a\,\text{ms}^{-2}$ is the deceleration.
But using $v^2 = u^2 + 2as$
$\quad\quad 8^2 = 12^2 - 2a(20) \Rightarrow a = 2$
$\quad \Rightarrow \mu mg = 2m$

$\quad\quad \Rightarrow \mu = \dfrac{2}{g} \approx 0{\cdot}2$ (taking $g = 10\,\text{ms}^{-2}$)

Coefficient of friction between puck and ice is $0{\cdot}2$

Worked example

A particle P, of mass $0{\cdot}2\,\text{kg}$, is placed on a rough plane inclined at $30°$ to the horizontal and rests in equilibrium. Show that the coefficient of friction between P and the plane is greater than $\dfrac{1}{3}\sqrt{3}$. Given that $\mu = 0{\cdot}6$, find the magnitude of the horizontal force which can be applied to P when P rests on the plane and is on the point of moving up the plane (Take $g = 10\,\text{ms}^{-2}$)

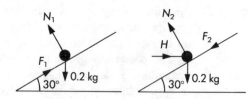

Fig. 18.5

Perpendicular to the plane, $N_1 = 0{\cdot}2g\cos 30°$
Along the plane, $F_1 = 0{\cdot}2g\sin 30°$ for equilibrium
Since equilibrium is not necessarily limiting
$\quad F_1 < \mu N_1$

Combining these equations gives $\mu > \tan 30° = \dfrac{1}{3}\sqrt{3}$.

The coefficient of friction between P and the plane is greater than $\dfrac{1}{3}\sqrt{3}$.

When the force of magnitude H N is applied, P is about to move up the plane and
$F_2 = 0{\cdot}6N_2$ \hfill (1)
Along the plane, $H\cos 30° = 0{\cdot}2g\sin 30° + F_2$ \hfill (2)
Perpendicular to plane, $N_2 = H\sin 30° + 0{\cdot}2g\cos 30°$ \hfill (3)
From (1), (2) and (3) we require H, so in (2) we have
$H\cos 30° = 0{\cdot}2g\sin 30° + 0{\cdot}6\,N_2$ [using (1)]
$H\cos 30° = 0{\cdot}2g\sin 30° + 0{\cdot}6\,(H\sin 30° + 0{\cdot}2g\cos 30°)$ [using (3)]

$H[\cos 30° - 0.6 \sin 30°] = 0.2g \sin 30° + 0.12g \cos 30°$
$$\Rightarrow H = 3.6$$
The horizontal force has magnitude $3.6\,\text{N}$

EXERCISES 18.1

1 A particle P moves in a straight line across a horizontal floor. The speed of P reduces uniformly from $8\,\text{ms}^{-1}$ to $6\,\text{ms}^{-1}$ in $4\,\text{s}$. Taking $g = 10\,\text{ms}^{-2}$, find the coefficient of friction between P and the floor

2 A particle P of mass $0.3\,\text{kg}$ is suspended by two inextensible strings whose other ends are tied to fixed points A and B at a horizontal distance of $2\,\text{m}$ apart. Given that PA and PB are of lengths $1.2\,\text{m}$ and $1.6\,\text{m}$ respectively find the tensions in the strings when P hangs in equilibrium. (Take $g = 10\text{ms}^{-2}$)

3 Two forces $\mathbf{F}_1 = (4\mathbf{i} + 7\mathbf{j})\,\text{N}$ and $\mathbf{F}_2 = (2\mathbf{i} - \mathbf{j})\,\text{N}$ act on a small body of mass $2\,\text{kg}$. Find the magnitude of the resultant force acting on the body and the distance covered from rest in $2\,\text{s}$ when these are the only forces acting on the body

4 A ring, of mass $m\,\text{kg}$, is threaded on a fixed straight wire inclined at $20°$ to the horizontal. Given that the ring is at rest in limiting equilibrium find, taking $g = 9.8\,\text{ms}^{-2}$
   a) the coefficient of friction between the ring and the wire, giving your answer to 2 decimal places
   b) the resultant force, in terms of m, exerted by the ring on the wire

5 A small box, of mass m, rests on a rough horizontal floor. The coefficient of friction between the box and the floor is μ. A force, of magnitude P, is applied to the box at an angle α to the horizontal and equilibrium is then limiting. Find P in terms of m, g, μ and α

6 A particle P, of mass m rests in limiting equilibrium on a rough plane which is inclined at an angle α to the horizontal (Fig. 18.6). Show that the coefficient of friction between P and the plane is $\tan\alpha$

   A force of magnitude T, acting at an angle θ to the inclined plane, is applied to P. The line of action of this force and the line of greatest slope of the inclined plane through P lie in the same vertical plane. Given that equilibrium is limiting with P on the point of moving up the plane, show that
   $$T = \frac{mg \sin 2\alpha}{\cos (\theta - \alpha)}$$
   Hence find in terms of m, g and α, the least value of T and the value of θ, in terms of α, for which this occurs

7 A particle, of mass m, rests on a rough horizontal plane. A force, of magnitude T making an angle θ with the horizontal, is applied to the particle and equilibrium is limiting. Given that λ is the angle of friction between the particle and the plane, show that
   $$T = \frac{mg \sin \lambda}{\cos (\theta - \lambda)}$$
   Deduce the least value of T and state the value of θ, in terms of λ, for which this occurs

Fig. 18.6

RIGID BODIES

Two new concepts are required in dealing with the equilibrium of a *rigid body*. When a rigid body is subjected to a system of coplanar forces, equilibrium occurs only if:

a) the vector sum of the forces acting on the body is zero,
b) there is no turning effect caused by the forces considered as a whole.

Condition a) can be fulfilled by showing that the sum of the resolved parts of the forces in two independent directions (often horizontal and vertical) are both zero. Condition b) requires us to define the moment of a force because moments are used to measure the turning effect of a system of forces.

The moment of a force of magnitude P about a point O is defined as Ph where h is the length of the perpendicular drawn from O to the line of action of the force (see Fig. 18.7).

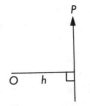

Fig. 18.7

Conventionally, we give a force which tends to turn a body in the *anti-clockwise sense a positive moment* and a force tending to turn the body in the *clockwise sense a negative moment*. Since the moment of a force is the product of force and distance, the unit of moment is the newton-metre (Nm).

Note: strictly, the moment of magnitude, Ph, is taken about an axis through O perpendicular to the plane in which the force system is acting.

PARALLEL FORCES AND COUPLES

The concept of moments is used to determine the line of action of the resultant of two or more parallel forces. When two forces act in the same direction, the magnitude of the resultant is equal to the sum of the magnitudes of the forces.

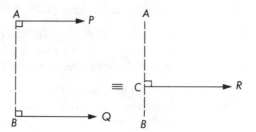

Fig. 18.8

Fig. 18.8 shows two parallel forces, of magnitudes P and Q, acting through points A and B, where AB is perpendicular to the lines of action of the forces. The resultant of these forces has magnitude R, where $R = P + Q$, and is parallel to the lines of action of the forces and acts through the point C in AB such that

$$\bigl(C \Rightarrow Q.CB - P.AC = 0 \qquad (1)$$

(where $\bigl(C$ is shorthand for 'moments about C')

$$\Rightarrow \frac{AC}{CB} = \frac{Q}{P}$$

and $AC = \left(\dfrac{Q}{P+Q}\right) AB$ and $CB = \left(\dfrac{P}{P+Q}\right) AB$

Note that in (1), we have used a very important result which is worth stating generally.

The sum of the moments of any system of forces about a point on the line of action of the resultant of the system is zero.

When the two forces of magnitudes P and Q, where $P > Q$, act in parallel lines in opposite directions, their resultant is of magnitude $P - Q$ and the line of action of this resultant is parallel to the line of action of the force of magnitude P (see Fig. 18.9)

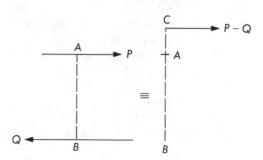

Fig. 18.9

$$\bigl(C, \Rightarrow P.CA - Q.CB = 0$$

$$\therefore CA = \left(\frac{Q}{P-Q}\right) AB \text{ and } CB = \left(\frac{P}{P-Q}\right) AB$$

When two unlike parallel forces are *equal in magnitude*, their effect is one of *turning only* and this combination of the two forces is called a **couple**. A couple is irreducible and cannot be replaced by any system which is simpler.

The couple shown in Fig. 18.10 has moment Ph and the sign of the moment is taken as positive because the turning effect of the couple is in an anti-clockwise sense. Further, a couple has the same moment about any point in its plane of action and may be replaced by any pair of equal and opposite parallel forces acting in opposite directions provided that their moment has the same magnitude and sense as the couple.

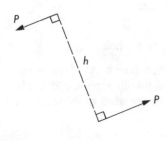

Fig. 18.10

THE CENTRE OF MASS OF A RIGID BODY

The second concept required in the study of forces acting on a rigid body is the centre of mass of the body.

The weight of the particles which make up a rigid body form a set of forces all directed towards the earth's centre and we assume that these forces are parallel. The resultant of the weights is equal in magnitude to the sum of the weights of the particles. This sum is called *the weight of the body* and it acts through a definite point called *the centre of mass of the body*. For uniform, symmetrical bodies, the position of the centre of mass coincides with the centre of symmetry, e.g. the centre of mass of a uniform straight rod is at the mid-point of the rod, the centre of mass of a thin, uniform square plate is at the centre of the square. Often, the position of the centre of mass of a body is determined by integration using the result that the sum of the moments of the weights of all the particles making up the body about any point is equal to the moment of the weight of the whole body about the same point.

You can find examples showing this method in Chapter 10 on the applications of integration. Some standard results for uniform bodies are often given in the formulae booklets issued by the examination boards and the following table gives some of these. Check carefully with your own booklet, but remember you may be asked to prove any result by integration at the start of a question.

66 Some standard results for uniform bodies. 99

BODY (UNIFORM)	CENTRE OF MASS LOCATION
Circular arc, angle 2θ, radius r	$\dfrac{r \sin \theta}{\theta}$ from centre
Circular sector, angle 2θ, radius r	$\dfrac{2r \sin \theta}{3\theta}$ from centre
Triangular lamina ABC, mid-point of BC is D	at G, intersection of medians, e.g. $\overrightarrow{AG} = \dfrac{2}{3}\overrightarrow{AD}$
Solid right circular cone, height h	$\dfrac{3}{4}h$ from vertex
Solid hemisphere, radius r	$\dfrac{3}{8}r$ from centre
Hemispherical shell, radius r	$\dfrac{1}{2}r$ from centre

Table 18.1 The centre of mass of some standard uniform laminas and bodies

EQUIVALENT SYSTEMS OF FORCES IN A PLANE

Two systems of forces acting in a plane on a body are defined to be equivalent when *both* of the following conditions are satisfied:

a) the vector sum of the forces in the first system is equal to the vector sum of the forces in the second system,

b) the sum of the moments of the forces in the first system is equal to the sum of the moments of the forces in the second system *about every point in the plane*.

In general, two forces acting on a body may be replaced by a single force, called their resultant. Any system of forces may be reduced successively using this technique until the simplest equivalent system has been obtained. This simplest equivalent system may be *a single force* or *a couple* or the special case when the system reduces to two equal and opposite forces acting in the same line when we have *equilibrium*.

A system of coplanar forces, in general, may be reduced to an equivalent system consisting of a single force acting through a specified point in the plane and a couple. This reduction is often of use as shown in the following example.

Worked example

Forces of magnitude $4P$, $5P$, $6P$, P, $2P$ and $3P$ act along $\overrightarrow{OA}$, $\overrightarrow{AB}$, $\overrightarrow{BC}$, $\overrightarrow{CD}$, $\overrightarrow{DE}$ and $\overrightarrow{EO}$ respectively of a regular hexagon $OABCDE$ of side $2a$. Reduce this system to a single force acting through O and a couple. By taking OA as the x-axis and OD as the y-axis, find an equation for the line of action of the resultant force on the system and find the coordinates of the point K at which this line of action meets OA.

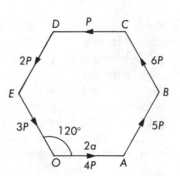

Fig. 18.11

Fig. 18.11 shows the given system of forces – always draw this if the question does not give a clear diagram.

Sum of forces parallel

to $\overrightarrow{OA} = 4P + 5P\cos 60° - 6P\cos 60° - P - 2P\cos 60° + 3P\cos 60°$
$= 3P$

Sum of forces parallel

to $\overrightarrow{OD} = 5P\sin 60° + 6P\sin 60° - 2P\sin 60° - 3P\sin 60°$
$= 3P\sqrt 3$

Moments about $O = 5Pa\sqrt 3 + 12Pa\sqrt 3 + 2Pa\sqrt 3 + 2Pa\sqrt 3$
$= 21Pa\sqrt 3$

We have reduced the given system to the system shown in Fig. 18.12

The force at O is $3P\mathbf{i} + 3P\sqrt 3\mathbf{j}$ and the couple has moment $21Pa\sqrt 3$ in the sense xOy

Now we wish to reduce this system to a single force acting through some point K in Ox

Consider the diagram (Fig. 18.13) and the forces added at K, which leave the system equivalent.

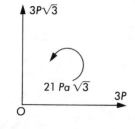

Fig. 18.12

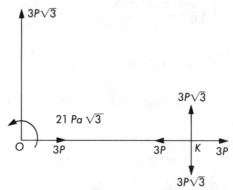

Fig. 18.13

Effectively we now have a force at K which is $3P\mathbf{i} + 3P\sqrt 3\mathbf{j}$, the couple of moment $+ 21Pa\sqrt 3$ and the couple of moment $- 3P\sqrt 3(OK)$. By choosing $OK = 7a$, we can eliminate the turning effect of both couples leaving just the force $3P\mathbf{i} + 3P\sqrt 3\mathbf{j}$ acting at K.

The original system of forces is equivalent to a single force $3P\mathbf{i} + 3P\sqrt 3\mathbf{j}$ acting through the point with position vector $7a\mathbf{i}$ relative to O.

In vector form an equation of the line of action of the resultant force is
$r = 7a\mathbf{i} + \lambda a(\mathbf{i} + \mathbf{j}\sqrt 3)$ or in cartesian form $y = (x - 7a)\sqrt 3$

Comment: notice how this solution is staged – first to reduce to a force at O and a couple – second to the single force at K and the line of action equation.

THE NATURE OF FORCES ACTING ON BODIES

Before concluding this chapter with some illustrative examples, we summarise the following important points:

a) the weight of a body acts vertically downwards through the centre of mass G of a body,

b) when a body is freely suspended from a fixed point O, the line OG is vertical when the body hangs in equilibrium,

c) the magnitude of the tension in a light inextensible string is unchanged when the string passes over smooth pegs or smooth pulleys,

d) when a body is resting against a smooth surface, the force exerted by the surface on the body acts at right-angles to the surface,

e) frictional forces at rough contact points only take their limiting value when equilibrium is limiting or when movement is occurring.

The strategy required to solve problems is:

i) draw a clear diagram, defining any forces you introduce,

> A useful strategy for finding solutions.

ii) explain each equation you write down,

iii) try to avoid writing down unnecessary equations,

iv) keep your processing well organised and tidy,

v) write a clear conclusion in words.

The following illustrative examples are typical of those set at AS- and A-level.

Worked example

A plane face of a uniform solid cylinder, of radius r and height h, is placed in contact with a rough plane whose inclination to the horizontal is gradually increased. Discuss the possible ways in which equilibrium will be eventually broken.

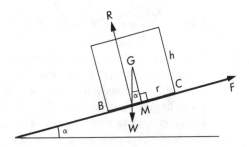

Fig. 18.14

The diagram (Fig. 18.14) shows a vertical section through the centre of mass G of the cylinder. Equilibrium is possible only if the vertical through G falls inside the base BC. Otherwise, unless friction is limiting, the cylinder will topple.

The forces acting are W, the weight of the cylinder, R the normal contact force, and F the frictional force.

Resolving horizontally $F \cos \alpha = R \sin \alpha \Rightarrow \dfrac{F}{R} = \tan \alpha$

The cylinder topples before it slides if:
$GM \tan \alpha > BM$

$GM = \dfrac{1}{2} h$ and $BM = r$, so we have

$\tan \alpha > \dfrac{2r}{h}$

Also, for sliding *not* to occur

$\dfrac{F}{R} < \mu \Rightarrow \mu > \tan \alpha$

Therefore the cylinder topples before it slides if $\mu > 2r/h$. The cylinder slides before it topples if the combined conditions $\tan \alpha < 2r/h$ and $\mu < \tan \alpha$ hold, that is, if $\mu < 2r/h$. If $\mu = 2r/h$ sliding and toppling would occur simultaneously.

Worked example

The end A of a uniform rod AB, of length $2a$ and mass m, is smoothly hinged at a fixed point to a vertical wall. A light inextensible string has one end tied to B and the other end is tied at a fixed point C in the wall vertically above A, where $AC = a$. Given that $AB = BC$ and that the rod is in equilibrium, calculate the tension in the string and the magnitude and direction of the force exerted by the hinge on the rod.

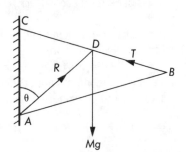

Fig. 18.15

With the notation in Fig. 18.15 and using given data, we have $AC = a$, $AB = BC = 2a$

$$\Rightarrow \angle ACB = \arccos\left(\frac{1}{4}\right).$$

$$\left(A \Rightarrow T . AC \sin \angle ACB = mg . a \sin \angle ACB\right.$$
$$\Rightarrow T = mg.$$

$\updownarrow$, $R \cos \theta = mg - T \cos \angle ACB$

$$= mg - \frac{1}{4}mg = \frac{3}{4}mg,$$

$\leftrightarrow$, $R \sin \theta = T \sin \angle ACB = mg \sqrt{(15)}/4$,

$$\Rightarrow R = \sqrt{\left[\left(\frac{3}{4}\right)^2 + \left(\frac{\sqrt{15}}{4}\right)^2\right]} mg = mg\sqrt{(3/2)},$$

$$\tan\theta = \frac{\sqrt{(15)}/4}{3/4} = \sqrt{(5/3)} \Rightarrow \theta = \arctan\sqrt{(5/3)} \approx 59°.$$

Note: the solution has used three equations obtained by moments and horizontal and vertical resolutions, to find the three unknowns T, R and θ.

Since there are only three forces, T, R and mg acting on the rod, the lines of action of these forces are concurrent. In particular, the line of action of R passes through the point D, the intersection of the lines of action of mg and T, both of whose directions are known. Using this fact, the angle θ can be determined by elementary trigonometry alone and forces R and T may then be found by using triangle ADC as a triangle of forces. This approach is more subtle and most students may prefer to use the method of solution given here.

Worked example

Show that the distance of the centre of mass of a uniform solid hemisphere is $\frac{3}{8}r$ from the centre of the plane face, r being the radius of the hemisphere

A uniform solid hemisphere, of weight W, is placed with its plane face uppermost and its curved surface in contact with rough horizontal ground and a smooth vertical wall. The points of contact and the centre of the hemisphere lie in the same vertical plane, at right-angles to the wall (see Fig. 18.16). Given that the coefficient of friction between the hemisphere and the ground is 1/5 and that the hemisphere is in equilibrium, find the set of possible values of θ, the angle made by the plane face of the hemisphere with the horizontal. Find also the magnitude of the force exerted on the hemisphere by the wall

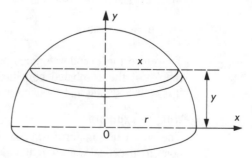

Fig. 18.16

Let the mass per unit volume of the hemisphere be σ

Mass of hemisphere $= \frac{2}{3}\pi r^3 \sigma$

A circular disc, radius x and thickness δy, is taken, as shown, at a distance y from the base.
Mass of disc $= \pi x^2 \sigma \delta y$
$$= \pi\sigma (r^2 - y^2)\, \delta y, \text{ since } y^2 + x^2 = r^2.$$
Taking the moments about the plane containing Ox which is perpendicular to Oy we have:
Moment of hemisphere = sum of moments of all discs over $(0, r)$ for y

$$\Rightarrow \frac{2}{3}\pi r^3 \sigma \bar{y} = \int_0^r \pi\sigma y(r^2 - y^2)\,dy$$

$$= \pi\sigma\left[\frac{1}{2}r^2y^2 - \frac{1}{4}y^4\right]_0^r$$

$$= \frac{1}{4}\pi\sigma r^4$$

$$\bar{y} = \frac{3\pi\sigma r^4}{8\pi\sigma r^3} = \frac{3}{8}r$$

Centre of mass of hemisphere is $\frac{3}{8}r$ from O

Fig 18.17 shows the forces acting on the hemisphere, where $OG = \frac{3}{8}r$

$\uparrow, \; N = W$
$\rightarrow, \; S = F$

The lines of action of N and S pass through O

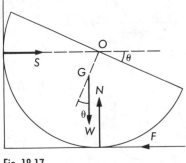

$$\left(O \Rightarrow Fr = W.(OG.\sin\theta) = \frac{3}{8}Wr\sin\theta\right.$$

$$\Rightarrow F = \frac{3}{8}W\sin\theta$$

For equilibrium $\dfrac{F}{N} \leqslant \dfrac{1}{5}$

Fig. 18.17

$$\Rightarrow \frac{3}{8}\sin\theta \leqslant \frac{1}{5}$$

$\Rightarrow \sin\theta \leqslant 8/15 \Rightarrow \theta \leqslant \arcsin(8/15) \approx 32°$
The least possible value of θ is zero
$\Rightarrow 0 \leqslant \theta \leqslant \arcsin(8/15)$

Also $S = F = \dfrac{3}{8}W\sin\theta$, the magnitude of the force exerted by the wall on the hemisphere

Note: the two critical stages of the solution are:
a) to get the diagram and forces correct,
b) to take moments about a convenient point to avoid having to use complicated trigonometry.

Once N and F have been found in terms of W, many students will prefer to find the greatest value of θ by taking $F = \mu N$, rather than using inequalities as in the solution above.

EXERCISES 18.2

1   A uniform straight bar, of length 5 m and of mass 250 kg, is simply supported at its ends A and B and rests horizontally. A man of mass 140 kg stands on the bar at the point C. Given that the force exerted by the support at A on the bar has twice the magnitude of the force exerted by the support at B on the bar, calculate the distance AC, giving your answer to the nearest cm

2   Use integration to show that the centre of mass of a uniform solid right circular cone of height h, is at a distance $\frac{3}{4}h$ from the vertex. The larger and smaller parallel plane faces of a frustum of a uniform right circular cone have radii $2a$ and a. The distance between the faces is H. Show that the distance of the centre of mass of the frustum from the larger plane face is $\dfrac{11H}{28}$.

(A frustum of a cone is the solid remaining when the top is cut off by a plane parallel to the circular base)

3   A light elastic string has natural length a and modulus of elasticity mg. The ends of the string are tied to points A and B, where AB is horizontal and of length $2a$. A particle P, of mass m, is tied at the mid-point of the string and gently lowered until it rests in

equilibrium. Given that each string is inclined at θ to the horizontal, show that

$$2\tan\theta = \sin\theta + \frac{1}{2}$$

4 A uniform rod AB, of length $2l$ and mass m, rests in equilibrium with the end A on rough horizontal ground and the end B against a smooth vertical wall. The vertical plane containing AB is at right angles to the wall. Given that AB is inclined at an angle θ to the horizontal, where $\tan\theta = \frac{4}{3}$, find the least possible value of the coefficient of friction between A and the ground. When the coefficient of friction has this value, calculate, in terms of m and g, the total force exerted by the rod on the ground at A

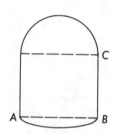

Fig. 18.18

5 A composite uniform body consists of a cylinder of radius $3r$ and height $6r$ and a hemisphere of radius $3r$, joined as shown in Fig. 18.18 with the plane face of the hemisphere and one circular face of the cylinder coincident. Find the distance of the centre of mass of the composite body from the base AB. A string is attached to C and the body is suspended by this string and is in equilibrium. Find the angle made by AB with the horizontal

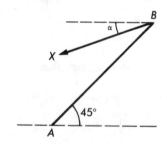

Fig. 18.19

6 A uniform smooth rod AB, of mass m and of length $2b$ rests in equilibrium, as shown in Fig. 18.19, in contact with a fixed smooth hemispherical bowl of inner radius c. Given that AB makes an angle θ with the horizontal, show that
$$2c\cos 2\theta = b\cos\theta$$
In this case, when $b = c$, show that θ is approximately $32\frac{1}{2}°$ and find, in terms of m and g, the magnitude of the force exerted by the hemisphere on the rod at A

7 Particles of masses m, $2m$, $3m$ and $4m$ are placed at the corners A, B, C and D respectively of a uniform square plate $ABCD$, of mass $5m$ and length of side $2a$. Find the distance of G, the centre of mass of the loaded plate, from i) AB, ii) BC. The loaded plate is smoothly hinged at D to a fixed point and hangs in equilibrium in a vertical plane. Calculate, to the nearest degree, the angle made by DB with the downward vertical

Fig. 18.20

8 The end of A of a uniform rod AB, of mass m, rests on a rough horizontal plane. The rod is inclined at $45°$ to the horizontal. A force of magnitude X is applied to the rod at B in the same vertical plane as A and B and at an angle α to the horizontal as shown in Fig. 18.20.

i) show that $X = \frac{1}{2}mg/(\cos\alpha - \sin\alpha)$

ii) find the horizontal and vertical components of the force exerted by the plane on the rod at A

iii) deduce that $\tan\alpha \leqslant 2 - \frac{1}{\mu}$, where μ is the coefficient of friction between the rod and the plane

NB: Examination questions on these topics can be found in Chapter 19.

ANSWERS TO EXERCISES

EXERCISES 18.1

1 $\dfrac{1}{20}$

2 $2\cdot4$N, $1\cdot8$N
3 $6\sqrt{2}$N, $6\sqrt{2}$m
4 a) $0\cdot36$;   b) $9\cdot8m$

5 $\dfrac{\mu mg}{\cos\alpha + \mu\sin\alpha}$

6 $mg\sin 2\alpha$, $\theta = \alpha$
7 $mg\sin\lambda$, $\theta = \lambda$

EXERCISES 18.2

1 18cm

4 $\dfrac{3}{8}$, $\dfrac{1}{8}mg\sqrt{73}$

5 $\dfrac{129}{32}r$, $56\cdot7°$

6 $0\cdot64\,mg$

7 i) $\dfrac{19a}{15}$;   ii) a; $9°$.

8 ii) $\dfrac{mg\cos\alpha}{2(\cos\alpha - \sin\alpha)}$, $\dfrac{mg(2\cos\alpha - \sin\alpha)}{2(\cos\alpha - \sin\alpha)}$

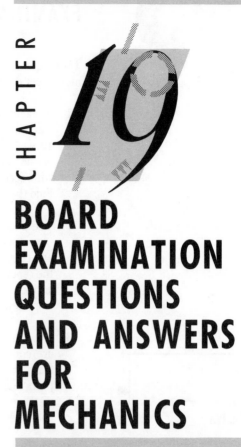

BOARD EXAMINATION QUESTIONS AND ANSWERS FOR MECHANICS

GETTING STARTED

When answering Examination Board Mechanics questions, try to employ the methods shown in Chapters 14–18. Define any symbols which you introduce, draw a clear sketch diagram and place your symbols and information from the question onto the diagram. *Check that you have included it all.*

Don't go rushing at a question *before* reading it through carefully and drawing your sketch, even if you have revised that topic thoroughly.

EXAMINATION QUESTIONS

These questions have been selected from recent examination papers and relate to the mechanics covered in Chapters 14–18.

1 A train travelling at $50\,\mathrm{m\,s^{-1}}$ applies its brakes on passing a yellow signal at a point A and decelerates uniformly, with a deceleration of $1\,\mathrm{m\,s^{-2}}$, until it reaches a speed of $10\,\mathrm{m\,s^{-1}}$. The train then travels for $2\,\mathrm{km}$ at the uniform speed of $10\,\mathrm{m\,s^{-1}}$ before passing a green signal. On passing the green signal the train accelerates uniformly, with acceleration $0\cdot2\,\mathrm{m\,s^{-2}}$, until it finally reaches a speed of $50\,\mathrm{m\,s^{-1}}$ at a point B. Find the distance AB and the time taken to travel that distance.   (AEB, June 1988)

2 The position vector of a particle at time t is given by
$$\mathbf{r} = (a\cos wt)\,\mathbf{i} + (a\sin wt)\,\mathbf{j} + bt\,\mathbf{k},$$
where a, b and w are constants. Find the velocity and acceleration vectors. Show that the speed of the particle is constant and that the magnitude of the particle's acceleration is always non-zero.
Comment briefly on these results.   (W)

3 Fig. 19.1 represents a particle A, of mass $3\,\mathrm{kg}$, and a particle B, of mass $2\,\mathrm{kg}$, which are connected by a light inextensible string. Particle A rests on a rough horizontal table, the string passes over a smooth fixed pulley at the edge of the table, and particle B hangs freely. The system is released from rest, and each particle moves a distance of $1\,\mathrm{m}$ in the first second of the motion. Find the tension in the string and the coefficient of friction between A and the table. (Take $g = 10\,\mathrm{m\,s^{-2}}$)   (C)

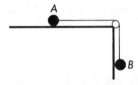

Fig. 19.1

4 The points O, A and B lie on a straight line OAB with $OA = 10$ metres and $OB = 40$ metres. A particle moving along the line with constant acceleration passes through O, A and B at times 0, 2 and 5 seconds, respectively. Find the speed of the particle when it is at B.   (JMB)

5 A particle of mass $1\,\mathrm{kg}$ falls vertically through a liquid. The motion of the particle is resisted by a force of magnitude $2v$ newtons, where $v\,\mathrm{m\,s^{-1}}$ is the speed of the particle at any time t seconds. Show that
$$\left(\frac{1}{5-v}\right)\frac{dv}{dt} = 2,$$
and hence find, correct to 3 significant figures, the value of v when $t = 1$, given that $v = 0$ when $t = 0$.   (C)
(Take $g = 10\mathrm{ms^{-2}}$)

6 A particle moves along Ox in the positive direction. At time $t = 0$, its speed is $8\,\mathrm{m\,s^{-1}}$. At time $t\,\mathrm{s}$, its acceleration is $3\mathrm{e}^{2t}\,\mathrm{m\,s^{-2}}$ in the positive direction. Calculate, in metres to 3 significant figures, the distance the particle moves in the first 2 seconds.   (L)

7 A car of mass $1500\,\mathrm{kg}$ is moving along a horizontal road. The resistance to the motion of the car is $750\,\mathrm{N}$. Assuming that the car's engine works at $15\,\mathrm{kW}$, find

i) the maximum constant speed at which the car can travel,
ii) the acceleration of the car when its speed is $8\,\mathrm{m\,s^{-1}}$.   (W)

8 A light inextensible string, of length $2\,\mathrm{m}$, passes through a small fixed smooth ring R and carries particle A, of mass $2M$, at one end and particle B, of mass M, at the other end. The particle A is at rest vertically below R. The particle B moves in a horizontal circle, the centre of which is vertically below R, with constant angular speed $4\ \mathrm{rad\,s^{-1}}$. Given that the string remains taut, calculate the distance AR. (Take g as $10\,\mathrm{m\,s^{-2}}$.)   (L)

9 Two particles of mass M and $4M$ respectively, are connected by a light inextensible string passing over a smooth fixed pulley. The particles are released from rest with the hanging parts of the string vertical. Find

i) the acceleration of the particles,
ii) the force exerted by the string on the pulley during the subsequent motion.   (W)

10   A particle P is projected, from a point O on horizontal ground, with speed V at an angle θ above the horizontal, where $\tan \theta = \dfrac{1}{3}$. The particle passes through the point with coordinates $\left(3a, \dfrac{3}{4}a\right)$ relative to horizontal and vertical axes at O in the plane of motion. Show that $V^2 = 20ga$.
A particle Q is projected from O at the instant when P is moving horizontally. It strikes the ground at the same place and at the same instant as P. Show that the speed of projection of Q is $\sqrt{\left(\dfrac{145ga}{2}\right)}$ and find the tangent of the angle of projection.

(C)

11   (In this question take $g = 10\,\mathrm{m\,s^{-2}}$)
With respect to a fixed origin O, the unit vectors $\mathbf{i}$ and $\mathbf{j}$ are directed horizontally and vertically upwards respectively. A particle P is projected with velocity $(u\mathbf{i} + v\mathbf{j})\,\mathrm{m\,s^{-1}}$ from a point A whose position vector is $50\mathbf{j}\,\mathrm{m}$. The particle moves freely under gravity and passes through the point B whose position vector is $200\mathbf{i}\,\mathrm{m}$.
Show that
$$u^2 + 4uv = 4000.$$
At time $1\cdot5\,\mathrm{s}$ after leaving A, the direction of motion of P is parallel to $\mathbf{i}$.

a) Find the value of v.
b) Find the value of u.
c) Hence find the time taken by P to reach B from A.
At the point C on the flight path of P between A and B, P is moving with speed $50\,\mathrm{m\,s^{-1}}$.
d) Determine the position vector of C.   (L)

12   From a point O on a horizontal plane a particle is projected in a direction which makes an angle of $45°$ with the plane. The particle moves freely under gravity and subsequently passes through a point which is at a horizontal distance a from O and a vertical distance b above the plane. Find, in terms of a, b and g, the speed of projection from O.   (JMB)

13   A particle P projected from a point O on level ground strikes the ground again at a distance of $120\,\mathrm{m}$ from O after a time $6\,\mathrm{s}$. Find the horizontal and vertical components of the initial velocity of P. The particle passes through a point Q whose horizontal displacement from O is $30\,\mathrm{m}$. Find

a) the height of Q above the ground,
b) the tangent of the angle between the horizontal and the direction of motion of P when it is at Q and also the speed of P at this instant,
c) the horizontal displacement from O of the point at which P is next at the level of Q.
(Take g to be $10\,\mathrm{m\,s^{-2}}$)   (AEB, Nov. 1985)

14   A particle is projected from a fixed point O with velocity $u\mathbf{i} + \lambda u\mathbf{j}$ and moves with acceleration $-g\mathbf{j}$, where u and λ are positive constants and $\mathbf{i}$ and $\mathbf{j}$ are unit vectors directed horizontally and vertically upwards respectively. At time t after leaving O, the particle is at the point P whose position vector relative to O is $x\mathbf{i} + y\mathbf{j}$. Show that
$$y = \lambda x - \frac{1}{2}g\frac{x^2}{u^2}.$$

Given that $x = a$ when $y = b$ and that $x = 2a$ when $y = 0$, find λ and u^2 in terms of a, b and g.
Find, in terms of a and b, the horizontal displacement of the particle from O at the instant when the velocity of the particle is at right angles to the velocity of projection.

(L)

15 In an experiment, observations of retardation r and speed v were made on a particle P which was moving in a straight line. A graph of retardation against speed produced a straight line. Given that two points on the line correspond to $v = 5\,\mathrm{ms}^{-1}$, $r = 13\,\mathrm{ms}^{-2}$ and $v = 2\,\mathrm{ms}^{-1}$, $r = 7\,\mathrm{ms}^{-2}$ respectively, show that the speed $v\,\mathrm{ms}^{-1}$ at time ts satisfies $\dfrac{\mathrm{d}v}{\mathrm{d}t} = -3 - 2v$.

Find the time taken for the speed to drop from $5\,\mathrm{ms}^{-1}$ to $2\,\mathrm{ms}^{-1}$ and the distance travelled by P during this period.

Given that P is of mass $0\cdot6\,\mathrm{kg}$ calculate

a) the work done by the forces acting on P during the period that the speed drops from $5\,\mathrm{ms}^{-1}$ to $2\,\mathrm{ms}^{-1}$,
b) the rate at which the forces are working when $v = 5\,\mathrm{ms}^{-1}$.

(Answers may, where relevant, be left in a form involving logarithms.)

(AEB, June 1985)

16 A car of mass $1500\,\mathrm{kg}$ has an engine which produces a maximum power of $75\,\mathrm{kW}$. The car is used to tow a trailer of mass $1000\,\mathrm{kg}$ along a level road. If the car experiences a constant resistance force of $1000\,\mathrm{N}$ and the trailer a constant resistance force of $1500\,\mathrm{N}$ find the power output from the engine when the car is travelling with a uniform speed of $20\,\mathrm{ms}^{-1}$.   *(5)*

Find the acceleration of the car and trailer when it is travelling at maximum power with a velocity of $25\,\mathrm{ms}^{-1}$. Hence show that the car cannot travel at maximum power without exceeding a speed limit of $25\,\mathrm{ms}^{-1}$.   *(6)*

If the safest minimum recommended speed for travel on the horizontal carriageway of the motorway is $15\,\mathrm{ms}^{-1}$, and the resistance on the trailer is $1\cdot5\,\mathrm{N}$ per kg of its mass, find the mass of the heaviest trailer that can safely be towed. When the trailer has a mass of $1000\,\mathrm{kg}$ find the work done by the engine in achieving the safest recommended speed in a distance of $100\,\mathrm{m}$ from rest.   (OXFORD)

17 Using Hooke's law, show by integration that the work done in stretching a light elastic string, of natural length l and modulus of elasticity λ, from length l to length $l + e$ is $\dfrac{\lambda e^2}{2l}$.

A light elastic string, of natural length 2 metres and modulus of elasticity $4mg$ newtons, has one end attached to a fixed point A. A particle P, of mass m kilograms, is attached to the other end of the string.

a) Show that, when P hangs in equilibrium, the length of AP is $2\frac{1}{2}\,\mathrm{m}$.
The particle P is released from rest at A and falls vertically. Use the work-energy principle to calculate
b) the speed, to 3 significant figures, of P at distance $2\frac{1}{2}\,\mathrm{m}$ below A,
c) the greatest length of the string.   (L)

18 A particle P of mass $0\cdot04\,\mathrm{kg}$ describes simple harmonic motion about a point O as centre. When P is at a distance of $3\,\mathrm{m}$ from O its acceleration is of magnitude $48\,\mathrm{ms}^{-2}$ and its kinetic energy is $5\cdot12\,\mathrm{J}$. Find the amplitude of the motion.   (AEB, June 1987)

19 Two fixed points A and B lie on a smooth horizontal table at a distance $6l$ apart. The mid-point of AB is O. A light elastic string of natural length $2l$ and modulus $16mln^2$ connects a particle P of mass m to A and an identical string connects P to B. At time $t = 0$, P is released from rest at the point in the line AB which is at a distance of $4l$ from A. At time t the displacement of P from O is x. Write down expressions for the tensions in the strings AP and BP at time t and hence show that

$$\frac{\mathrm{d}^2x}{\mathrm{d}t^2} = -16n^2x.$$

Write down the period of the motion of P.
Find
  i) the maximum speed of P,
 ii) the distance of P from O when its speed is half the maximum speed,
iii) the value of t at which P first reaches half its maximum speed.   (JMB)

20 A particle is describing simple harmonic motion in a straight line about a point O as centre. At a particular instant its displacement from O, its speed and the magnitude of its acceleration are $3\,\mathrm{cm}$, $6\,\mathrm{cm\,s^{-1}}$ and $12\,\mathrm{cm\,s^{-2}}$ respectively.
Find
a) the greatest speed of the particle, and
b) the period of its motion.   (AEB, Nov. 1987)

21 A particle of mass $2\,\mathrm{kg}$ is acted upon by a variable force which makes it move in simple harmonic motion about O on the straight line Ox. Given that the maximum speed attained is $0\cdot5\,\mathrm{m\,s^{-1}}$ and the maximum acceleration is $0\cdot1\,\mathrm{m\,s^{-2}}$, find the period of the motion. Show that the amplitude of the motion is $2\cdot5\,\mathrm{m}$.
The particle passes through O at time $t = 0$.

i) Write down the displacement of the particle from O at time t. Hence find the velocity and acceleration at time t.
ii) Find the force acting on the particle at time t. Show that the force generates a power of $-0\cdot05\sin(0\cdot4t)$ watts during the motion and find the values of t for which the power is a maximum.   (W)

22 At noon a boat A is $9\,\mathrm{km}$ due west of another boat B. To an observer on B the boat A always appears to be moving on a bearing of $150°$ (i.e. S30°E) with constant speed $2\cdot5\,\mathrm{m\,s^{-1}}$. Find the time at which the boats are closest together and the distance between them at this time. Find also, to the nearest minute, the length of time for which the boats are less than $8\,\mathrm{km}$ apart. [No credit will be given for methods using scale drawing].   (AEB, June 1987)

23 A port A is $20\,\mathrm{km}$ due north of another port B. Steamers P and Q leave A and B respectively at 12 noon. P travels at $15\,\mathrm{km}$ per hour due east and Q at $10\,\mathrm{km}$ per hour in a direction $\theta°$ east of north, where $\tan\theta° = \frac{3}{4}$. Define $\hat{\mathbf{i}}$ and $\hat{\mathbf{j}}$ to be unit vectors in east and north directions respectively. Find in terms of $\hat{\mathbf{i}}$ and $\hat{\mathbf{j}}$ the velocity of Q relative to P and show this velocity in a diagram. Find at what time P is nearest to Q and their distance apart. What will be the bearing of P from Q 1 hour after leaving port?   (OXFORD)

24 Two particles, A and B, of masses $2m$ and $3m$ respectively, are moving in a straight line in the same direction on a smooth horizontal plane. The particles collide and, *after* the collision, A and B continue to move in the same straight line and in the same direction with speeds u and $3u/2$ respectively. Given that the coefficient of restitution between A and B is $1/5$, find, in terms of u, the speed of A and the speed of B *before* their collision. Find also, in terms of m and u, the magnitude of the impulse of the force exerted by B on A during the collision.
After the collision, the particle B continues to move with constant speed $3u/2$. At a given instant, B is moving towards a fixed point O in the line of motion and is at a distance b from O. At this same instant a particle C, moving on the same horizontal plane, passes through O with constant velocity of magnitude $9u/8$ at right angles to the line of motion of B.
By considering the direction of the velocity of C relative to B, or otherwise, calculate in terms of b, the least distance between B and C in their ensuing motions.   (L)

25 Two small smooth spheres A and B of equal radii but of masses m and $4m$ respectively are moving towards each other on a smooth horizontal table and collide directly. The speeds of A and B before collision are $2u$ and $4u$ respectively. After collision the direction of motion of A is reversed and it moves with speed $3u$. Find the speed of B after collision and also the coefficient of restitution between A and B.
After collision, sphere A moves with constant speed $3u$ until it catches up and collides directly with a third sphere C which is identical to A and moving in the same direction as A with speed u. Given that the kinetic energy lost in this second collision is $\frac{1}{2}mu^2$ find the speed of A immediately after this collision. [The coefficient of restitution between spheres A and B is not the same as that between spheres A and C.]   (AEB, June 1987)

26   A car of mass M is travelling up a straight road inclined at an angle α to the horizontal. The engine of the car is working at a rate P. At an instant when the speed is v and the acceleration is f the resistance to motion due to frictional forces is kMg, where k is a positive constant. Express P in terms of M, g, v, f, k and α.   (C)

27   A small bead of mass m is threaded on a smooth wire in the form of a circle of radius a. The wire is fixed in a vertical plane and a light elastic string of natural length a and modulus $2mg$ joins the bead to the highest point of the wire. The bead is released from rest in a position where the string is just taut, and in the subsequent motion it passes through the lowest point of the wire with speed U. Using the principle of conservation of energy, or otherwise, find U in terms of g and a.   (C)

28   The foot of a uniform ladder of length 1m and mass M kg rests on rough horizontal ground and the top of the ladder rests against a smooth vertical wall. The ladder is inclined at $30°$ to the horizontal and the ground is sufficiently rough to prevent slipping. Find the frictional force and hence the magnitude of the total force exerted by the ground on the ladder.   (W)

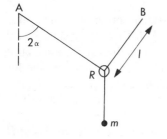

Fig. 19.2

29   Fig. 19.2 shows a particle of mass m suspended from a fixed point A by a light inextensible string which passes through a small smooth *light* ring R. The ring is attached to a fixed point B, at the same horizontal level as A, by a light inextensible string of length l (where $l < AB$). The system hangs at rest in equilibrium. Denoting the angle between AR and the vertical by 2α, show that angle $ABR = \alpha$ and find, in terms of m, g and α, the tension in the string BR.   (C)

30   A straight uniform rod AB, of mass M, rests in equilibrium with the end A in contact with horizontal ground and the end B against a smooth vertical wall. The vertical plane containing AB is at right angles to the wall and AB is inclined at $60°$ to the horizontal. The coefficient of friction between the rod and the ground is μ.

a) Find, in terms of M and g, the magnitude of the force exerted by the wall on the rod.

b) Show that $\mu \geqslant \dfrac{1}{6}\sqrt{3}$.

A load of mass kM is attached to the rod at B, where k is a positive constant.

c) Given that $\mu = \dfrac{1}{5}\sqrt{3}$ and that equilibrium is limiting, find the value of k.   (L)

31   A smooth horizontal rail is fixed at a height $3h$ above rough horizontal ground. A uniform rod AB, of mass M and length $6h$, is placed in a vertical plane perpendicular to the rail with the end A resting on the ground. The distance $AC = 5h$, where C is the point of contact between the rail and the rod. Show that the force exerted by the rail on the rod is of magnitude $12Mg/25$.

Given that equilibrium is limiting, find the coefficient of friction between the rod and the ground and show that the force exerted by the ground on the rod is of magnitude $17Mg/25$.

Find, in terms of M and g, the greatest magnitude of the horizontal force which could be applied to the rod at A without disturbing equilibrium.   (L)

32   Two particles P and Q of masses m and $2m$, respectively, are connected to a fixed point O by light inextensible strings OP and OQ, each of length l. The particle Q hangs in equilibrium vertically below O and P is held so that the string OP is taut and horizontal and is then released. The coefficient of restitution between P and Q is e. Given that P is brought to rest by its impact with Q, show that $e = \frac{1}{2}$. Given also that Q rises until OQ makes an angle α with the downward vertical, find $\cos \alpha$.

Subsequently Q strikes P. If P rises until OP makes an angle β with the downward vertical, show that $\beta = \alpha$.   (JMB)

33   A smooth cylinder of radius a is fixed on a rough horizontal table with its axis parallel to the table. A uniform rod ACB of length $6a$ and mass M rests in limiting equilibrium with the end A on the table and the point C touching the cylinder. The vertical plane containing the rod is perpendicular to the axis of the cylinder and the rod makes an angle 2θ with the table.

a) Show that the magnitude of the force exerted by the cylinder on the rod is
$3Mg \cos 2\theta \tan \theta$.

b) Show also that μ, the coefficient of friction between the rod and the table, is given by
$$\mu(\cot \theta - 3 \cos^2 2\theta) = 3 \sin 2\theta \cos 2\theta. \qquad \text{(L)}$$

34 Fig. 19.3 shows a framework consisting of seven equal smoothly jointed light rods AB, BC, DC, DE, AE, EB and EC. The framework is in a vertical plane, with AE, ED and BC horizontal and is simply supported at A and D. It carries vertical loads of $20\,\text{N}$ and $100\,\text{N}$ at B and C respectively.

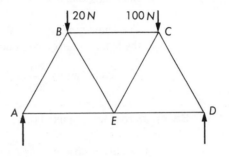

Fig. 19.3

a) Calculate the reactions at A and D.
b) Find the forces in AB, AE and BC.
Show also that, for all possible values of the vertical loads at B and C, the force in BC is proportional to the sum of those loads.   (AEB, Nov 1988)

35 A simple suspension bridge consists of a uniform plane horizontal rectangular roadway of mass M, freely supported by two sets of four equally spaced light vertical cables. Each set of four cables lies in the same vertical plane and is attached to one edge of the roadway. These cables are in turn attached to a single light cable ABCDEF as shown in Fig. 19.4. A, B, C are on the same horizontal level as F, E, D respectively.

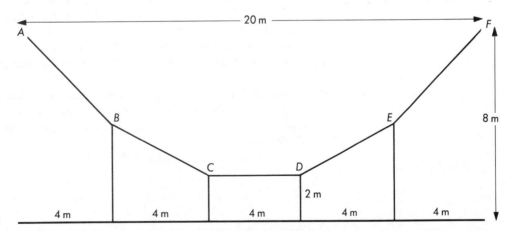

Fig. 19.4

A and F are 8 m and C and D are 2 m above the level of the roadway which has length 20 m. AF = 20 m. Given that the tensions in all vertical cables are the same, show that the tension in CD is $\frac{1}{4}Mg$ and find the tensions in each of the parts AB, BC, DE, EF of the connecting cable.   (OXFORD)

36 A particle moves along a straight line with acceleration k/v, where k is a constant and v is the velocity of the particle. When the particle is at a point O its velocity is $1\,\text{m}\,\text{s}^{-1}$, and when its displacement from O is 13 metres its velocity is $3\,\text{m}\,\text{s}^{-1}$. Show that $k = \frac{2}{3}$. Find the velocity of the particle when it is at the point A whose displacement from O is 62 metres.
Find also the time taken for the particle to travel from O to A.   (JMB)

37 Two identical small smooth spheres A and B, each of mass M kg, are connected by a light inextensible string AB of length 1 m. They are placed on a smooth horizontal table with AB just taut. The line AB is at right angles to the plane of a vertical wall, B being the sphere nearer to the wall and distant 10 m from it. A is projected towards B with speed $2\,\mathrm{ms}^{-1}$. The coefficient of restitution for collisions between the spheres, and for collisions between the spheres and the wall is $e(e > 0)$.

i) Show that the speeds of A and B after their first collision are $(1 - e)\,\mathrm{ms}^{-1}$ and $(1 + e)\,\mathrm{ms}^{-1}$ respectively.

ii) Show that if the string does not become taut before B collides with the wall then $e < \dfrac{1}{19}$.

iii) Assuming that $e < \dfrac{1}{19}$, find the speed of B after its collision with the wall.

The kinetic energy lost as a result of the first collision between A and B is k times the kinetic energy lost due to the collision of B with the wall. Show that the value of k lies between $\dfrac{361}{200}$ and 2. (W)

38 With respect to a fixed origin O, the position vector $\mathbf{r}$ of a particle P at time t is given by
$$\mathbf{r} = a(wt - \sin wt)\mathbf{i} + a(wt + \cos wt)\mathbf{j},$$
where $0 \leqslant t \leqslant \pi/w$ and w and a are positive constants.

a) Determine the velocity $\mathbf{v}$ and the acceleration $\mathbf{f}$ of P at time t.
b) Show that $|\mathbf{f}|$ is constant.
Prove that at the instant when $\mathbf{v}.\mathbf{f} = 0$,
c) the speed of P is $aw(\sqrt{2} - 1)$,
d) the distance of P from O is $a(1 + \pi^2/8)^{\frac{1}{2}}$. (L)

39 Using integration, show that the centre of mass of a uniform thin hemispherical bowl of radius a is at a distance $\frac{1}{2}a$ from the centre C of the circular rim of the bowl.

A lid is attached to the bowl to form a closed composite body B. The lid is a thin circular disc of radius a and centre C. The lid is made of the same uniform material as the bowl. Show that the centre of mass of B is at a distance $\frac{1}{3}a$ from C.

The body B has mass M. A particle P, also of mass M, is attached at a point on the circumference of the plane circular face of B. The body is placed with a point of its curved surface in contact with a horizontal plane and rests in equilibrium. Find, to the nearest half degree, the angle made by the line PC with the horizontal. (L)

40 A particle P of mass 2 kg is attached to the end of a light inextensible string OP of length 1·5 m. The end O is fixed and P describes a horizontal circle with uniform angular speed w radians/second, with OP inclined at an angle of $60°$ to the vertical. Show that
i) the tension in the string is $4g$ N,

ii) $w = 2\sqrt{\dfrac{g}{3}}$. (W)

41 A cricket ball was thrown by one player and caught at the same height from which it was thrown by another player, 30 m away. The ball moved freely under gravity. The greatest height reached by the ball above the point from which it was thrown was 10 m.

a) Show that the vertical component of the initial velocity of the ball was $14\,\mathrm{ms}^{-1}$.
Calculate:
b) the time of flight, in s, of the ball,
c) the speed, in ms^{-1}, with which the ball left the thrower's hand. (L)

42 A particle A, of mass 2 kg, has a velocity $(16\mathbf{i} - 12\mathbf{j})\,\mathrm{ms}^{-1}$ and collides directly with another particle B, of mass 3 kg. Before the collision B is at rest. After the collision B has speed $10\,\mathrm{ms}^{-1}$.

Find

a) the unit vector in the direction in which B begins to move,
b) the velocities, in ms^{-1}, of A and B after the collision,
c) the impulse, in Ns, received by A as a result of the collision.   (L)

43 The two forces $(4\mathbf{i} - 6\mathbf{j})\,\mathrm{N}$ and $(6\mathbf{i} + 2\mathbf{j})\,\mathrm{N}$ act on a particle P, of mass $2\,\mathrm{kg}$.

a) Show that the acceleration of P is $(5\mathbf{i} - 2\mathbf{j})\,\mathrm{ms}^{-2}$.
At time $t = 0$, P is at a point with position vector $(5\mathbf{i} + 3\mathbf{j})\,\mathrm{m}$ relative to a fixed origin O and has velocity $(-7\mathbf{i})\,\mathrm{ms}^{-1}$. Calculate, at time $t = 2$ seconds,
b) the position vector of P relative to O,
c) the velocity, in ms^{-1}, of P
d) the kinetic energy, in J, of P.   (L)

44 The points A, B and C of a horizontal plane have coordinates $(4, 3)$, $(-4, 0)$ and $(4, -3)$, respectively, these dimensions being in metres. A particle P on the plane is subject to three forces which are directed towards A, B and C.

a) When P is at the origin the forces directed towards A, B and C have magnitudes 4N, 2N and 4N, respectively, as shown in Fig. 19.5.

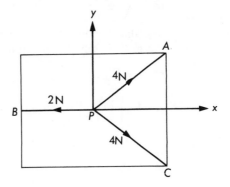

Fig. 19.5

Calculate the magnitude of the resultant of the three forces, and state its direction.
i) Given that the plane is rough, and that P is in equilibrium at the origin, state the magnitude and direction of the frictional force on P.
ii) Given that the plane is smooth and that the mass of P is $0.1\,\mathrm{kg}$, calculate the acceleration of P when it is at the origin.
b) When P is at the point $(-4, 3)$ the force directed towards B is zero, and the forces directed towards A and C have magnitudes 10N and 14N, respectively. Calculate the magnitude and direction of the resultant of the two non-zero forces.   (C)

45 (In this question take g to be $10\,\mathrm{ms}^{-2}$.)
A long light inextensible string passes over a smooth pulley, and particles of masses m and $4m$ are fixed to the two ends of the string. The system is released from rest with the string taut and with each particle at a height $1.2\,\mathrm{m}$ from the floor.

i) Show that the acceleration of either particle is $6\,\mathrm{ms}^{-2}$.
ii) Calculate the time taken for the heavier particle to reach the floor, and the speed on impact.
iii) Assuming the heavier particle does not rebound, calculate the greatest height above the floor attained by the lighter particle in the subsequent motion.   (C)

ANSWERS TO EXAMINATION QUESTIONS

1   $9 \cdot 2$ km, $7\frac{1}{3}$ min

2   $(-aw\sin wt)\mathbf{i} + (aw\cos wt)\mathbf{j} + b\mathbf{k}$;
   $(-aw^2\cos wt)\mathbf{i} + (-aw^2\sin wt)\mathbf{j}$;
   speed $= \sqrt{a^2w^2 + b^2}$,
   acceleration magnitude $= aw^2 \neq 0$

3   16N, $\frac{1}{3}$

4   $13 \mathrm{ms}^{-1}$

5   $4 \cdot 32$

6   $53 \cdot 2$ m

7   $20 \mathrm{ms}^{-1}$, $0 \cdot 75 \mathrm{ms}^{-2}$

8   $0 \cdot 75$ m

9   $\dfrac{3}{5}$g, $\dfrac{16}{5}Mg$

10   $\dfrac{1}{12}$

11   a) 15;   b) 40;   c) 5s: $180\mathbf{i} + 16 \cdot 25\mathbf{j}$

12   $a\sqrt{\dfrac{g}{a-b}}$

13   $20 \mathrm{ms}^{-1}$, $30 \mathrm{ms}^{-1}$;   a) $33 \cdot 8$ m;
     b) $0 \cdot 75$, $25 \mathrm{ms}^{-1}$;   c) 90 m

14   $\lambda = \dfrac{2b}{a}$, $u^2 = \dfrac{ga^2}{2b} \cdot \dfrac{a}{4b^2}(a^2 + 4b^2)$

15   $\dfrac{1}{2}\ln\left(\dfrac{13}{7}\right)$s, $\left(\dfrac{3}{2} - \dfrac{3}{4}\ln\dfrac{13}{7}\right)$m;
     a) $6 \cdot 3$J;   b) 39W

16   50kW, $0 \cdot 2 \mathrm{ms}^{-2}$; $\dfrac{8000}{3}$ kg, 531250J

17   b) $6 \cdot 64 \mathrm{ms}^{-1}$ c) 4m

18   5m

19   Period $= \dfrac{\pi}{2n}$;   i) $4ln$;   ii) $\dfrac{1}{2}l\sqrt{3}$;
     iii) $\dfrac{\pi}{24n}$

20   a) $6\sqrt{2} \mathrm{cms}^{-1}$;   b) πs

21   10πs;   i) $2 \cdot 5\sin\dfrac{t}{5}$, $0 \cdot 5\cos\dfrac{t}{5} - 0 \cdot 1\sin\dfrac{t}{5}$;
     ii) $-0 \cdot 2\sin\dfrac{t}{5}$, $\dfrac{5\pi}{4}$ $(2n+1)$

22   1230, $7 \cdot 79$ km; 47 min

23   $-9\mathbf{i} + 8\mathbf{j}$, 1306, $14 \cdot 95$ km; $036 \cdot 9°$

24   $\dfrac{14u}{5}$, $\dfrac{3u}{10}$, $\dfrac{18mu}{5}$, $\dfrac{3b}{5}$

25   $\dfrac{11}{4}u$, $\dfrac{1}{24}$; $\dfrac{1}{2}u[4 + \sqrt{2}]$

26   $P = v(kMg + Mg\sin\alpha + Mf)$

27   $U = \sqrt{ga}$

28   $\dfrac{1}{2}Mg\sqrt{3}$, $\dfrac{1}{2}Mg\sqrt{7}$

29   $2mg\sin\alpha$

30   a) $\dfrac{1}{6}Mg\sqrt{3}$;   c) $0 \cdot 25$

31   $\dfrac{36}{77}$, $72Mg/125$

32   $\dfrac{3}{4}$

34   a) 40N, 80N;   b) $80/\sqrt{3}$ N, $40/\sqrt{3}$ N, $20\sqrt{3}$N

35   $\dfrac{\sqrt{5}}{2}Mg$ (AB and EF),

     $\dfrac{\sqrt{2}}{4}Mg$ (BC and DE)

36   $5 \mathrm{ms}^{-1}$, 18s

37   iii) $e(1 + e) \mathrm{ms}^{-1}$

38   a) $aw[(1 - \cos wt)\mathbf{i} + (1 - \sin wt)\mathbf{j}]$,
     $aw^2[(\sin wt)\mathbf{i} + (-\cos wt)\mathbf{j}]$

39   $71 \cdot 5°$

41   b) $2\dfrac{6}{7}$s;   c) $17 \cdot 5 \mathrm{ms}^{-1}$

42   a) $0 \cdot 8\mathbf{i} - 0 \cdot 6\mathbf{j}$;
     b) $(4\mathbf{i} - 3\mathbf{j}) \mathrm{ms}^{-1}$, $(8\mathbf{i} - 6\mathbf{j}) \mathrm{ms}^{-1}$;
     c) $(24\mathbf{i} - 18\mathbf{j}) \mathrm{N}$s

43   b) $(\mathbf{i} - \mathbf{j})$;   c) $(3\mathbf{i} - 4\mathbf{j})$;   d) 25J

44   a) $4 \cdot 4$N in positive x-direction;
     i) $4 \cdot 4$N in negative x-direction.
     ii) $44 \mathrm{ms}^{-1}$;   b) $22 \cdot 8$N, $-21 \cdot 6°$

45   ii) $0 \cdot 6325$, $3 \cdot 79 \mathrm{ms}^{-1}$;   iii) $1 \cdot 92$m

DESCRIPTIVE STATISTICS

GETTING STARTED

In statistics we study data obtained from observation. In this chapter we consider the collection and the display of data and then proceed to look at some of the measures we can calculate, e.g. central tendency and spread, for such sets of data.

SOME STANDARD TERMS

The characteristic that we are observing and recording is called the variable, e.g. the number of vehicles passing by a school entrance each minute or the height (in cm) of a certain species of plant growing in a local park. A variable is said to be discrete if it can take only a finite number of values, e.g. the set of integers. We have a discrete variable whenever we are counting the number of times some particular thing happens. A variable is called continuous when it can take any real value in a particular interval. We are concerned with variables that can fluctuate in a random way, and we call these **random variables**. From the illustrations above, the number of vehicles is a *discrete random variable* and the height of a plant is a *continuous random variable*.

SOURCES OF DATA

Statistical data may be collected from direct observation, from a series of experiments, from interviews, from questionnaires or from published sources. When collecting data you should appreciate that it is rare to take into account the whole population under consideration. If we wish to find out properties of a whole population, it is often impossible to test every member as it is too expensive or too time consuming, or the population may be destroyed by testing (e.g. if light-bulbs are tested to see how long they last). We take a sample of the population and obtain data from this. If we are to draw conclusions about the population from the sample data, the sample must be chosen with great care. We assume that the sample under consideration is a **random sample**. In effect this means that every member of the whole population has an equal chance of being selected in the sample. This simple idea is often very difficult to achieve in practice and you must always bear this in mind when collecting data for your own project work. Unless stated otherwise, we shall consider samples containing a fairly large number of members. These samples then have similar properties to the whole population and we can avoid taking the special precautions that would be necessary for a small sample from a large (or infinite) population.

THE CLASSIFICATION AND TABULATION OF DATA

MEASURES OF CENTRAL TENDENCY

MEASURES OF DISPERSION (OR SPREAD)

COMBINED MEAN AND VARIANCE

INDEX NUMBERS

THE CLASSIFICATION AND TABULATION OF DATA

ESSENTIAL PRINCIPLES

Once the data have been collected for a statistical investigation, the next vital steps are to classify and then to tabulate these data. Although to a large extent this is done for you in what follows in the examples and exercises, the importance of these steps should not be underestimated. Many employees in their daily work spend much of their time recording data for future analysis. The classification is usually presented in tabular form in both manual and computerised record keeping and the following comments should prove helpful for you in your project work:

" Properties of tables. "

i) the tables should be self-explanatory with a clear title, telling the reader what the data are about, where the data come from and when the data were collected,
ii) each column and row in the table should be headed clearly,
iii) the degree of accuracy of any approximate figures in the data should be given.

FREQUENCY DISTRIBUTIONS

Raw readings (or data) are condensed into a **frequency distribution** so that salient points can be easily seen and any underlying patterns are more easily identified.

Worked example

A photographic competition is divided into a number of sections. In the wild-life section there were 84 entries and the judge awarded marks for each entry as follows:

```
 5   4   10   7   2   8   5   7   6   6
 6  10    7   2   9   4   6  10   8   5
 9   7    1   9   6  10   5   4  10   4
10   5    5   6   4   8   7   9   6   7
 7   8    2   7   5   1   6   4   9   5
 6   4   10   5   9   6   9   5   8   6
 7   9    6   9   2  10   6   5  10   7
 5   6    8   4  10   6   7   5   6   8
 7   4    8   6
```

Classify the data and give your results in tabular form

MARK		FREQUENCY	MARK		FREQUENCY
1	II	2	6	THL THL THL II	17
2	IIII	4	7	THL THL II	12
3		0	8	THL III	8
4	THL IIII	9	9	THL IIII	9
5	THL THL III	13	10	THL THL	10
					84

Table 20.1 Frequency distribution to show the marks awarded to 84 entries in the wild-life section

Here we are dealing with a discrete random variable. As the marks take integer values only, the data are best displayed graphically by a **line graph**, often called a *bar chart* as in Fig. 20.1.

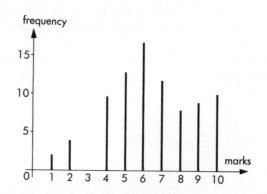

Fig. 20.1 Graphical representation of the marks of 84 entries in the wild-life section

The length of each line is proportional to the frequency of occurrence of the corresponding mark.

Worked example

From the records at a garage it was found that 80 cars were given a short-service in one week. The time, in minutes, to complete each service was recorded on the time-sheets and the following frequency distribution was prepared from these data.

Time (min)	45 –	50 –	55 –	60 –	65 –	70 –	75 –	80 –	85 –
Frequency	2	4	9	19	15	13	8	6	4

The notation 45 – means that the time in this interval is at least 45 minutes, but less than 50 minutes and in the last interval the time is at least 85 minutes, but less than 90 minutes.

A continuous variable such as time, which is illustrated here, requires us to choose class intervals with no gaps. The most widely used method of illustrating these data graphically is with a **histogram** in which the *areas* of the rectangles represent the frequencies, as shown in Fig. 20.2.

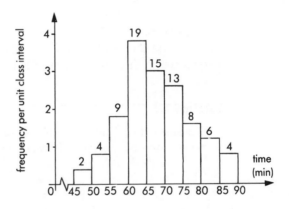

Fig. 20.2 Graphical representation of the times to give 80 cars a short-service

Several points are worth noting about the histogram:

> **Properties of the histogram.**

i) the actual frequency in each class interval is written over the top of the corresponding rectangle.

ii) the height of the rectangle is obtained by dividing the length of class interval (5 min) by the area, which is of course, also the frequency or some constant factor times the frequency; for example, the height of the 55 – interval is $\frac{9}{5}$ units = 1·8 units,

iii) the same method can be used when a histogram with unequal class intervals needs to be drawn.

The data in this exercise could also be displayed using what is called a **frequency polygon**. The polygon is drawn by joining the mid-points of the tops of each rectangle in the histogram, as shown in Fig. 20.3.

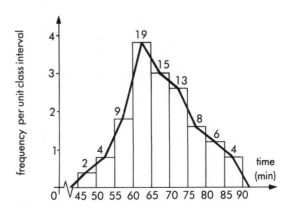

Fig. 20.3 Frequency polygon showing the times taken to give 80 cars a short-service

**MEASURES OF
CENTRAL
TENDENCY**

Various averages are used to distinguish a particular feature of a collection of data. An average is some central value of the data, although it may not always be a member of the set from which the variable is taken. For example, we could say that the average number of children per family when considering all the families in a large block of flats is 2·6, although it is obvious that the number of children in a particular family cannot be anything but a positive integer or zero.

The three averages most commonly used are the *mean*, the *median* and the *mode*.

For a list of numbers such as the marks awarded to the 84 entries in the wild-life section of the photographic competition we determine the mode, median and mean in the following ways.

THE MODE

This average is the mark which occurs most often. From the frequency distribution we see that this is a mark of 6. The **mode** is therefore 6.

THE MEDIAN

The median mark is found by arranging the list of 84 marks in either ascending or descending order. When the list contains an odd number of marks in total, the **median** is the middle mark. In ascending order here, with 84 marks, the middle two marks, that is the 42nd and 43rd marks, are found. The median is then taken to be half the sum of these two marks. In this case both the 42nd and 43rd marks are 6 and the median is also 6.

THE MEAN

The **mean** is found by adding all 84 marks and dividing by 84. The following tabulated working shows one method of finding the mean.

Mark (x)	Frequency (f)	fx
1	2	2
2	4	8
3	0	0
4	9	36
5	13	65
6	17	102
7	12	84
8	8	64
9	9	81
10	10	100
	84	542

The mean is

$$\frac{\Sigma fx}{\Sigma f} = \frac{542}{84} = 6{\cdot}45 \text{ (2 decimal places)}$$

The Σ notation provides us with a concise formula for the mean, which is used frequently in later calculations. For this reason we make the following generalisation at this stage.

The n numbers $x_1, x_2, x_3, \ldots, x_n$ occurring with frequencies $f_1, f_2, f_3, \ldots, f_n$ respectively have mean $\bar{x}$ defined by

$$\bar{x} = \frac{\displaystyle\sum_{r=1}^{n} f_r x_r}{\displaystyle\sum_{r=1}^{n} f_r}$$

Consider the linear relation

$$x_r = at_r + b$$

where a and b are constants and x_r represents a typical member of the set $x_1, x_2, x_3, \ldots, x_n$.

Another corresponding set t_1, t_2, t_3, ..., t_n is produced.

We have $\bar{x} = \dfrac{\Sigma f_r x_r}{\Sigma f_r} = \dfrac{\Sigma f_r(at_r + b)}{\Sigma f_r}$

$$\bar{x} = \dfrac{a\Sigma f_r t_r}{\Sigma f_r} + \dfrac{b\Sigma f_r}{\Sigma f_r}$$

$$\bar{x} = a\bar{t} + b$$

> This is an important result.

We see that $\bar{t}$ is the mean of the n numbers t_1, t_2, t_3, ..., t_n occurring with frequencies f_1, f_2, f_3, ..., f_n respectively.

This important result is used throughout statistical work because it can dramatically reduce numerical processing and later, it enables us to use one set of values for a standard distribution, such as the normal distribution.

The relation $x = at + b$ contains two basic steps:

Look at $t = \dfrac{1}{a}(x - b)$, which is the same relation with t in terms of x.

i) $x - b$ effectively translates the origin from $(0, 0)$ to $(b, 0)$ without changing the units of the x variable,

ii) $\dfrac{1}{a}$ alters the scale by effectively making a units of the x variable be 1 unit for the t variable.

Worked example

We consider now how to estimate the mean of the frequency distribution for the time taken to give a short-service to 80 cars. The mid-interval value of each class interval is taken as the x values, as shown in the following table.

Time (min)	Mid-interval value x	Frequency f	$t = \dfrac{x - 62\cdot5}{5}$	ft
45 –	47·5	2	–3	–6
50 –	52·5	4	–2	–8
55 –	57·5	9	–1	–9
60 –	62·5	19	0	0
65 –	67·5	15	1	15
70 –	72·5	13	2	26
75 –	77·5	8	3	24
80 –	82·5	6	4	24
85 –	87·5	4	5	20
		$\Sigma f = 80$		$\Sigma ft = 109{-}23$ $= 86$

Estimate of mean $= 62\cdot5 + 5\left(\dfrac{86}{80}\right) \approx 67\cdot9$ (1 decimal place)

Explanation: the linear transformation $x = 62\cdot5 + 5t$ has been used in the form

$$t = \dfrac{1}{5}(x - 62\cdot5)$$

to compute values of t from the corresponding values of x.
The estimate of the mean has been found by using

$$\bar{x} = 5\bar{t} + 62\cdot5 \text{ where } \bar{t} = \dfrac{\Sigma ft}{\Sigma f} \text{ (see table)}$$

You should convince yourself that the choice of the linear transformation is arbitrary. Equally well the transformation $x = 5t + 67\cdot5$ or the transformation $x = 5t + 72\cdot5$ could be used here. Try these for yourself and this should convince you about the effectiveness and simplicity of the calculations.

The following work illustrates the methods used to find estimates for a) the mode and b) the median of this frequency distribution.

ESTIMATE OF THE MODE FROM THE HISTOGRAM

For this we use the histogram drawn to display the data about short-service times and shown in Fig. 20.4.
The mode is read off the horizontal axis at the point marked M.

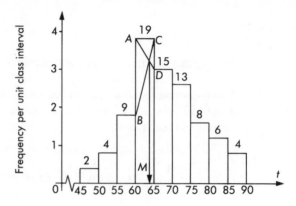

Fig. 20.4

Explanation: the class interval 60 – contains the largest number of members and all class intervals are of the same length. This is called the *modal class*. The rectangles on either side of the modal class are also considered and the lines AD and BC are joined. The estimate of the mode is then taken as the horizontal axis coordinate of the intersection of these lines.

Estimate of mode = 63·6

ESTIMATION OF THE MEDIAN FROM THE HISTOGRAM

A little preliminary investigation reveals that the median falls in the 65 – interval because 34 times fall below this and 31 times are above this. For the median we require the line, $t = d$, say, drawn at right angles to the base which divides the area of the whole histogram into two equal parts, that is an area of 40 units^2 on either side of $t = d$.

This gives for the left hand side $34 + (d - 65)\dfrac{15}{5} = 40$

giving $d = 65 + 2 = 67$
The estimate of the median is 67

The second method of estimating the median of a frequency distribution is obtained from a drawing of a cumulative frequency curve, sometimes called the ogive. The cumulative frequency table for short-service times is:

Time (min)	<45	<50	<55	<60	<65	<70	<75	<80	<85	<90
Cumulative frequency	0	2	6	15	34	49	62	70	76	80

Percentiles. These are a useful device for locating particular positions in the cumulative frequency distribution. For example the 50th percentile is the median. The Kth percentile is the value of t corresponding to a cumulative frequency of

$\dfrac{80K}{100}$ i.e. $\dfrac{K}{100} \times$ (total frequency)

The median is then read off at N as 67, as before (see Fig. 20.5).

MEASURES OF DISPERSION (OR SPREAD)

> We are also interested in the spread of the data.

Once a central measure has been established for a frequency distribution, we often wish to investigate the dispersion of the distribution. The ogive provides one way of measuring dispersion and **the inter-quartile range** is often used. This range is found by reading off the 25th and 75th percentiles from the ogive, as shown, at L and U.

The interval $LU = 74·2 - 61·3 = 12·9$ is the inter-quartile range.

Any other inter-percentile range can be obtained from the ogive in a similar way.

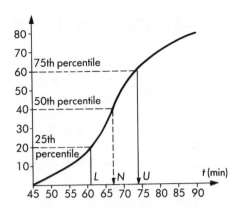

Fig. 20.5

Other measures of spread used in descriptive statistics for a frequency distribution are the **range** and the **absolute mean deviation from the mean** (*or* from the *median*). Like inter-percentile ranges though, these are not used in further work because they are either too simplistic or they do not lend themselves well to calculation. The measures widely used are the **variance** and its square root called the **standard deviation**.

We use again the data from our two previous examples to illustrate how to calculate these important measures of dispersion.

Worked example

For the distribution of marks awarded by the judge in the wild-life section of the photographic competition, we found the mean to be 6·45. The variance, in Σ notation, is defined to be:

$$\frac{\Sigma f(\bar{x}-x)^2}{\Sigma f}$$

and we know that $\Sigma f = 84$.

We show two methods for calculating the variance, the first by direct calculation from the above formula and the second by processing this formula using some properties of the mean of the distribution:

Method 1

$$\begin{aligned}
\Sigma f(\bar{x}-x)^2 &= 2(6·45-1)^2 + 4(6·45-2)^2 + 9(6·45-4)^2 + 13(6·45-5)^2 + 17(6·45-6)^2 \\
&\quad + 12(6·45-7)^2 + 8(6·45-8)^2 + 9(6·45-9)^2 + 10(6·45-10)^2 \\
&= 59·405 + 79·210 + 54·023 + 27·333 + 3·443 + 3·630 + 19·220 + 58·523 + 126·025 \\
&= 430·812
\end{aligned}$$

The variance $= \dfrac{430·812}{84} = 5·13$ (2 decimal places)

Method 2

First, we note that
$$\begin{aligned}
\Sigma f(\bar{x}-x)^2 &= \Sigma f(\bar{x}^2 - 2\bar{x}x + x^2) \\
&= \bar{x}^2 \Sigma f - 2\bar{x} \Sigma fx + \Sigma fx^2
\end{aligned}$$

But we know that $\Sigma fx = \bar{x}\Sigma f$ from the definition of $\bar{x}$

This gives $\Sigma f(\bar{x}-x)^2 = \bar{x}^2 \Sigma f - 2\bar{x}^2 \Sigma f + \Sigma fx^2$
$$= \Sigma fx^2 - \bar{x}^2 \Sigma f$$

$$\text{Variance} = \frac{\Sigma f(\bar{x}-x)^2}{\Sigma f} = \frac{\Sigma fx^2}{\Sigma f} - \frac{\bar{x}^2 \Sigma f}{\Sigma f}$$

$$\text{Variance} = \frac{\Sigma fx^2}{\Sigma f} - \bar{x}^2$$

This formula makes the calculations much easier because we only need to evaluate Σfx^2 as Σf and $\bar{x}$ are known

$$\begin{aligned}
\Sigma fx^2 &= 2 + 16 + 144 + 325 + 612 + 588 + 512 + 729 + 1000 \\
&= 3928
\end{aligned}$$

$$\text{Variance} = \frac{3928}{84} - \left(\frac{542}{84}\right)^2 = 5·13$$

The standard deviation of this distribution $= \sqrt{5·13}$
$$\approx 2·26$$

THE STANDARD DEVIATION (SD) AND VARIANCE FOR TRANSFORMED DATA

The linear transformation given by the relation $x = at + b$, translates the origin from $(0, 0)$ to $(b, 0)$ and changes the scale so that a units of the x variable become 1 unit for the t variable.

The translation does not affect the standard deviation and we have

   Standard deviation of x data $= a$ (standard deviation of t data)

and

   Variance of x data $= a^2$ (variance of t data)

Worked example

Find the variance and standard deviation of the times taken to give a short-service to 80 cars

We set out the workings in full and strongly advise students to follow this practice always in their own work.

Mid-interval value x	Frequency f	$\left(t = \dfrac{x - 62 \cdot 5}{5}\right)$ t	ft	ft^2
47·5	2	−3	−6	18
52·5	4	−2	−8	16
57·5	9	−1	−9	9
62·5	19	0	0	0
67·5	15	1	15	15
72·5	13	2	26	52
77·5	8	3	24	72
82·5	6	4	24	96
87·5	4	5	20	100
	$\Sigma f = 80$		$\Sigma ft = 86$	$\Sigma ft^2 = 378$

$$\text{Estimate of variance} = a^2 \left[\frac{\Sigma ft^2}{\Sigma f} - \left(\frac{\Sigma ft}{\Sigma f} \right)^2 \right]$$

$$= 25 \left[\frac{378}{80} - \left(\frac{86}{80} \right)^2 \right]$$

$$= 89 \cdot 23 \text{ (2 decimal places)}$$

$$\text{Estimate of standard deviation} = \sqrt{\text{variance}}$$
$$= 9 \cdot 45 \text{ (2 decimal places)}$$

> Descriptive statistics are a vital part of many projects.

Final Note: at A-level, in mathematics subjects containing either a paper or parts of a paper with statistics questions, you are unlikely to get full questions on the material contained in this chapter. Our experience suggests however, that students need to master and understand this work as a foundation for future success. Many students will be undertaking project work as a necessary component for the final assessment. In this work, the collection, tabulation, display and processing of your own data will be essential. Biology, geography, social sciences, economics, accountancy and psychology are all subjects with sources of data requiring interpretation and analysis. The following questions should be worked through and you can then investigate sets of data from associated studies or from your own statistical project work.

COMBINED MEAN AND VARIANCE

We sometimes need to combine results for the means and variances from two or more samples of data arising from the same experiment. This can be done without calculating the mean and variance from the original readings although the saving of work in evaluating the variance of the combined sample is not very great.

Consider: Sample 1, size n, values x_1, x_2, x_3, ..., x_n. Mean $\bar{x}$, SD s_x

   Sample 2, size m, values y_1, y_2, y_3, ..., y_m. Mean $\bar{y}$, SD s_y

$$\text{Combined mean is } \bar{X} = \frac{\displaystyle\sum_{r=1}^{n} x_r + \sum_{r=1}^{m} y_r}{m + n} = \frac{n\bar{x} + m\bar{y}}{m + n} \text{ since}$$

$$\frac{\sum_{r=1}^{n} x_r}{n} = \bar{x} \text{ and } \frac{\sum_{r=1}^{m} y_r}{m} = \bar{y}$$

Variance of combined sample $\Rightarrow s^2 = \dfrac{\sum_{r=1}^{n}(x_r - \bar{X})^2 + \sum_{r=1}^{m}(y_r - \bar{X})^2}{m+n}$,

and since it can be shown that

$$\sum_{r=1}^{n}(x_r - \bar{X})^2 = \sum_{r=1}^{n}(x_r - \bar{x})^2 + \sum_{r=1}^{m}(\bar{x} - \bar{X})^2 = n[s_x^2 + (\bar{x} - \bar{X})^2]$$

and similarly $\sum_{r=1}^{m}(y_r - \bar{X})^2 = m[s_y^2 + (\bar{y} - \bar{X})^2]$

we have $\quad s^2 = \dfrac{n[s_x^2 + (\bar{x} - \bar{X})^2] + m[s_y^2 + (\bar{y} - \bar{X})^2]}{m+n}$

Worked example

Calculate the mean and variance of each of the two samples given. Use your results to calculate the mean and variance of the combined sample

Sample 1: 7·4, 8·2, 7·5, 8·1, 7·8

Sample 2: 6·8, 7·6, 8·1, 7·3

$$\text{Sample 1} \quad \text{Mean} = \frac{7·4 + 8·2 + 7·5 + 8·1 + 7·8}{5} = 7·8$$

$$\text{Variance} \Rightarrow s_1^2 = \frac{0·4^2 + 0·4^2 + 0·3^2 + 0·3^2 + 0}{5} = \frac{0·5}{5} = 0·1$$

$$s_1 = 0·316$$

$$\text{Sample 2} \quad \text{Mean} = \frac{6·8 + 7·6 + 8·1 + 7·3}{4} = 7·45$$

$$\text{Variance} \Rightarrow s_2^2 = \frac{0·65^2 + 0·15^2 + 0·65^2 + 0·15^2}{4} = \frac{0·89}{4} = 0·2225$$

$$\Rightarrow s_2 = 0·472$$

$$\text{Combined mean } \bar{X} = \frac{5 \times 7·8 + 4 \times 7·45}{9} = 7·64$$

$$\text{Combined variance} \Rightarrow s^2 = \frac{5 \times 0·1 + 4 \times 0·2225 + 5(7·8 - 7·64)^2 + 4(7·45 - 7·64)^2}{9}$$

$$= 0·185$$

INDEX NUMBERS

You may already be familiar with the retail price index which measures the change in the cost of certain household commodities each month, and in particular the cost now compared with the cost some few years ago, the base cost, when the exercise first started. A simple example will show how the calculations are undertaken and the difference between the three main indices.

Generally n basic items are considered, the average quantity of each item purchased, both in the base year and in the year under consideration, together with the prices of each item in the two years.

> Be familiar with these three index numbers.

The **Paasche index** is then given by:

$$\frac{\sum_{\text{all items}} (\text{Present price}) . (\text{Average quantity now consumed})}{\sum_{\text{all items}} (\text{Base year price}) . (\text{Average quantity consumed in base year})}$$

The **Laspeyres index** is given by:

$$\frac{\sum_{\text{all items}} (\text{Present price}) . (\text{Average quantity consumed in base year})}{\sum_{\text{all items}} (\text{Base year price}) . (\text{Average quantity consumed in base year})}$$

The **Fisher index** is then the geometric mean of the Paasche and Laspeyres indices, i.e.

Fisher index $= \sqrt{[(\text{Paasche index})(\text{Laspeyres index})]}$

Worked example

The table below shows the average price in pence per kilogram of seven staple food items for the years 1984 and 1989, together with the average weight in kilograms consumed by a family of four in those years. Calculate the following indices a) Paasche, b) Laspeyres, c) Fisher.

ITEM	1984		1989	
	PRICE $(\text{p}\,\text{kg}^{-1})$	WEIGHT (kg)	PRICE $(\text{p}\,\text{kg}^{-1})$	WEIGHT (kg)
Bread	55	200	62	192
Butter	192	64	208	50
Margarine	96	56	116	78
Potatoes	28	248	30	260
Sugar	52	80	58	76
Tea	140	59	154	54
Coffee	1260	10	1490	16

Paasche index

$$= \frac{62 \times 192 + 208 \times 50 + 116 \times 78 + 30 \times 260 + 58 \times 76 + 154 \times 54 + 1490 \times 16}{55 \times 200 + 192 \times 64 + 96 \times 56 + 28 \times 248 + 52 \times 80 + 140 \times 59 + 1260 \times 10}$$

$$= \frac{75\,716}{60\,628} = 1 \cdot 249$$

Laspeyres index

$$= \frac{62 \times 200 + 208 \times 64 + 116 \times 56 + 30 \times 248 + 58 \times 80 + 154 \times 59 + 1490 \times 10}{60\,628}$$

$$= \frac{68\,274}{60\,628} = 1 \cdot 126$$

Fisher index

$$= \sqrt{(1 \cdot 249 \times 1 \cdot 126)}$$

$$= 1 \cdot 186$$

Thus taking 1984 as 100% then the 1989 price index is:

Paasche $\approx 125\%$, Laspeyres $\approx 113\%$, Fisher $\approx 119\%$

The figures may appear quite disparate, but to a large extent this is due to a 60% increase in the consumption of coffee between 1984 and 1989. The Paasche index shows this more clearly by taking into account the increase in the consumption of coffee, the Laspeyres index does not take it into account whilst the Fisher index attempts to balance the two.

EXERCISES 20.1

1 Find a) the mean, b) the standard deviation of the numbers
2, 3, 4, 4, 6, 7, 7, 8, 9, 10

2 The numbers
3, 6, 4, 12, 10, 4, 12, a, b
have mode 4 and mean 7. Calculate:
i) values for a and b
ii) the median of these nine numbers
iii) the variance of these numbers

3   An examination in statistics was taken by 59 students and the marks they obtained are shown in the following table

45	36	26	50	56	34	39	32	49	67
30	46	48	37	23	47	60	34	44	54
65	43	28	41	52	40	64	34	32	45
62	19	40	57	31	33	38	35	43	42
55	28	71	58	41	36	60	54	41	25
47	33	68	38	52	65	37	49	33	

i) Find the median mark

ii) Using a suitable frequency table, estimate the mean and the standard deviation of the marks

4   A survey of the mass, in kg, of 97 girls in a college final year was taken and the results are shown in the table.

Mass (kg)	40 –	45 –	50 –	55 –	60 –	65 –	70 –	75 –	80 –	85 – 90
Frequency	3	9	16	21	17	14	5	7	4	1

a) Display the information using a histogram

b) Estimate the mode

c) Estimate the median and find the inter-quantile range

d) Calculate estimates for the mean, variance and standard deviation

5   The times (in min) taken by each of 100 students to prepare and write a set essay in economics are given in the frequency table

Time (min)	– 80	– 100	– 120	– 140	– 160	– 180	– 200	– 220
Frequency	3	7	15	19	28	16	11	1

a) Estimate the mode and the median

b) Calculate estimates for the mean, variance and standard deviation

NB: Examination questions on these topics can be found in Chapter 27.

ANSWERS TO EXERCISES

EXERCISES 20.1

1   a) 6;   b) 2·53

2   i) 4, 8;   ii) 6;   iii) $11\dfrac{5}{9}$

3   i) 42;   ii) 43·9, 12·4

4   b) 58;   c) 59, 13·8;   d) 61·2, 101·3, 10·1

5   a) 149, 144;   b) 141·8, 969, 31·12

CHAPTER 21

PROBABILITY

SOLUTIONS TO PERMUTATIONS

COMBINATIONS (SELECTION)

PROBABILITY

CONDITIONAL PROBABILITY

USE OF TREE DIAGRAMS

GETTING STARTED

In order to be successful in solving problems on probability, you first need to have clear, logical ideas about how to count the total different possible events occurring in a sample space. We start therefore with some work on permutations and combinations.

PERMUTATIONS (ARRANGEMENTS)

Suppose we have a set containing 5 children and we wish to find the number of different ways of arranging a) all of them, b) just 3 of them in a line.

a) First, we think of 5 spaces, and count like this: any one of the 5 can fill the first space, any 4 are left to fill the next space and so on. In all then we have $5 \times 4 \times 3 \times 2 \times 1$ ways of seating the children. It is convenient to have a notation for numbers like $5.4.3.2.1$ and we write this as 5! (called 5 factorial) and you may have a button on your calculator for this. So the answer here is 120 different ways.

b) Using a similar approach to that described in a) we now have only three spaces and 5 possible children can fill the first space, 4 the second space and 3 the third space. You can see therefore that this time there are $5 \times 4 \times 3$ different ways $= 60$ ways of arranging any 3 children from the 5 children.

It is useful to note that we can write $5.4.3 = \dfrac{5.4.3.2.1}{2.1} = \dfrac{5!}{2!}$

Generalising we can say that the number of different arrangements, called **permutations**, of n unlike items taken k at a time is

$$\frac{n!}{(n-k)!}$$

When items are repeated, we adopt the following method. Suppose we wish to find the number of permutations of the 4 letters of the word NONE. If the letter Ns were different, then the answer would be 4! However the Ns are identical and can be arranged in 2! ways, so NONE has $\dfrac{4!}{2!}$ permutations using all the letters.

The answer is 12 and if you need to convince yourself write them out in just this simple case (They are NNOE, NNEO, NEON, NOEN, NONE, NENO, OENN, EONN, ONEN, ENON, ENNO, ONNE)

All the letters of the word BEETLE can be permutated in $\dfrac{6!}{3!} = 120$ ways

using a similar argument.

Notation: the following useful shorthand is used for permutations

$$^nP_r = \frac{n!}{(n-r)!}$$

where ^nP_r is a number standing for the number of permutations of n different items taken r at a time.

SOLUTIONS TO PERMUTATIONS

ESSENTIAL PRINCIPLES

The following worked examples illustrate the methods used:

Worked example
How many four-digit numbers can be made from the set 1, 2, 3, 4, 5 when
i) no integer is used more than once
ii) an integer can be used any number of times?

In i), how many of the numbers are odd?

i) $^5P_4 = \dfrac{5!}{(5-4)!} = 5! = 120$

ii) $5^4 = 625$, because any of the 5 numbers can now fill each place.

In i) the number of odd numbers is
$3 \cdot {}^4P_3 = 3 \cdot 4! = 72$, because each number must end in either 1 or 3 or 5 each time.

Worked example
In how many ways can 4 people from a party of 7 people be arranged around a circular table?
Consider seats A, B, C and D around a circular table (see Fig. 21.1)

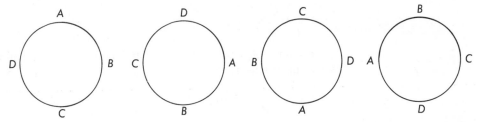

Fig. 21.1

Seat A can be occupied by any one of the 7 people; $B \Rightarrow$ by any one of the remaining 6, $C \Rightarrow$ by 5, $D \Rightarrow$ by 4.
Seats A, B, C, $D \Rightarrow 7 \cdot 6 \cdot 5 \cdot 4$ ways. However, we are only interested in who sits next to who and the above selection includes duplications due to the possibility of moving chairs around the table as shown in Fig. 21.1, without changing the seating order.

Hence there are $\dfrac{7 \cdot 6 \cdot 5 \cdot 4}{4} = 210$ different arrangements.

You should note that if the problem had been one of arranging coloured beads on a ring then a similar argument could have been followed, but it would be necessary to make a further division by 2 because of the duplications due to the possibility of turning the ring over, see Fig. 21.2.

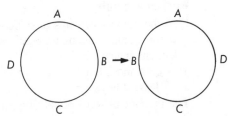

Fig. 21.2

EXERCISES 21.1

1 Evaluate $^6P_4$, $^5P_5$, $^7P_2$, $^4P_1$

2 Find the number of permutations of all the letters of the word SOLDIER.
   a) How many of these permutations start with SO and end in R?
   b) How many permutations of 5 letters can be made from the 7 letters in SOLDIER?

3 A girl has 4 different cookery books and 3 different needlework books. If the 7 books are arranged in any order on a shelf, what is the number of arrangements possible? If, on the other hand, she arranges them on the shelf with the cookery books next to each other and the needlework books next to each other, how many permutations are now possible?

4   MISSISSIPPI has 11 letters. In how many ways can all of these letters be permutated? (Leave your answer in factorials). Repeat the question for the word ASSASSINATIONS.

5   In how many different ways can a football team of 11 players be photographed if there are to be two rows, with 5 seated on the front row and 6 standing in the back row? In how many of these photographs would the captain be sitting in the centre of the front row?

COMBINATIONS (SELECTIONS)

> Here the order does not matter.

Suppose we require a committee of three boys from A, B, C, D, E. We know already that we could permutate these 3 from 5 in $^5P_3$ ways, that is, in

$$\frac{5!}{(5-3)!} = 60 \text{ ways}$$

But we require only a selection, so the order within each group of 3 does not matter. As any group of 3 can be permutated in $3! = 6$ ways, we divide 60 by 6 to obtain 10, which is the possible number of committees.

To convince yourself, write them out (just this once!) (ABC, ABD, ABE, ACD, ADE, BCD, BCE, BDE, ACE, CDE)

Generalising, we can write that a selection of k items from n different items can be chosen in

$$\frac{n!}{(n-k)!\,k!} \text{ ways.}$$

From the illustration above we have, $n = 5$ and $k = 3$ which gives the required number of combinations as

$$\frac{5!}{(5-3)!\,3!} = 10$$

Notation: the following useful shorthand is used for combinations

$$^nC_r = \frac{n!}{(n-r)!\,r!}$$

where ^nC_r is a number standing for the number of combinations of n items taken r at a time.

Note also that some examiners use $\binom{n}{r}$ to stand for ^nC_r

The following worked examples illustrate the methods used:

Worked example

Find the number of ways of selecting five children from a group of 12 children
    If the group contains an equal number of boys and girls, find the number of different selections which contain
i) 2 girls and 3 boys
ii) at least 3 boys
The number of ways of selecting five children from 12 children is

$$\binom{12}{5} = \frac{12!}{7!\,5!} = \frac{12.11.10.9.8}{1.2.3.4.5} = 792$$

i) If the selection contains 2 girls and 3 boys, it is necessary to select 2 girls from 6 girls and 3 boys from 6 boys
    The number of ways of selecting the 2 girls is

$$\binom{6}{2} = \frac{6.5}{1.2} = 15$$

    The number of ways of selecting the 3 boys is

$$\binom{6}{3} = \frac{6.5.4}{1.2.3} = 20$$

Any of the selections of girls can be taken with any of the selections of boys
⇒ total number of selections containing 2 girls and 3 boys is

$$\binom{6}{2} \cdot \binom{6}{5} = 15 \times 20 = 300.$$

ii) If the selection contains at least 3 boys it could contain a) 3 boys and 2 girls,
b) 4 boys and 1 girl, c) 5 boys
Number of selections containing 3 boys and 2 girls is

$$\binom{6}{3} \cdot \binom{6}{2} = 300$$

Number of selections containing 4 boys and 1 girl is

$$\binom{6}{4} \cdot \binom{6}{1} = \frac{6.5}{1.2} \cdot \frac{6}{1} = 90$$

Number of selections containing 5 boys is

$$\binom{6}{5} = \frac{6}{1} = 6$$

Total number of selections containing at least 3 boys is $300 + 90 + 6 = 396$

Worked example

In how many ways can 8 different books be divided into
i) two groups of 5 books and 3 books;   ii) two groups of 4 books and 4 books

i) Two groups of 5 and 3 ⇒ $\binom{8}{5} = \binom{8}{3} = \frac{8.7.6}{1.2.3} = 56$ ways

ii) Two groups of 4 and 4 ⇒ $\frac{1}{2}\binom{8}{4} = \frac{1}{2} \cdot \frac{8.7.6.5}{1.2.3.4} = 35$ ways

Worked example

A bag contains 12 coloured balls, of which 5 are red, 4 blue, 2 green and 1 yellow. Find the number of different selections of 3 balls

Care must be taken with questions such as this one to differentiate between the various possibilities. They can all be listed, for instance:

i) All 3 of the same colour ⇒ 3R or 3B   ⇒ 2 selections
ii) Two of the same colour ⇒ 2R + 1B or 1G or 1Y ⇒ 3 selections
                                         ⇒ 2B + 1R or 1G or 1Y ⇒ 3 selections
                                         ⇒ 2G + 1R or 1B or 1Y ⇒ 3 selections
iii) All different colours   ⇒ 1R + 1B + 1G ⎤
                                              1R + 1B + 1Y ⎥
                                              1R + 1G + 1Y ⎬ ⇒ 4 selections
                                              1B + 1G + 1Y ⎦

Total number of different selections is 15
However the questions could be answered as follows:

i) All 3 of the same colour ⇒ $\binom{2}{1} = 2$

ii) Two of the same colour ⇒ $\binom{3}{1} \cdot \binom{3}{1} = 3 \times 3 = 9$

iii) All different colours   ⇒ $\binom{4}{1} = 4$

Total number of different selections = 15

EXERCISES 21.2

1 Evaluate $\binom{6}{4}, \binom{5}{5}, \binom{7}{2}, \binom{8}{3}$. Evaluate also $\binom{14}{4}\binom{22}{10}$, giving your answers in factorial form.

2   A committee of 4 is to be selected from 7 people. In how many ways can this be done? How many selections are possible when the oldest member of this group of 7 people is always excluded?

3   Show that $^7C_3 = {}^7C_4$ by direct calculation.
    Prove generally that $^nC_r = {}^nC_{n-r}$

4   Find the number of different committees, containing 2 staff and 3 students, that could be made up from 5 staff and 7 students

5   A sixth-form social committee is to made up so that there are 3 members from the first year and 4 members from the second year. There are 80 students in the first year and 75 in the second year. Find, to 3 significant figures, the number of different committees which could be formed given that all students are eligible for selection

6   Two car loads containing 4 passengers in each are to be made up from 8 people. One car is a Ford, the other is a Vauxhall. Find the total number of selections possible for this

7   A secretary writes 12 letters, but only has enough stamps in her desk for just 4 letters. She does in fact send 4 after choosing the 4 at random. In how many ways could she choose the letters for stamping?

8   A youth club has just 18 members and 7 of these are athletes. In how many ways can 10 members be chosen so as to include
    a) exactly 5 athletes
    b) at least 3 athletes

PROBABILITY

Probability is an attempt to use mathematics to estimate by means of a numerical answer the chance that some event will happen. We are for instance all used to the idea of a raffle where one buys a number of tickets in the hope of winning a prize. If, for instance, we buy 3 of the raffle tickets of which a total of 1000 are sold, then assuming that any one of these 1000 tickets sold is as likely to be drawn as any other, we can say with confidence that we have a 3 in 1000 chance of winning a prize. We say we have a probability of $\dfrac{3}{1000}$ of winning a prize. Had we not bought a ticket the probability would be $\dfrac{0}{1000}$, i.e. impossibility, whilst on the other hand had we purchased all of the tickets the probability would have been $\dfrac{1000}{1000} = 1$, i.e. certainty. The probability that it is either impossible or certain for something to happen is very rare and generally we are not interested in such cases. Consequently, we shall concern ourselves with problems where the probability lies between 0 and 1.

Most GCE A-level examination questions on probability involve the use of games of chance; throwing dice, selecting cards, tossing coins. We define the act of say, throwing a six with a die, drawing an Ace from a pack of cards or winning a raffle as a simple *event*. The set of all such events concerned with an experiment is known as the *sample space*. Thus, for the throwing of a die the sample space is the numbers 1, 2, 3, 4, 5 and 6. Consequently, the probability that the event of throwing a six with a die is $\dfrac{1}{6}$ i.e.:

$$\frac{\text{No. of ways in which a 6 occurs}}{\text{No. of ways in which equally likely events occur}}$$

Hence we define the probability that an event E occurs as:

$$P(E) = \frac{\text{No. of ways in which } E \text{ can occur}}{\text{Total no. of ways in which events such as } E \text{ can occur}}$$

or simply as:

$$P(E) = \frac{\text{No. of successful outcomes}}{\text{Total no. of possible outcomes}}$$

Worked example

Five children, two of whom are Betty and Alan, sit at random in a row on a bench. Find the probability that Betty and Alan sit next to each other

The 5 children can sit on the bench in $^5P_5 = 120$ ways; the sample space. When Betty and Alan sit next to each other the number of different ways is $2.\,^4P_4 = 48$; the 2 being required since Betty and Alan can be interchanged. This represents the number of successful outcomes. Hence

$$\text{P (Alan and Betty sit next to each other)} = \frac{48}{120} = \frac{2}{5} = 0\cdot4$$

Worked example

Two books are selected simultaneously and at random from a shelf containing 5 biology books and 3 chemistry books. Find the probability that one and only one biology book is selected.

$$\text{No. ways of selecting any two books from eight} = {}^8C_2 = \frac{8!}{6!2!} = 28$$

No. ways of selecting one biology book from 5 biology books = 5
No. ways of selecting one chemistry book from 3 chemistry books = 3
$\Rightarrow$ No. of ways of selecting one biology and one chemistry book is $5 \times 3 = 15$

$$\Rightarrow \text{P (selecting one and only one biology book)} = \frac{15}{28} = 0\cdot54$$

LAWS AND FORMULAE ASSOCIATED WITH PROBABILITY

Set theory proves very useful in building up a sound approach to probability since the elements in a set are analogous to the events in a sample space.

Consider the following worked example.

Worked example

Find the probability of drawing an Ace or a Heart from a normal pack of 52 cards.

The problem can be illustrated by using a Venn diagram as follows.

The sample space consists of 52 simple events – this can be represented by the total area of the rectangular box

There are 13 Hearts in the pack which can be represented by a set H and 4 Aces which can be represented by a set A, but as the Ace of Hearts is contained in each of these sets we draw the sets intersecting as shown in Fig. 21.3

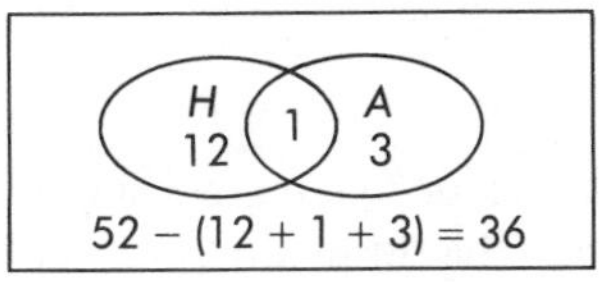

$52 - (12 + 1 + 3) = 36$

Fig. 21.3

$$\Rightarrow \text{P (Heart)} = \frac{13}{52} = \frac{1}{4} = \text{P}(H)$$

$$\text{P (Ace)} \quad = \frac{4}{52} = \frac{1}{13} = \text{P}(A)$$

$$\text{P (Heart } or \text{ an Ace)} = \frac{12 + 1 + 3}{52} = \frac{16}{52} = \frac{4}{13} = \text{P}(H \cup A) \text{ where } \cup \equiv \text{or, i.e. } \text{P}(H \text{ } or \text{ } A).$$

The event $H \cap A$ stands for a Heart *and* an Ace. Thus $H \cap A$ is represented by the intersection of the sets H and A and hence

$$\text{P (Ace of Hearts)} = \frac{1}{52}$$

Note: $\text{P}(H \cup A) = \text{P}(H) + \text{P}(A) - \text{P}(H \cap A)$. It is known as the addition law for probabilities.
The probability of *not* drawing a Heart or an Ace is given by

$$\Rightarrow \text{P (not Heart or Ace)} = \frac{36}{52} = \frac{9}{13}$$

It is known as the complement of the event (Heart or Ace) and is written as (Heart or Ace)′

$$\text{Thus P (not Heart)} = \text{P}(H)' = \frac{39}{52} = \frac{3}{4} = 1 - \text{P}(H)$$

$$\text{P (not Ace)} \quad = \text{P}(A)' = 1 - \text{P}(A) = 1 - \frac{1}{13} = \frac{12}{13}$$

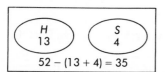

Fig. 21.4

In this example the sets H and A intersected. When we are concerned with sets which do not intersect, e.g. the probability of drawing a Heart or the Jack, Queen, King or Ace of Spades, then our two sets would be represented as shown in Fig. 21.4.

$$P(H \text{ or } S) = \frac{13+4}{52} = \frac{17}{52} = P(H) + P(S)$$

and we say the event H and S are mutually exclusive.

Note: two events H and S are mutually exclusive when $P(H \cap S) = 0$

Worked example

A group of 20 people consists of 8 men, 6 women, 4 boys and 2 girls. A person is chosen at random from the group. Find the probability that:
a) the person is female
b) the person is either a man, woman or a girl
The Venn diagram of this situation can be represented as shown in Fig. 21.5

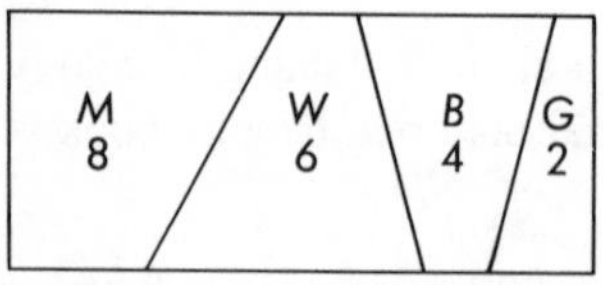

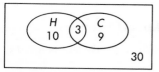

Fig. 21.5

Note: as the four events M, W, B, G cover the whole of the sample space they are said to be exhaustive.

a) $P(\text{Female}) = P(W \cup G) = P(W) + P(G) = \frac{6}{20} + \frac{2}{20} = \frac{8}{20} = \frac{2}{5} = 0{\cdot}4$

b) $P(M \cup W \cup G) = P(M) + P(W) + P(G) = \frac{8}{20} + \frac{6}{20} + \frac{2}{20} = \frac{16}{20} = \frac{4}{5} = 0{\cdot}8$

Worked example

Find the probability of drawing at random a Heart or a picture card from a well-shuffled pack of cards
The Venn diagram for this problem is as shown in Fig. 21.6 where
$H \equiv$ Hearts and $C \equiv$ picture card
You will notice the 3 in the overlap of the sets covers the Jack, Queen and King of Hearts.

Fig. 21.6

$$P(\text{Heart or picture card}) = P(H \cup C) = \frac{10+3+9}{52} = \frac{22}{52} = \frac{11}{26}$$

CONDITIONAL PROBABILITY

The question could have asked for the probability of drawing a Jack or Queen or King of Hearts given that it was known that a picture card was drawn. This brings us into the realms of **conditional probability**. A little thought will show that this question reduces to finding the probability that one of three particular cards is chosen out of 12 known picture cards, i.e. the sample space is reduced to the set C, and so:

$$P(\text{Jack, Queen or King of Hearts}) = \frac{3}{12} = \frac{1}{4}$$

We write this as $P(H|C)$ and see from the Venn diagram that

$$P(H|C) = \frac{P(H \cap C)}{P(C)} = \frac{3}{12} = \frac{1}{4}$$

i.e. $P(A|B) = \dfrac{P(A \cap B)}{P(B)}$ = the probability that event A occurs given that it is known that event B has occurred.

INDEPENDENT EVENTS

Two events A and B are said to be **independent** if the occurence of one event, say A, does not depend upon the occurrence of B, and conversely. E.g. the tossing of a coin to get a head and the throwing of a die to obtain a six. In such an instance

$$P(A) = P(A|B) \Rightarrow P(A) = \frac{P(A \cap B)}{P(B)}$$

or $P(A \cap B) = P(A).P(B)$ where $P(A) \neq 0$ and $P(B) \neq 0$

MULTIPLICATION LAW

When events A and B are independent then $P(A \cap B) = P(A) \cdot P(B)$ i.e. the probability of both event A and event B happening is given by $P(A \cap B) = P(A) \cdot P(B)$

This is known as the **multiplication law** in probability.

Worked example

A bag contains 5 white balls and 3 red balls. Find the probability of drawing in succession 2 white balls when:
a) the first ball is replaced in the bag
b) the first ball is not replaced in the bag

a) $P(\text{white ball}) = P(W_1) = \dfrac{5}{8}$

If the ball is replaced the probability of drawing a second white ball is still $P(W_2) = \dfrac{5}{8}$

$$\Rightarrow P(W_1, W_2) = \frac{5}{8} \cdot \frac{5}{8} = \frac{25}{64}$$

b) If the first white ball is not replaced the bag will now only contain 4 white balls

$$\Rightarrow \text{probability of second white ball } P(W_2) = \frac{4}{7}$$

$$\Rightarrow P(W_1, W_2) = \frac{5}{8} \cdot \frac{4}{7} = \frac{5}{14}$$

SUMMARY OF RULES

"Note these rules."

1. $A \cup B \equiv$ Either A or B
2. $A \cap B \equiv$ Both A and B
3. $A | B \equiv A$ given that B has already occurred
4. Generally, $P(A \cup B) = P(A) + P(B) - P(A \cap B)$
5. For mutually exclusive events, since $P(A \cap B) = 0$
   $\Rightarrow P(A \cup B) = P(A) + P(B)$
6. $P(A | B) = \dfrac{P(A \cap B)}{P(B)}$
7. Independent events $\Rightarrow P(A \cap B) = P(A) \cdot P(B)$
   and $P(A | B) = P(A)$, $\qquad P(B | A) = P(B)$

Worked example

Find the least number of throws of a die that will ensure a 60% chance of obtaining at least one six

$$\text{Probability of not throwing a six} = \frac{5}{6}$$

$\Rightarrow$ Probability of not obtaining a six with n throws of a die $= \left(\dfrac{5}{6}\right)^n$

$\Rightarrow$ Probability of obtaining at least one six with n throws of a die $= 1 - \left(\dfrac{5}{6}\right)^n = \dfrac{3}{5}$ for a 60% chance

$$\Rightarrow \left(\frac{5}{6}\right)^n = 1 - \frac{3}{5} = \frac{2}{5} \quad \text{or} \quad \left(\frac{6}{5}\right)^n = \left(\frac{5}{2}\right). \text{ Taking logs} \Rightarrow n \lg(1 \cdot 2) = \lg(2 \cdot 5)$$

$$n = \frac{\lg(2 \cdot 5)}{\lg(1 \cdot 2)} = 5 \cdot 03$$

$\Rightarrow$ least number of throws is 6

Worked example

Three boys Alan, Brian and Colin are each to have one shot at a target using long-bows.

The independent probabilities that Alan, Brian and Colin will hit the target are $\dfrac{1}{3}$, $\dfrac{1}{4}$ and $\dfrac{1}{6}$

respectively. Find the probability that a) no boy will hit the target, b) one and only one boy will hit the target.

The events A, B and C are taken as "Alan hits the target", "Brian hits the target" and "Colin hits the target" respectively. We have then

$$P(A) = \frac{1}{3},\; P(A') = \frac{2}{3},\; P(B) = \frac{1}{4},\; P(B') = \frac{3}{4},\; P(C) = \frac{1}{6},\; P(C') = \frac{5}{6}$$

a) P(no boy hits target) $= P(A')\,P(B')\,P(C') = \dfrac{2}{3} \times \dfrac{3}{4} \times \dfrac{5}{6} = \dfrac{5}{12}$

b) P(one and only one hits target) $= P(A)\,P(B')\,P(C') + P(A')\,P(B)\,P(C')$
$$+ P(A')\,P(B')\,P(C)$$

$$= \frac{1}{3} \times \frac{3}{4} \times \frac{5}{6} + \frac{2}{3} \times \frac{1}{4} \times \frac{5}{6} + \frac{2}{3} \times \frac{3}{4} \times \frac{1}{6}$$

$$= \frac{31}{72}$$

Worked example

Two dice are thrown. The event A is "the sum of the numbers on the two dice is a multiple of 3" and the event B is "the difference between the numbers on the two dice is greater than 3". Calculate the values of $P(A)$, $P(B)$, $P(A \cap B)$, $P(A \cup B)$, $P(A|B)$ and $P(B|A)$ Determine whether the events A and B are independent

When two dice are thrown, the sample space consists of 36 simple events which can be written as number pairs (x, y) where x is the number on the first and y the number on the second and both x and y can take any of the values 1, 2, 3, 4, 5, 6. The whole sample space is then shown on Fig. 21.7 as 36 points, where each point represents a simple event.

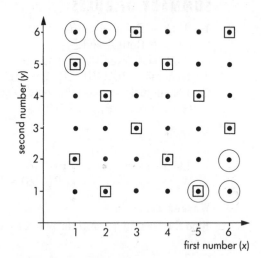

Fig. 21.7

Each point with a square around it is a member of A and each point with a circle around it is a member of B. From this count we see that

$$P(A) = \frac{12}{36} = \frac{1}{3} \quad \text{and} \quad P(B) = \frac{6}{36} = \frac{1}{6}$$

Further, the event $A \cap B$ can be identified as having just the simple events $(5, 1)$ and $(1, 5)$.

Therefore $P(A \cap B) = \dfrac{2}{36} = \dfrac{1}{18}$

Again by direct counting, we see that $P(A \cup B) = \dfrac{16}{36} = \dfrac{4}{9}$

As a check on our work to date, we see that
$$P(A) + P(B) = P(A \cup B) + P(A \cap B)$$

Finally, $P(A|B) = \dfrac{P(A \cap B)}{P(B)} = \dfrac{1/18}{1/6} = \dfrac{1}{3}$

and $P(B|A) = \dfrac{P(A \cap B)}{P(A)} = \dfrac{1/18}{1/3} = \dfrac{1}{6}$

To test whether or not A and B are independent, we calculate $P(A).\,P(B) = \dfrac{1}{3} \times \dfrac{1}{6} = \dfrac{1}{18}$

Since $P(A \cap B) = \dfrac{1}{18}$, we have $P(A).\,P(B) = P(A \cap B)$ and the events are independent

Worked example

The events A and B are members of a sample space such that
$P(A) = 0{\cdot}2$, $P(B|A) = 0{\cdot}7$ and $P(B|A') = 0{\cdot}4$
Calculate the value of a) $P(B)$, b) $P(A \cup B)$, c) $P(A|B)$

a) $P(A) = 0{\cdot}2$ and so $P(A') = 1 - 0{\cdot}2 = 0{\cdot}8$
$$\begin{aligned}
P(B) &= P(A \cap B) + P(A' \cap B) \\
&= P(B|A)\,P(A) + P(B|A')\,P(A') \\
&= 0{\cdot}7 \times 0{\cdot}2 + 0{\cdot}4 \times 0{\cdot}8 \\
&= 0{\cdot}46
\end{aligned}$$

b) $\begin{aligned}[t]
P(A \cap B) &= P(B|A)\,P(A) \\
&= 0{\cdot}7 \times 0{\cdot}2 = 0{\cdot}14
\end{aligned}$
Using $P(A) + P(B) = P(A \cup B) + P(A \cap B)$, we have
$\qquad 0{\cdot}2 \;\; + 0{\cdot}46 \; = P(A \cup B) + 0{\cdot}14$
We have then $P(A \cup B) = 0{\cdot}52$

c) $\begin{aligned}[t]
P(A|B) &= \frac{P(A \cap B)}{P(B)} \\[2mm]
&= \frac{0{\cdot}14}{0{\cdot}46} = 0{\cdot}304 \text{ (3 decimal places)}
\end{aligned}$

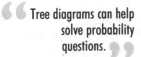

USE OF TREE DIAGRAMS

A very useful notation in solving certain probability questions is that of a **tree diagram** as will be seen in the following example.

Worked example

A bag contains 5 red balls (R), 7 black balls (B) and 4 white balls (W). Two balls are drawn from the bag, one after the other, and without replacement. Find the probability of drawing:
a) 2 balls of the same colour
b) 1 red and 1 black ball

> Tree diagrams can help solve probability questions.

The different ways in which the balls can be drawn from the bag are represented by "the branches of a tree," each branch showing the probability of drawing the ball (R, B or W) indicated at the end of the branch.

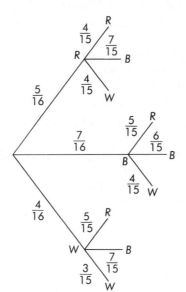

$$P(2R \text{ or } 2B \text{ or } 2W) = \frac{5}{16} \cdot \frac{4}{15} + \frac{7}{16} \cdot \frac{6}{15} + \frac{4}{16} \cdot \frac{3}{15}$$

$$= \frac{20 + 42 + 12}{16.15} = \frac{37}{120}$$

$$P(1R+1B \text{ or } 1B+1R) = \frac{5}{16} \cdot \frac{7}{15} + \frac{7}{16} \cdot \frac{5}{15}$$

$$= \frac{70}{16.15} = \frac{7}{24}$$

Fig. 21.8

EXERCISES 21.3

1   There are 30 men and 40 women at a party. Two people are to be selected at random without replacement. Calculate the probability that the two people will be of the same sex

2   A box contains 9 marbles of which 4 are red and the rest are blue. Three marbles are drawn one at a time and without replacement from this box. Find the probability that 3 marbles of each colour are left in the box

3   Given that A and B are events in a sample space such that

$$P(A) = \frac{1}{3}, \ P(B) = \frac{3}{5} \text{ and } P(A \cup B) = \frac{11}{15}$$

show that A and B are independent
Calculate the values of $P(A|B)$ and $P(B|A)$

4   Two events A and B of a sample space are such that
$P(A) = 0 \cdot 3$, $P(A' \cap B) = 0 \cdot 2$ and $P(A \cap B) = 0 \cdot 25$
Find the values of $P(A \cap B')$, $P(A|B)$ and $P(B|A)$.

5   The events C and D are members of a sample space such that

$$P(C) = x, \ P(D) = \frac{2}{5} \text{ and } P(C \cup D) = \frac{7}{10}. \text{ Find the value of } x \text{ when}$$

i) C and D are independent events
ii) C and D are mutually exclusive events

6   The independent probabilities that Luke, Mark and Nigel can solve a particular puzzle are $\frac{1}{3}$, $\frac{1}{4}$ and $\frac{3}{4}$ respectively. If all three boys try to solve the puzzle, find the probability that:

i) only Nigel solves the puzzle
ii) two, and only two boys, solve the puzzle

7   Three cards are drawn at random, one at a time without replacement, from a pack of 52 playing cards. Find the probability that
i) three Spades will be drawn
ii) three Kings will be drawn
iii) one Heart, one Diamond and one Club will be drawn

8   The bag A contains just 9 cards, numbered 1, 2, ..., 9
The bag B contains just 5 cards, numbered 1, 2, ..., 5
A bag is selected at random and a card is selected at random from the bag. The event E is "bag A is selected" and the event F is "the card selected has an odd number on it". Calculate the values of $P(E)$, $P(F)$, $P(E \cap F)$, $P(E \cup F)$, $P(E|F)$, $P(F|E)$

9   A box contains just 16 balls, of which 7 are green, 5 are blue and 4 are red. Three balls are drawn from the bag, one after another, at random and without replacement. Find the probability that:
a) one ball of each colour will be drawn
b) just one of the balls drawn is blue
c) no red balls are drawn
d) the second ball is green, given that the first drawn was green

10   The events A and B are independent and the events A and C are mutually exclusive.

Given that $P(A) = \frac{2}{5}$, $P(B) = \frac{3}{10}$, $P(C) = \frac{1}{2}$ and $P(A \cup B \cup C) = 1$, find the values of

a) $P(A \cap B)$,   b) $P(B \cap C)$,   c) $P(B|C)$

NB: Exam Board Questions can be found in Chapter 27.

ANSWERS TO EXERCISES

EXERCISES 21.1

1   360, 120, 42, 4
2   5040;   a) 24;   b) 2520
3   5040, 288
4   $\dfrac{11!}{(4!)^2 2!}$, $\dfrac{14!}{5!\,3!\,(2!)^2}$
5   11!, 10!

EXERCISES 21.2

1   15, 1, 21, 56, $\dfrac{14!}{10!\,4!}$, $\dfrac{22!}{10!\,12!}$
2   35, 15
4   350
5   9.99×10^{10}
6   140
7   495
8   a) 9702;   b) 39 897

EXERCISES 21.3

1   $\dfrac{81}{161}$
2   $\dfrac{10}{21}$
3   $\dfrac{1}{3}$, $\dfrac{3}{5}$
4   0.05, $\dfrac{5}{9}$, $\dfrac{5}{6}$
5   i) $\dfrac{1}{2}$;   ii) $\dfrac{3}{10}$
6   i) $\dfrac{3}{8}$;   ii) $\dfrac{1}{3}$
7   i) $\dfrac{11}{850}$;   ii) $\dfrac{1}{5525}$;   iii) $\dfrac{169}{1700}$
8   $\dfrac{1}{2}$, $\dfrac{26}{45}$, $\dfrac{5}{18}$, $\dfrac{4}{5}$, $\dfrac{25}{52}$, $\dfrac{5}{9}$
9   a) $\dfrac{1}{4}$;   b) $\dfrac{55}{112}$;   c) $\dfrac{11}{28}$;   d) $\dfrac{6}{15}$
10   a) $\dfrac{3}{25}$;   b) $\dfrac{2}{25}$;   c) $\dfrac{4}{25}$

PROBABILITY FUNCTIONS OF A DISCRETE RANDOM VARIABLE

GETTING STARTED

Many, but not all, of the events in problems of probability result in a sample space S that is a set of numerical values. E.g. the throwing of a die results in a sample space of 1, 2, 3, 4, 5 and 6. The drawing of a ball from a bag containing black, white and red balls however does not. It would though be possible by assigning numbers, say 1 to black, 2 to white and 3 to red, to make the outcome numerical.

This is particularly useful in probability. By assigning a real number x_r to each event E_r in the sample S we then have a function X defined at all points of the sample space. (i.e. X takes the value x_r when event E_r occurs.) The function X is called a random variable. When the number of possible values of X is finite then we define X to be a discrete random variable.

Consider the discrete random variable X which takes values x_1, x_2, ..., x_n when events E_1, E_2, ..., E_n occur respectively. Then if the probability that $X = x_r$, i.e. $P(X = x_r) = p(x_r)$, we define $p(x)$ where $P(X = x) \equiv p(x)$ as the probability function of the discrete random variable X.

We further define $F(x_r) = P(X \leqslant x_r) = \sum_{r=1}^{r} p(x_r)$ as the distribution function $F(x)$ of the discrete random variable X – it is the cumulative probability function of X from the lowest value of X, i.e. x_1 up to and including the value x_r.

ESSENTIAL PRINCIPLES

SOLUTIONS TO PROBABILITY FUNCTION QUESTIONS

Worked example

Show graphically the probability function of the discrete random variable X where X is the number of heads appearing when an unbiased coin is tossed twice in succession.

The sample space has 4 events $\{TT, TH, HT, HH\}$

$\Rightarrow$ numerically the number of heads obtained being $\{0, 1, 2\}$

and the probability function of the discrete random variable X, the number of heads obtained is therefore given in the table.

$X = x$	0	1	2
$P(X = x)$	$\dfrac{1}{4}$	$\dfrac{1}{2}$	$\dfrac{1}{4}$

It can be displayed graphically as shown in Fig. 22.1

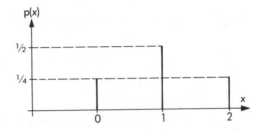

Fig. 22.1

Note: the sum of the length of the bars in the graph represents the sum of the probabilities of all the possible outcomes $= 1$, (certainty), i.e. $\displaystyle\sum_{r=1}^{3} \mathrm{p}(x_r) = 1$

Worked example

The probability function of a discrete random variable X is given by

$$\mathrm{p}(x) = \frac{K}{3^x} \quad x = 1, 2, 3 \ldots$$

Find a) the value of K b) $P(X \leqslant 4)$

a) $\displaystyle\sum_{x=1}^{\infty} \mathrm{p}(x) = 1 \Rightarrow 1 = \sum_{x=1}^{\infty} \frac{K}{3^x} = K\left(\frac{1}{3} + \frac{1}{3^2} + \frac{1}{3^3} + \ldots\right),$

an infinite geometrical progression.

Hence $\displaystyle\sum_{x=1}^{\infty} \frac{K}{3^x} = \frac{K \cdot \frac{1}{3}}{1 - \frac{1}{3}} = 1 = \frac{K}{2} \Rightarrow K = 2$

b) $P(X \leqslant 4) = 2 \cdot \left(\dfrac{1}{3} + \dfrac{1}{3^2} + \dfrac{1}{3^3} + \dfrac{1}{3^4}\right) = 2 \cdot \dfrac{(27 + 9 + 3 + 1)}{81} = \dfrac{80}{81}$

MEAN AND VARIANCE OF A DISCRETE PROBABILITY FUNCTION

For a discrete random variable X the mean, μ, sometimes called the *expected value* or the expectation of X, is defined to be:

$$\mu = \mathrm{E}(X) = \sum_{r=1}^{n} x_r \mathrm{p}(x_r) \text{ i.e. summed over all possible value of } X$$

The variance σ^2, of X is defined to be:

$$\sigma^2 = \mathrm{Var}(X) = \sum_{r=1}^{n} (x_r - \mu)^2 \, \mathrm{p}(x_r) \text{ which can be shown to be}$$

$$= \sum_{r=1}^{n} x_r^2 \mathrm{p}(x_r) - \mu^2$$

The standard deviation of the probability is the square root of the variance, i.e. σ.

Worked example

A discrete random variable X takes values x only from the set N, where $N = \{0, 1, 2, 3\}$. The probability that X takes the value x is given by

$$P(X = x) = k(x + 1)$$

where k is a constant.

a) Determine the value of k

b) Find the mean and the variance of X

From the data given, we can make up a table:

$X = x$	0	1	2	3
$P(X = x)$	k	$2k$	$3k$	$4k$

a) Since $\sum P(X = x) = 1$, we have

$$k + 2k + 3k + 4k = 1 \Rightarrow k = \frac{1}{10}$$

b) Mean $= \mu = \sum_{i=0}^{4} x_i\, p(x_i)$

$$= 0 + 2k + 6k + 12k = 20k$$

But $k = \dfrac{1}{10}$ and therefore $\mu = 2$

$$\text{Var}(X) = \sigma^2 = \sum_{i=1}^{4} x_i^2\, p(x_i) - \mu^2$$
$$= 0 + 2k + 12k + 36k - 2^2$$
$$= 50k - 4 = 5 - 4 = 1$$

Worked example

Consider another discrete random variable Y, where $Y = 3X + 2$ and X is the random variable with mean 2 and variance 1, as given in the last example.

As X takes values 0, 1, 2, 3, Y takes values 2, 5, 8, 11 respectively and the probability distribution for Y in tabular form is

Y	2	5	8	11
$P(Y)$	0·1	0·2	0·3	0·4

The mean of Y, $\mu_Y = 0{\cdot}2 + 1{\cdot}0 + 2{\cdot}4 + 4{\cdot}4 = 8$

The variance of Y, $\sigma_Y^2 = 4(0{\cdot}1) + 25(0{\cdot}2) + 64(0{\cdot}3) + 121(0{\cdot}4) - 8^2$
$$= 73 - 64 = 9$$

In this example we note that $\mu_Y = 3(\mu_X) + 2$ and $\sigma_Y^2 = 3^2 \sigma_X^2$

Generally, if $Y = aX + b$, where a and b are constants, it is true that

i) Mean of $Y = a(\text{Mean of } X) + b$

ii)   Var $(Y) = a^2\, \text{Var}(X)$

Result i) is also often written as $E(Y) = a E(X) + b$

Note also that the standard deviation of $Y = a.(\text{the standard deviation of } X)$.

EXERCISES 22.1

1   From prolonged observation at a snack-bar it was recorded that the probabilities of finding 0, 1, 2, 3, 4, 5 customers waiting to be served were 0·12, 0·18, 0·29, 0·23, 0·15, 0·03 respectively.

a) Draw a graph to illustrate this discrete probability distribution

b) Calculate the expected number of customers waiting to be served

c) Calculate the variance of the number of customers waiting to be served

2   A discrete random variable X takes values x, where $x \in N$ and $N = \{1, 2, 3, \ldots, n.\}$
The probability that X takes the value x is given by

$$P(X = x) = \frac{1}{n}$$

Show that the expected value of X is $\dfrac{1}{2}(n + 1)$ and that the variance

of X is $\dfrac{1}{12}(n^2 - 1)$

Note: this discrete probability distribution is sometimes called the rectangular distribution or the uniform distribution. In your solution, you need to use the following formulae from your work in pure mathematics and you may quote these as required:

$$\sum_{r=1}^{n} r = \frac{1}{2}n(n+1) \text{ and } \sum_{r=1}^{n} r^2 = \frac{1}{6}n(n+1)(2n+1)$$

3   The probability distribution of the discrete random variable X is given in the following table:

X	−1	0	1	2	3	4
$P(X)$	0·2	0·3	0·2	0·1	0·1	0·1

Find the expected value of X, denoted by $E(X)$ and the variance of X, denoted by $Var(X)$.

The random variables Y and Z are related to X such that

$$Y = \frac{1}{2}X \text{ and } Z = 4X - 3$$

Determine values for $E(Y)$, $E(Z)$, $Var(Y)$ and $Var(Z)$

4   The probability distribution of the discrete random variable X is given in the table

X	1	2	3
$P(X)$	a	b	c

Given that $E(X) = 1\cdot8$ and $Var(X) = 0\cdot76$, find the values of a, b, c

We now look at a number of examples of discrete probability distributions.

RECTANGULAR DISTRIBUTION

The tossing of an unbiased die is an example of a discrete rectangular distribution since the probability of throwing a 1, 2, 3, 4, 5 or 6 is in each case $\frac{1}{6}$. The graph of the probability distribution is therefore *rectangular* in appearance.

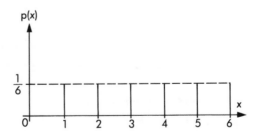

Fig. 22.2

The mean of the distribution is $\mu = E(X) = \frac{1}{6}(1+2+3+4+5+6)$

$$\Rightarrow E(X) = \frac{21}{6} = 3\cdot5$$

The variance is $\sigma^2 = Var(X) = \frac{1}{6}(1^2+2^2+3^2+4^2+5^2+6^2) - 3\cdot5^2 = \frac{35}{12}$

Generally if $p(x_r) = \frac{1}{n}$ a constant, and $X = r$ for $r = 1, 2 \ldots n$

then $\mu = E(X) = \frac{n+1}{2}$

and $\sigma^2 = Var(X) = \frac{n^2-1}{12}$

BINOMIAL DISTRIBUTION

> *Here there are only two possible outcomes.*

In the *binomial distribution* we are concerned with an experiment which requires repetition of the same test a specific number of times, say, n. Each test is independent of all of the other tests, but each test only has two possible outcomes, one which we will call a success (S), the other which we will call a failure (F). We will let $P(S) = p$ then $P(F) = q = 1 - p$ since the sum of the two probabilities must be 1.

Let us now consider the probability of the n tests resulting in r successes and $n - r$ failures.

The probability of r tests with r successes $= p^r$

The probability of $n - r$ tests with $n - r$ failures $= q^{n-r}$

$\Rightarrow$ Probability of r successes followed by $n - r$ failures $= p^r q^{n-r}$

However if the r successes can occur randomly within the group of n tests then there will be ^nC_r or $\binom{n}{r}$ ways in which we can choose the r places for the successes within the n tests. Hence this demonstrates that the probability of r successes in a set of n tests is $^nC_r p^r q^{n-r}$

i.e. $P(X = r) = {}^nC_r p^r q^{n-r} = \binom{n}{r} p^r q^{n-r} = \binom{n}{r} p^r (1-p)^{n-r}$

$$\text{for } r = 0, 1, 2, \ldots, n$$

You should note that these probabilities are the $(n+1)$ consecutive terms of the binomial expansion of $(p+q)^n$; a fact which confirms that the total sum of the probabilities is 1 since $(p+q)^n = 1^n = 1$.

This distribution is therefore called the binomial distribution. It is defined formally as follows.

The discrete random variable X having a probability function:

$$P(X = r) \equiv P(r) = \binom{n}{r} p^r (1-p)^{n-r}, \text{ where } 0 \leqslant p \leqslant 1 \text{ and } r = 0, 1, 2, \ldots, n,$$

is said to have a binomial distribution $B(n, p)$. It is often written as $X \sim B(n, p)$ i.e. X is distributed binomially with n independent tests, each test having a constant probability p of success, and $(1-p)$ of failure.

The mean, μ, or expected value of X, $E(X)$ is np

The variance $\sigma^2 = \text{Var}(X)$ is $npq = np(1-p)$

Worked example

Calculate the probability of throwing either a 3 or a 4 six times in ten throws of an unbiased die

$P(3 \text{ or } 4) = \dfrac{2}{6} = \dfrac{1}{3} = p \Rightarrow q = 1 - p = 1 - \dfrac{1}{3} = \dfrac{2}{3}$

$P(X = 6) = \binom{10}{6} \cdot p^6 q^4 = \binom{10}{6} \left(\dfrac{1}{3}\right)^6 \left(\dfrac{2}{3}\right)^4 \approx 0.057$

Worked example

Out of a large batch of components it is known that 20% are defective. Ten components are to be selected at random. Estimate the probability that, of the 10 components,
a) none will be defective
b) at most two will be defective
We take X as the random variable representing the number of defective components and we then write $X \sim B(10, 0.2)$

a) $P(X = 0) = p(0) = \binom{10}{0}(0.2)^0(0.8)^{10} = (0.8)^{10} \approx 0.107$

b) $P(\text{at most 2 defective}) = P(X \leqslant 2) = p(0) + p(1) + p(2)$

$$= (0.8)^{10} + \binom{10}{1}(0.8)^9(0.2)^1 + \binom{10}{2}(0.8)^8(0.2)^2$$
$$= (0.8)^{10} + 10(0.8)^9(0.2) + 45(0.8)^8(0.2)^2$$
$$\approx 0.68$$

Note: when we are told that the number of components in a population is large, we assume that each time we select a component, the probability that it will be defective remains constant from trial to trial.

We take the probability, on any trial, for a defective item to be selected as $0 \cdot 2$ and the probability for a non-defective item to be selected as $0 \cdot 8$. Further, these are the only possible outcomes, which we call "a success" and "a failure" respectively. We set up the probability function of X as a binomial distribution, as shown, and we are then able to make estimates for the probabilities of specific events by choosing relevant values for X.

Worked example

A certain binomial distribution has mean 4 and variance 3. Find, for this distribution, the probability of obtaining exactly 2 successes

Taking n as the size of the sample and p as the probability of success in a single trial, we have

$$np = 4 \qquad\qquad (1)$$
$$np(1-p) = 3 \qquad\qquad (2)$$

Solving (1) and (2) we have $p = 0 \cdot 25$ and $n = 16$

$$P(2 \text{ successes}) = \binom{n}{2} p^2 (1-p)^{14} = \binom{16}{2}(0 \cdot 25)^2 (0 \cdot 75)^{14}$$

$$= 120 (0 \cdot 25)^2 (0 \cdot 75)^{14} \approx 0 \cdot 13$$

EXERCISES 22.2

1   A certain game can be lost or won only. The probability of a win for player A when playing player B at this game is $0 \cdot 6$. They plan to play a series of 5 games. Calculate the probability that
   a) A will win 4 games
   b) B will win 3 games

2   From a large batch of light-bulbs the probability that a bulb will fail in the first month of use is $\dfrac{1}{15}$. A random sample of 6 bulbs is to be selected. Calculate that in the first month of use of these 6 bulbs
   a) none will fail
   b) one will fail
   c) at least 2 will fail

3   An insurance agent sells life policies to 6 young women who are all aged 25 years and in good health. According to the records held by the agent's company, the probability of a women aged 25 years being alive in 30 years time is $0 \cdot 8$. Find the probability that in 30 years time for the group of 6 women
   a) all will be alive
   b) exactly 4 will be alive
   c) at least 4 will be alive

4   It is known that 20% of the potatoes in a large stock are rotten. A random sample of 12 potatoes is to be taken. Find the probability that this sample will contain
   a) 3 rotten potatoes
   b) at most 3 rotten potatoes
   A larger random sample of 400 potatoes is to be taken. For this sample estimate
   c) the expected number of rotten potatoes
   d) the standard deviation

5   A large number of clay pots is moulded and fired. After firing the pots are checked and 15% are found to have flaws. A random sample of 8 pots is taken from a large number of pots produced under identical conditions. Estimate the probability that this sample will contain
   a) at least 2 pots with flaws
   b) at most 2 pots with flaws

POISSON DISBRIBUTION

The arithmetical calculations involved in probability problems concerned with the binomial distribution $B(n, p)$ is very considerable when n is large. It can be eased by using the **Poisson distribution** which gives a good approximation to the binomial distribution for large values of n and small values of p. E.g. $n \geqslant 20$, $p \leqslant 0 \cdot 05$ – the larger the value of n the closer the approximation.

> **You can use this for large values of n and small values of p.**

The Poisson distribution states that for a discrete random variable X, the probability function is

$$P(X = r) = p(r) = \frac{\lambda^r e^{-\lambda}}{r!} \quad r = 0, 1, 2, \ldots, \text{ where } \lambda \text{ is a positive constant.}$$

We write it as $X \sim \text{Poi}(\lambda)$

The mean μ of the distribution is $\mu = E(X) = \lambda = np$

The variance σ^2 of the distribution is $\sigma^2 = \text{Var}(X) = \lambda$

Worked example

It is believed that on average 0·8% of the screws being produced by a machine are defective. Estimate the probability, to 3 decimal places, that a sample of 250 screws will contain a) no defective screws, b) not more than 3 defective screws

$p = 0{\cdot}008 \quad n = 250 \Rightarrow np = \lambda = 2$

$X \sim B(250, 0{\cdot}008)$ which can be approximated by $X \sim \text{Poi}(2)$

$\Rightarrow P(r = 0) = e^{-2} = 0{\cdot}135$ (to 3 decimal places)

i.e. P (No defective screws in the sample) $= 0{\cdot}135$ (to 3 decimal places)

$$P(X \leqslant 3) = p(0) + p(1) + p(2) + p(3) = e^{-2} + \frac{2e^{-2}}{1!} + \frac{2^2 e^{-2}}{2!} + \frac{2^3 e^{-2}}{3!}$$

$$= e^{-2}\left(1 + 2 + 2 + \frac{4}{3}\right) = \frac{19}{3}e^{-2} = 0{\cdot}857 \text{ (to 3 decimal places)}$$

We have just shown how a Poisson distribution can be used as an approximation to a binomial distribution, $B(n, p)$, provided n is large and p is small. However the Poisson distribution may, under certain conditions represent the distribution of a random variable X, where X is the number of events that happen in an interval of fixed length, e.g. the number of red cars passing a fixed point on a road in a 5-minute interval. The conditions which are necessary for the Poisson distribution to model X are that the events should happen singly, independently, uniformly and at random in a continuous space or time.

Worked example

The telephone calls received in a busy office follow a Poisson distribution with a mean of 2·1 per minute. Find the probability that in a given minute there will be

a) 2 calls

b) at least three calls

a) $\lambda = 2{\cdot}1 \Rightarrow P(X = 2) = \frac{2{\cdot}1^2 e^{-2{\cdot}1}}{2!} = 0{\cdot}2700$

b) $P(X = 0) = e^{-2{\cdot}1}$

$P(X = 1) = \frac{2{\cdot}1 e^{-2{\cdot}1}}{1!}$

$\Rightarrow P(X \geqslant 3) = 1 - P(X = 0) - P(X = 1) - P(X = 2)$

$= 1 - e^{-2{\cdot}1}\left(1 + 2{\cdot}1 + \frac{2{\cdot}1^2}{2}\right)$

$= 1 - (0{\cdot}1225 + 0{\cdot}2572 + 0{\cdot}2700) = 0{\cdot}3503$

GEOMETRIC DISTRIBUTION

If we are concerned with an experiment which requires repetition of the same test a specific number of times, and that each test results in failure except for the last one which is a success, then we obtain a **geometric probability distribution**. As with the binomial distribution each test is independent of the others and the probability of a successful outcome is constant and equal to p.

$\Rightarrow$ Probability of $(r-1)$ tests failing in succession $= q^{r-1} = (1-p)^{r-1}$

$\Rightarrow$ Probability of $(r-1)$ tests failing in succession followed by a test which is successful $= (1-p)^{r-1}p$

Hence defining X as the number of tests required up to and including a success, $P(X = r) = (1-p)^{r-1}p$.

Thus a discrete random variable X with probability function $P(X = r) = (1-p)^{r-1}p$ where $0 \leqslant p \leqslant 1$, $r = 1, 2, \ldots$, is known as a geometric distribution with parameter p. It is often written as $X \sim G(p)$

The mean μ or expected value of X, $E(X)$ is $\dfrac{1}{p}$

The variance $\sigma^2 = \text{Var}(X) = \dfrac{1-p}{p^2}$

Worked example

A fair die is to be thrown repeatedly until a 6 appears. Find
a) the probability that a 6 will appear on the fourth throw
b) the minimum number of throws n required in order to ensure that the probability that a 6 will appear on or before the nth throw is greater than $\dfrac{3}{4}$

a) Probability of obtaining a 6 with a throw $= \dfrac{1}{6}$

$$\Rightarrow P(X=4) = \left(\frac{5}{6}\right)^3 \frac{1}{6} = 0{\cdot}0965$$

b) $P(X \leqslant n) = P(X=1) + P(X=2) + P(X=3) \ldots + P(X=n)$

$$= \frac{1}{6} + \frac{5}{6}\cdot\frac{1}{6} + \left(\frac{5}{6}\right)^2\left(\frac{1}{6}\right) + \ldots + \left(\frac{5}{6}\right)^{n-1}\left(\frac{1}{6}\right)$$

$$= \frac{1}{6}\left[1 + \left(\frac{5}{6}\right) + \left(\frac{5}{6}\right)^2 + \ldots + \left(\frac{5}{6}\right)^{n-1}\right]$$

$$= \frac{1}{6}\, \frac{\left[1-\left(\frac{5}{6}\right)^n\right]}{1-\left(\frac{5}{6}\right)} = 1 - \left(\frac{5}{6}\right)^n$$

$$\Rightarrow 1 - \left(\frac{5}{6}\right)^n > \frac{3}{4} \Leftrightarrow \left(\frac{5}{6}\right)^n < 0{\cdot}25 \text{ or } \left(\frac{6}{5}\right)^n > 4$$

$n\lg(1{\cdot}2) > \lg 4$, $n > 7{\cdot}604$

i.e. a minimum of 8 throws is required

EXERCISES 22.3

1 A pair of unbiased dice are thrown 20 times. Find
   i) by using a binomial distribution,
   ii) by using a Poisson approximation
   the possibility that "double 6" will be obtained 3 times out of the 20. Explain why this data is appropriate for using a Poisson distribution as an approximation of the binomial distribution

2 Over a long period it is found that 0·2% of micro-chips produced by a small factory are defective. A random sample of 200 micro-chips are taken. Find out the probability that all the 200 are non-defective. Find also the probability that less than 3 micro-chips are non-defective

3 All the adult sheep at a large farm are immunised and it is known that 1·5% of the sheep are likely to become very ill as a result. There are 120 adult sheep in one field. Estimate the probability that at least 2 sheep in the field are likely to become very ill as a result of immunisation. Estimate also the probability that more than 4 of these sheep become very ill as a result of immunisation.

4 The number of minor accidents in a school year of 200 days was recorded.

Accidents per day	0	1	2	3	4	5 or more
Frequency	128	52	14	4	2	0

Calculate the mean and the variance of the number of accidents per day. Taking your mean as the parameter of a Poisson distribution, calculate the probability for 0, 1, 2, 3, 4, 5 accidents on a random day at the school

5   The independent discrete random variables X and Y have Poisson distributions with parameters a_1 and a_2 respectively. Show that $X + Y$ has a Poisson distribution with parameter $a_1 + a_2$

6   A company has two buildings, each having an external independent telephone switchboard, A and B. A security officer, in each building at night, answers calls to A and B. The number of calls to A is a Poisson distribution with parameter 5 and independently the number of calls to B is a Poisson distribution with parameter 3. Find, to 3 significant figures, the probability that on a night chosen at random there will be:
a) just 4 calls to A
b) at least 3 calls to B
c) 5 calls and only 5 calls to A and B together

7   A discrete random variable X has a geometric distribution defined by
$$P(X = x) = \left(\frac{3}{4}\right)^{x-1}\left(\frac{1}{4}\right) \quad x = 1, 2, 3 \dots$$
a) Write down the expected value and the variance of X
b) Calculate the values of i) $P(X = 4)$,
                            ii) $P(X < 4)$

8   A discrete random variable X has a geometric distribution with expected value 5. Write down the variance of X and the probability distribution of X in the form
$$P(X = x) = (1 - \lambda)^{x-1}\lambda$$
stating the value of λ
Find the value of
a) $P(X = 3)$, b) $P(X \geqslant 4)$

NB: Examination questions on these topics can be found in Chapter 27.

ANSWERS TO EXERCISES

EXERCISES 22.1

1   b) 2·2; c) 1·72
3   0·9, 2·49; 0·45, 0·6, 0·623, 39·84
4   0·5, 0·2, 0·3

EXERCISES 22.2

1   a) 0·259; b) 0·230
2   a) 0·661; b) 0·283; c) 0·056
3   a) 0·262; b) 0·246; c) 0·901
4   a) 0·236; b) 0·794; c) 80; d) 8
5   a) 0·343; b) 0·895

EXERCISES 22.3

1   i) 0·015; ii) 0·016
2   0·67, 0·992
3   0·537, 0·036
4   0·5, 0·63; 0·607, 0·303, 0·076, 0·013, 0·0016, 0·00016
6   a) 0·175; b) 0·577; c) 0·092
7   a) $E(X) = 4$, $Var(X) = 12$; b) 0·105, 0·578

8   $Var(X) = 20$, $\lambda = \dfrac{1}{5}$, $\dfrac{1}{5}\left(\dfrac{4}{5}\right)^{x-1}$;

    a) 0·128; b) 0·512

PROBABILITY FUNCTIONS OF A CONTINUOUS RANDOM VARIABLE

CHAPTER

SOLUTIONS TO CONTINUOUS RANDOM VARIABLE QUESTIONS

DISTRIBUTION FUNCTIONS OF A CONTINUOUS RANDOM VARIABLE

NORMAL DISTRIBUTION

GETTING STARTED

For a continuous probability distribution, the random variable may take any value between specified limits or just any real value. The number of possible values that could be taken by the random variable is infinite, and we cannot speak of the probability that the variable will take some specific value. Instead we consider ways of finding the probability that the value of the random variable will be in a particular interval: e.g. if we are measuring the lengths of leaves to the nearest mm and we wish to find the probability that a given leaf will measure 8mm, we are really asking "what is the probability of the leaf measuring between 7.5mm and 8.5mm?" since length is a continuous variable and we are measuring to the nearest mm.

The probability density function (pdf) of a continuous variable X is defined as the function f that satisfies the conditions

i) $f(x) \geq 0$ for all $x \in S$, the sample space,

ii) $\int_S f(x)dx = 1$

where $\int_S$ means integration over the sample space

iii) $P(a \leq X \leq b) = \int_a^b f(x)dx$ for any $a < b$ in S,

and hence $P(a \leq X \leq b)$ is represented by the area of the region under the graph of the pdf $f(x)$, between limits a and b. When we are given a problem in which X takes values only in a finite interval, we assume that the pdf is zero everywhere else. We can then write

$$\int_{-\infty}^{+\infty} f(x)dx = 1 \text{ (the certain event)}$$

You should note that if we let a approach b, then, in the limit

$$\int_a^b f(x)dx$$

becomes zero. This means that, where we can speak of the probability that the value of a discrete random variable is exactly b, there is no equivalent to this for a continuous variable.

The mean μ, often called the expected value or the expectation of X, is defined to be

$$E(X) = \mu = \int_{-\infty}^{+\infty} xf(x)dx$$

The variance σ^2 of X is defined to be

$$Var(X) = \int_{-\infty}^{+\infty} (x-\mu)^2 f(x)dx$$

$$= \int_{-\infty}^{+\infty} x^2 f(x)dx - 2\mu \int_{-\infty}^{+\infty} xf(x)dx + \mu^2 \int_{-\infty}^{+\infty} f(x)dx$$

$$= \int_{-\infty}^{+\infty} x^2 f(x)dx - \mu^2, \text{ from the results above}$$

(Compare this last result with the analogous formula for discrete probability functions in the last chapter.)

ESSENTIAL PRINCIPLES

Worked example

A continuous random variable X takes all real values x in the interval $0 \leqslant x \leqslant 3$. The probability density function f of X is given by

$$f(x) = \frac{1}{9}(3 - 4x + 2x^2) \text{ for } 0 \leqslant x \leqslant 3$$

$f(x) = 0$, otherwise

Sketch the graph of f and calculate the mean and the variance of X. Find the probability that X lies in the interval $0 \leqslant x \leqslant 1$.

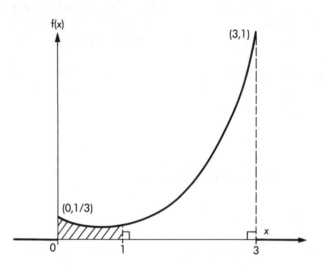

Fig. 23.1

The graph of f is shown in Fig. 23.1 as the curve $(0 \leqslant x \leqslant 3)$, the x-axis to the right of $(3, 0)$ and the whole negative x-axis to the left of O

$$\text{Mean of } X = \int_0^3 \frac{x}{9}(3 - 4x + 2x^2)\mathrm{d}x$$

$$= \left[\frac{x^2}{6} - \frac{4x^3}{27} + \frac{x^4}{18}\right]_0^3$$

$$= (3/2 - 4 + 9/2) = 2$$

$$\text{Variance of } X = \int_0^3 \frac{1}{9}(x - 2)^2(3 - 4x + 2x^2)\mathrm{d}x$$

$$= \int_0^3 \frac{1}{9}(12 - 28x + 27x^2 - 12x^3 + 2x^4)\mathrm{d}x$$

$$= \frac{1}{9}\left[12x - 14x^2 + 9x^3 - 3x^4 + 2x^5/5\right]_0^3$$

$$= \frac{1}{9}(36 - 126 + 243 - 243 + 486/5)$$

$$= 0 \cdot 8$$

$$P(0 \leqslant x < 1) = \int_0^1 \frac{1}{9}(3 - 4x + 2x^2)\mathrm{d}x \quad \text{(this is the area of the shaded region shown)}$$

$$= \left[\frac{1}{9}(3x - 2x^2 + 2x^3/3)\right]_0^1$$

$$= \frac{1}{9}(3 - 2 + 2/3) = \frac{5}{27} \approx 0 \cdot 185$$

FURTHER NOTES ON THIS EXAMPLE

The graph of f is sketched as for an algebraic function, but remember that no part of the curve for a probability density function can lie below the x-axis. The calculation for σ^2 can be much shortened by using the result

$$\sigma^2 = \int x^2 f(x) dx - \mu^2$$

$$= \int_0^3 \frac{1}{9} x^2 (3 - 4x + 2x^2) dx - 2^2$$

$$= \left[\frac{1}{9} x^3 - \frac{1}{4} x^4 + \frac{2}{45} x^5 \right]_0^3 = (3 - 9 + \frac{54}{5}) - 4 = 0{\cdot}8$$

Worked example

A random variable X has probability density function f given by

f$(x) = Kx^a$, $0 \leqslant x \leqslant 1$

f$(x) = 0$, otherwise,

where K and a are constants.

Calculate, giving your answers in terms of a

a) K

b) the expected value of X

a) Since f is a probability density function

$$\int_0^1 Kx^a dx = 1$$

$$\Rightarrow \left[\frac{K}{a+1} x^{a+1} \right]_0^1 = 1$$

$$\frac{K}{a+1} = 1 \quad \Rightarrow \quad K = a+1$$

b) E$(X) \int_0^1 x(Kx^a) dx = (a+1) \int_0^1 x^{a+1} dx$

$$= (a+1) \left[\frac{x^{a+2}}{a+2} \right]_0^1 = \frac{a+1}{a+2}$$

<table>
<tr><td>

DISTRIBUTION FUNCTIONS OF A CONTINUOUS RANDOM VARIABLE

</td><td>

Consider again the continuous random variable X with probability density function f given by

$$f(x) = \frac{1}{9} (3 - 4x + 2x^2), \ 0 \leqslant x \leqslant 3$$

f$(x) = 0$, otherwise

</td></tr>
</table>

The function F given by

$$F(x) = 0, \ x < 0$$

$$F(x) = \frac{1}{9} \left[3x - 2x^2 + \frac{2}{3} x^3 \right], 0 \leqslant x \leqslant 3$$

$$F(x) = 1, \ x > 3$$

is related to f by the equation

$$\frac{dF}{dx} = f$$

In this specific case we could also write

$$F(x) = \int_0^x f(x) dx \text{ for } 0 \leqslant x \leqslant 3$$

F is called the *distribution function* (or the cumulative probability function) of the random variable X

We now define F, (using the same convention that we had previously, that the pdf is zero where it is not defined, thus:

$$F(b) = P(X \leqslant b) = \int_{-\infty}^b f(x) dx$$

This is represented by the area of the graph of the pdf from $x = -\infty$ to $x = b$. Hence F(x) obviously increases from zero at $x = -\infty$ to unity at $x = +\infty$

From this definition we see that

$$\frac{dF(x)}{dx} = f(x)$$

since $F(b)$ is the integral of $f(x)$ from $x = -\infty$ to $x = b$

When writing solutions you will find the following useful:

a) When f is defined and F is asked for, use integration taking particular care over the limits

b) When F is defined and f is asked for, use differentiation

c) Read every question very carefully. Many students mix up the facts given in a) and b).

Worked example

The continuous random variable X has distribution function F given by

$$F(x) = \begin{cases} 0 & \text{for } x < 0 \\ 4x^{\frac{3}{2}} - 3x^2 & \text{for } 0 \leqslant x \leqslant 1, \\ 1 & \text{for } x > 1 \end{cases}$$

a) Sketch the graph of F

b) Find the probability density function f of X and sketch its graph

c) Determine the modal value of X

d) Show that m, the median value of X, satisfies the equation

$$4m^{\frac{3}{2}} - 3m^2 = 0 \cdot 5 \text{ and verify that } m \approx 0 \cdot 38.$$

e) Find E (X) and Var (X)

a)

F(x)

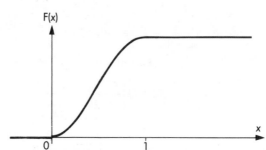

Fig. 23.2

The graph of F in Fig. 23.2 consists of the negative x axis, the curve shown between $x = 0$ and $x = 1$ and the line from $(1, 1)$ parallel to $0x$. The gradient of the curve at $(0, 0)$ and $(1, 1)$ is zero.

b) The probability density function f of X is given by

$$f(x) = \begin{cases} 0 & \text{for } x < 0 \\ 6(x^{\frac{1}{2}} - x) & \text{for } 0 \leqslant x \leqslant 1, \\ 0 & \text{for } x > 1. \end{cases}$$

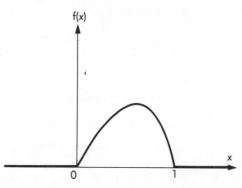

$$\left(\text{Check for yourself that this makes } \frac{dF}{dx} = f \right)$$

f(x)

Fig. 23.3

The graph of f in Fig. 23.3 consists of the negative x-axis, that part of the curve $y = 6(x^{\frac{1}{2}} - x)$ between $(x = 0$ and $x = 1)$ and the positive x-axis for $x > 1$.

c) From the graph of f, we can see that the *modal* value of X occurs between $x = 0$ and $x = 1$ for the value of x at which $f'(x) = 0$

That is $f'(x) = 6\left(\frac{1}{2}x^{-\frac{1}{2}} - 1\right) = 0 \Rightarrow x = \frac{1}{4}$

The modal value of x is $\frac{1}{4}$

d) The median value of X occurs between $x = 0$ and $x = 1$ for the value of x at which $f(x) = \frac{1}{2}$. That is, the line parallel to $x = 0$ which bisects the total area in the graph of f. Taking the equation of this line as $x = m$, we see that

$4m^{\frac{3}{2}} - 3m^2 = \frac{1}{2}$

By direct verification $m \approx 0.38$

e) $E(X) = \int_0^1 xf(x)dx = \int_0^1 6x(x^{\frac{1}{2}} - x)dx$

$= \int_0^1 (6x^{\frac{3}{2}} - 6x^2)dx$

$= \left[\frac{12}{5}x^{\frac{5}{2}} - 2x^3\right]_0^1 = 0.4$

$Var(X) = \int_0^1 x^2 f(x)dx - [E(X)]^2$

$= \int_0^1 (6x^{\frac{5}{2}} - 6x^3)dx - (0.4)^2$

$= \left[\frac{12}{7}x^{\frac{7}{2}} - \frac{3}{2}x^4\right]_0^1 - (0.4)^2$

$= \frac{12}{7} - \frac{3}{2} - (0.4)^2 = 0.054$

EXERCISES 23.1

1 The continuous random variable X has probability density function f given by

$f(x) = \begin{cases} 0 & \text{for } x<0 \\ ax^2 + \frac{3}{4}x & \text{for } 0 \leqslant x \leqslant 2 \\ 0 & \text{for } x>2 \end{cases}$

Calculate the value of a, the expected value of X and $Var(X)$

2 The continuous random variable X has probability density function f given by

$f(x) = \begin{cases} 0 & \text{for } x<0 \\ \frac{1}{9}(ax - x^2) & \text{for } 0 \leqslant x \leqslant 3 \\ 0 & \text{for } x>3 \end{cases}$

a) Show that $a = 4$
b) Calculate $E(X)$
c) Sketch the graph of f
d) Define and sketch the graph of the distribution function F of X

3 The continuous random variable X has probability density function f defined by

$f(x) = \begin{cases} \alpha x + \beta, & 0 \leqslant x \leqslant \frac{3}{2} \\ 0, & \text{otherwise.} \end{cases}$

a) Given that $E(X) = \frac{5}{8}$, find the values of the constants α and β

b) Find the value of $Var(X)$
c) Define the distribution function F of X
d) Calculate the value of $P\left(\frac{1}{2} \leqslant X \leqslant 1\right)$.

4 The continuous random variable X has distribution function F given by

$F(x) = \begin{cases} 0 & \text{for } x<0 \\ Kx^2 & \text{for } 0 \leqslant x \leqslant 6 \\ 1 & \text{for } x>6 \end{cases}$

a) Determine the value of the constant K

b) Find the probability that X will take a value in the interval $(2, 4)$

c) Determine the median value of X and calculate

   i) $E(X)$,   ii) $\operatorname{Var}(X)$

5   The continuous random variable X has probability density function f given by

$$f(x) = \lambda e^{-\lambda x}$$

where λ is a positive constant (Known as the *exponential* distribution).

a) Sketch the graph of f

b) Show that $E(X) = \dfrac{1}{\lambda}$ and $\operatorname{Var}(X) = \dfrac{1}{\lambda^2}$

6   The duration t minutes of a telephone call to the complaints section of a large manufacturing company follows an exponential distribution given by

$$f(t) = \frac{1}{3}e^{-\frac{1}{3}t}, \; t \in \mathbb{R}^+$$

Find the probability that a call will last:

a) at least 2 min

b) at most 3 min

c) between 1 and 4 min

7   The random variable X can take all values x in the interval $(a \leqslant x \leqslant b)$ and its probability density function f is given by

$$f(x) = \begin{cases} \dfrac{1}{b-a} & \text{for } a \leqslant x \leqslant b, \\ 0 & \text{otherwise} \end{cases}$$

(This distribution is often called the *uniform* or *continuous rectangular distribution*.) Show that:

a) $E(X) \quad = \dfrac{1}{2}(a+b)$

b) $\operatorname{Var}(X) = \dfrac{1}{12}(b-a)^2$

Define the distribution F of this random variable

NORMAL DISTRIBUTION

Many sets of observations in disciplines such as science, geography, economics, agriculture and medicine produce data which are said to be **normally distributed**. If a frequency curve were drawn to illustrate a set of data known to be normally distributed, the curve would take the bell-shape shown in Fig. 23.4.

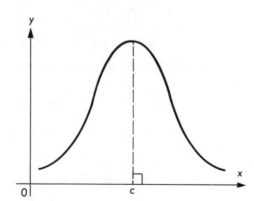

Fig. 23.4

Each set of normally distributed data has a curve of this distinctive bell-shape whose equation is of the form

$$y = ae^{-b(x-c)^2}$$

where a, b and c are constants. Notice that the curve has the following properties

> **Properties of the curve of the normal distribution.**

1) the line $x = c$ is the axis of symmetry

2) $y > 0$ for all values of x

It is also possible, but outside our scope mathematically, to show that the area of the region enclosed between the curve and the x-axis is finite.

In order to facilitate calculations and estimates concerning normal populations a standardised normal distribution is used. By selecting specific values for the constants a, b and c, the curve with equation

$$y = a\mathrm{e}^{-b(x-c)^2}$$

can be transformed so that it becomes the graph of the probability density function of a continuous random variable Z which is normally distributed. Rather more formally, we write:

A continuous random variable Z takes values z where $z \in \mathbb{R}$

The probability density function ϕ of z is given by

$$\phi(z) = \frac{1}{\sqrt{(2\pi)}}\,\mathrm{e}^{-z^2/2}, \ z \in \mathbb{R}$$

This continuous probability function is called **the standardised normal probability function.**

The mean of the distribution is 0. The variance of the distribution is 1. When referring to this standard normal distribution we write N(0,1).

N(μ, σ^2) is written for a normal distribution with mean μ and variance σ^2.

The graph of ϕ is shown in Fig. 23.5.

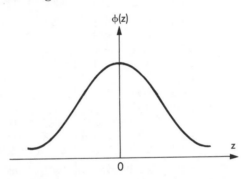

Fig. 23.5 Graph of the normal probability function ϕ

The curve has three special properties which tie up with the choice of

$$a = \frac{1}{\sqrt{(2\pi)}}, \ b = \frac{1}{2} \text{ and } c = 0$$

the constants in the general equation. These properties are:

1) the curve is symmetrical about the line $z = 0$
2) the area enclosed by the region between the curve and the z-axis is 1
3 the positive square root of the variance, called the standard deviation, is the unit used on the z-axis

Since the probability density function ϕ cannot be integrated directly, we have a table of values provided for the cumulative probability function $\Phi(z)$, where

$$\Phi(z) = \mathrm{P}(X \leqslant z) = \int_{-\infty}^{z} \phi(t)\,\mathrm{d}t$$

In geometrical terms, $\Phi(z)$ represents the shaded area shown in Fig. 23.6.

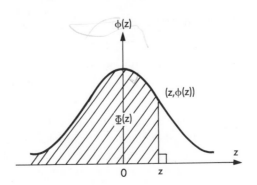

Fig. 23.6

When we write $Z \sim N(0, 1)$ it is shorthand for the "continuous random variable Z is normally distributed with mean 0 and variance 1".

If $Y \sim N(\mu, \sigma^2)$, then we have the simple linear relation $Z = \dfrac{Y-\mu}{\sigma}$ to link any value of Y with the corresponding value of Z. A table of values of the cumulative probability function $\Phi(z)$, usually called the distribution function $\Phi(z)$ is now given.

z	·00	·01	·02	·03	·04	·05	·06	·07	·08	·09
0·0	·5000	·5040	·5080	·5120	·5160	·5199	·5239	·5279	·5319	·5359
0·1	·5398	·5438	·5478	·5517	·5557	·5596	·5636	·5675	·5714	·5753
0·2	·5793	·5832	·5871	·5910	·5948	·5987	·6026	·6064	·6103	·6141
0·3	·6179	·6217	·6255	·6293	·6331	·6368	·6406	·6443	·6480	·6517
0·4	·6554	·6591	·6628	·6664	·6700	·6736	·6772	·6808	·6844	·6879
0·5	·6915	·6950	·6985	·7019	·7054	·7088	·7123	·7157	·7190	·7224
0·6	·7257	·7291	·7324	·7357	·7389	·7422	·7454	·7486	·7517	·7549
0·7	·7580	·7611	·7642	·7673	·7704	·7734	·7764	·7794	·7823	·7852
0·8	·7881	·7910	·7939	·7967	·7995	·8023	·8051	·8078	·8106	8133
0·9	·8159	·8186	·8212	·8238	·8264	·8289	·8315	·8340	·8365	·8389
1·0	·8413	·8438	·8461	·8485	·8508	·8531	·8554	·8577	·8599	·8621
1·1	·8643	·8665	·8686	·8708	·8729	·8749	·8770	·8790	·8810	·8830
1·2	·8849	·8869	·8888	·8907	·8925	·8944	·8962	·8980	·8997	·9015
1·3	·9032	·9049	·9066	·9082	·9099	·9115	·9131	·9147	·9162	·9177
1·4	·9192	·9207	·9222	·9236	·9251	·9265	·9279	·9292	·9306	·9319
1·5	·9332	·9345	·9357	·9370	·9382	·9394	·9406	·9418	·9429	·9441
1·6	·9452	·9463	·9474	·9484	·9495	·9505	·9515	·9525	·9535	·9545
1·7	·9554	·9564	·9573	·9582	·9591	·9599	·9608	·9616	·9625	·9633
1·8	·9641	·9649	·9656	·9664	·9671	·9678	·9686	·9693	·9699	·9706
1·9	·9713	·9719	·9726	·9732	·9738	·9744	·9750	·9756	·9761	·9767
2·0	·97725	·97778	·97831	·97882	·97932	·97982	·98030	·98077	·98124	·98169
2·1	·98214	·98257	·98300	·98341	·98382	·98422	·98461	·98500	·98537	·98574
2·2	·98610	·98645	·98679	·98713	·98745	·98778	·98809	·98840	·98870	·98899
2·3	·98928	·98956	·98983	·99010	·99036	·99061	·99086	·99111	·99134	·99158
2·4	·99180	·99202	·99224	·99245	·99266	·99286	·99305	·99324	·99343	·99361
2·5	·99379	·99396	·99413	·99430	·99446	·99461	·99477	·99492	·99506	·99520
2·6	·99534	·99547	·99560	·99573	·99585	·99598	·99609	·99621	·99632	·99643
2·7	·99653	·99664	·99674	·99683	·99693	·99702	·99711	·99720	·99728	·99736
2·8	·99744	·99752	·99760	·99767	·99774	·99781	·99788	·99795	·99801	·99807
2·9	·99813	·99819	·99825	·99831	·99836	·99841	·99846	·99851	·99856	·99861
3·0	·99865	·99869	·99874	·99878	·99882	·99886	·99889	·99893	·99896	·99900

z	3·1	3·2	3·3	3·4	3·5	3·6	3·7	3·8	3·9	4·0
Φ	·99903	·99931	·99952	·99966	·99977	·99984	·99989	·99993	·99995	·99997

Table 23.1 Values of the distribution function $\Phi(z)$, of the normal probability function $\phi(z)$

Worked example

The standardised normal variable Z takes values z, where $z \in \mathbb{R}$. Use tables to find the probability that

a) $z > 0·7$

b) $-0·51 \leqslant z \leqslant 0·51$

a) $\quad P(z \leqslant 0·7) = \displaystyle\int_{-\infty}^{0·7} \phi(t)dt \approx 0·758$ (from tables)

$\Rightarrow P(z > 0·7) \approx 1 - 0·758 = 0·242$

b) $\quad P(z < 0·51) = \displaystyle\int_{-\infty}^{0·51} \phi(t)\,dt \approx 0·695$ (from tables)

$\Rightarrow P(-0·51 \leqslant z \leqslant 0·51) = \displaystyle\int_{-0·51}^{0·51} \phi(t)\,dt \approx 2(0·695 - 0·5)$

$= 0·39$

Note: in the solution to b) how we make use of the symmetrical property of the normal curve to obtain the probability. In both a) and b), draw a sketch of the curve and shade in the region required because this will help you to understand the solution given and when you are writing your own solutions.

Worked example

A continuous random variable X is normally distributed and takes values x, where $x \in \mathbb{R}$. The variance of X is 100. Use tables to estimate the mean of X, given that the probability of X taking a value less than 85 is 0·33

The standard deviation of X is $\sqrt{100} = 10$, so that each unit of X is 10 times that of the standardised normal variable Z.

If $z = -c$ when $x = 85$, we have

$$\int_{-\infty}^{-c} \phi(t)\,dt = 0{\cdot}33$$

$$\Rightarrow \int_{-\infty}^{c} \phi(t)\,dt = 1 - 0{\cdot}33 = 0{\cdot}67$$

$$\Rightarrow c = 0{\cdot}44 \text{ (from using the tables in reverse)}$$

Taking μ_1 as the mean of X we have

$$\frac{\mu_1 - 85}{10} = 0{\cdot}44$$

$$\Rightarrow \mu_1 = 89{\cdot}4$$

The mean of X is 89·4

Worked example

A javelin thrower finds that the distances he throws form a normal distribution with mean 50 m. The probability that he throws a distance greater than 60 m is 0·01. Find, to 2 decimal places, the standard deviation of the distribution of distances

$$\Phi\left(\frac{60 - 50}{\sigma}\right) = 0{\cdot}99$$

$$\Rightarrow \frac{10}{\sigma} = 2{\cdot}327 \text{ (from tables in reverse)}$$

$$\Rightarrow \sigma = 4{\cdot}30$$

The standard deviation of the distribution of distances is 4·30 m

Worked example

In an examination, 2% of the candidates scored more than 80 marks and 5% of the candidates scored less than 30 marks. Given that the marks of the candidates taking this examination are approximately normally distributed, find estimates for the mean and the standard deviation of the mark distribution, giving your answers to 3 significant figures

Let μ and σ be the mean and the standard deviation of the mark distribution. Since marks are usually awarded as integers, we introduce what is known as a continuity correction at both the limits given, because a normal random variable is of course continuous. At the upper end we consider all those scoring less than 80·5 marks and at the lower end all those scoring more than 29·5 marks.

The probability of scoring less than 80·5 marks is 0·98, (since 2% score more)

$$\Rightarrow 80{\cdot}5 - \mu = 2{\cdot}054\sigma \text{ (from tables in reverse)}$$

The probability of scoring more than 29·5 marks is 0·95, (since 5% score less)

$$\Rightarrow \mu - 29{\cdot}5 = 1{\cdot}64\sigma \text{ (from tables in reverse)}$$

Solving these equations simultaneously gives $\mu = 52{\cdot}1$, $\sigma = 13{\cdot}8$ (to 3 significant figures)

USING A NORMAL DISTRIBUTION AS AN APPROXIMATION TO A BINOMIAL DISTRIBUTION

Remember the need for a continuity correction.

Finding the probability of the occurrence of a given number of events in the binomial distribution $B(n, p)$ may be tedious to calculate and under certain conditions good approximations may be obtained by considering the normal distribution $N(\mu, \sigma^2)$ where $\mu = np$ and $\sigma^2 = npq$.

In general, for good approximations we need n to be fairly large and we need p not to be too near 0 or 1. A good working rule is to use the approximation when both np and nq are greater than 5.

As binomial distributions are concerned with discrete random variables and normal distributions with continuous random variables, a *continuity correction* is needed also in these approximations.

Worked example

An unbiased penny is to be tossed 80 times. Estimate the probability that:
a) between 35 and 50 heads will result
b) fewer than 30 heads will result

a) the binomial distribution under consideration is $B(80, \frac{1}{2})$ and we take the normal distribution $N(40, 20)$ as the approximation. Taking x to be the number of heads, we require the value of $P(34 \cdot 5 \leqslant x \leqslant 50 \cdot 5)$ in $N(40, 20)$, (where we are taking x as a continuous random variable). Standardizing to the continuous random variable z in $N(0, 1)$ we have

$$z = \frac{x - 40}{\sqrt{20}}$$

$$\begin{aligned} P(34 \cdot 5 \leqslant x \leqslant 50 \cdot 5) &= P(-1 \cdot 230 \leqslant z \leqslant 2 \cdot 348) \\ &= 0 \cdot 99056 - (1 - 0 \cdot 8907) \text{ (from tables)} \\ &\approx 0 \cdot 881 \end{aligned}$$

b) $\begin{aligned}[t] P(x < 29 \cdot 5) &= P(z < -2 \cdot 348) \\ &\approx \cdot 0094 \text{ (from tables)} \end{aligned}$

Note carefully how the continuity corrections have been used in each part of the solution.

THE NORMAL DISTRIBUTION USED AS AN APPROXIMATION TO THE POISSON DISTRIBUTION

The Poisson distribution $Poi(\mu)$, with mean and variance μ, can, under certain circumstances be aproximated by the normal distribution $N(\mu, \mu)$. The required condition is that μ should be "large"; this is usually taken to mean $\mu > 20$, although the approximation is quite good when $\mu > 10$. Since we are approximating a discrete variable distribution by a continuous distribution, we must use a continuity correction, just as we did when approximating the binomial by a normal distribution.

Worked example

The number, X, of traffic-light failures in a large city in a week can be assumed to be distributed $Poi(28)$. Estimate, to 3 decimal places the probabilities that there will be, in a coming week, a) at most 25, b) more than 20, traffic-light failures

We can approximate $Poi(28)$ by $N(28, 28)$. Using a continuity correction, we have

a) $\begin{aligned}[t] P(X < 25 \cdot 5) &= \Phi\left(\frac{25 \cdot 5 - 28}{\sqrt{28}}\right) \\ &= \Phi(-0 \cdot 4725) \\ &= 1 - 0 \cdot 6817 \\ &\approx 0 \cdot 318 \end{aligned}$

b) $\begin{aligned}[t] P(X > 20) &= 1 - \Phi\left(\frac{20 \cdot 5 - 28}{\sqrt{28}}\right) = 1 - \Phi(-1 \cdot 417) \\ &= \Phi(1 \cdot 417) \\ &= 0 \cdot 922 \end{aligned}$

EXERCISES 23.2

1 The continuous random variable z is $N(0, 1)$.
Calculate the values of
a) $P(z > 2 \cdot 12)$,
b) $P(z < -1 \cdot 18)$,
c) $P(-1 \cdot 18 < z < 2 \cdot 12)$
Show, in each case, on a sketch the area of the region under the standard normal curve associated with your answer

2 The continuous random variable X is $N(14, 16)$
Calculate the values of:
a) $P(X > 15 \cdot 5)$
b) $P(X > 19)$
c) $P(14 < X < 22)$

3   The lengths of a species of fish are normally distributed with mean length 35 cm and standard deviation 3 cm. Find the percentage of these fish having lengths:
   a) greater than 40 cm
   b) between 30·5 cm and 39·5 cm

4   The masses of a large number of cats are normally distributed. It is known that 10% of the cats have masses exceeding 1·8 kg and that 15% of the cats have masses less than 1·35 kg. Calculate the mean mass and the standard deviation, in kg to 2 decimal places, of this distribution

5   The marks of a large number of candidates in a physics examination are normally distributed with mean 45 marks and standard deviation 20 marks.
   a) Given that the pass-mark is 52, estimate the percentage of candidates who passed the examination
   b) If 8% of candidates gained a grade A by scoring x marks or more, estimate the value of x

6   An unbiased die is to be thrown 240 times. Using the normal distribution approximation to the binomial distribution, estimate, to 2 decimal places, the probability that between 36 and 43 sixes, inclusive, will be obtained

7   It is estimated that 40% of the adult population of Wales will watch the Welsh Rugby Union Cup Final on television. A random sample of 100 adults is taken. Estimate the probability that between 34 and 44, inclusive, watch the final

8   The heights of a large number of young men are normally distributed with mean 176 cm and standard deviation 6·5 cm.
   a) Find the probability that the height of a young man selected at random will exceed 180 cm
   b) Two young men are selected at random. Find the probability that:
      i) the height of each young man exceeds 180 cm
      ii) the sum of their heights exceeds 360 cm

9   Explain briefly the circumstances under which a normal distribution may be used as an approximation to a binomial distribution. In each question of a chemistry multiple-choice examination there are 4 possible answers given, only one of which is correct. A student selects an answer to each question at random. Find, giving your answers to 3 decimal places, the probability that this student will obtain:
   a) just 1 correct answer when 6 questions are attempted
   b) more than 13 correct answers when 40 questions are attempted

10   A large mixture of cornflower seeds contains just two types, blue-flowering and white-flowering in the ratio 5 : 1 repsectively. Seeds are chosen at random from the mixture and planted in plastic strips with 12 seeds in each strip. Assuming that the seeds germinate and produce plants, find the mean and the variance of the number of white-flowering plants per strip. Find the probability that a strip will contain 3 or more white-flowering plants. Estimate the probability that in a total of 20 strips there will be more than 44 white-flowering plants.

11   The number of children absent each day in a comprehensive school can be assumed to have a Poisson distribution with mean 25. Using the normal approximation to the Poisson distribution calculate approximate values for the probability that on any day there will be:
   a) less than 23 children absent
   b) 30 or more children absent

12   In a factory the average number of electric light-bulbs failing each day is 3. Find the probability that 7 or more bulbs fail on two consecutive days. Estimate the probability that, in a 2-week period of 10 working days, more than 25 bulbs fail.

NB: Examination questions on these topics can be found in Chapter 27.

ANSWERS TO EXERCISES

EXERCISES 23.1

1 $-\dfrac{3}{16}$, $1\cdot25$, $\dfrac{19}{80}$

2 b) $1\dfrac{3}{4}$;

d) $F(x) = \begin{cases} 0, & x<0 \\ \dfrac{1}{9}\left(2x^2-\dfrac{x^3}{3}\right), & 0\leqslant x\leqslant3 \\ 1, & x>3 \end{cases}$

3 a) $-\dfrac{4}{9}$, 1; b) $\dfrac{11}{64}$;

c) $F(x) = \begin{cases} 0, & x<0, \\ x-\dfrac{2}{9}x^2, & 0\leqslant x\leqslant\dfrac{3}{2}; \\ 1, & x>\dfrac{3}{2} \end{cases}$

d) $\dfrac{1}{3}$

4 a) $\dfrac{1}{36}$; b) $\dfrac{1}{3}$; c) i) $4\cdot24$; ii) 4; iii) 2

6 a) $0\cdot513$; b) 0.632; c) $0\cdot453$

EXERCISES 23.2

1 a) $0\cdot017$; b) $0\cdot119$; c) $0\cdot864$
2 a) $0\cdot354$; b) $0\cdot106$; c) $0\cdot477$
3 a) $4\cdot8\%$; b) $86\cdot6\%$
4 $1\cdot55\,\text{kg}$, $0\cdot19\,\text{kg}$
5 a) $37\cdot2\%$; b) $73\cdot6$
6 $0\cdot51$
7 $0\cdot73$
8 a) $0\cdot269$; b) $0\cdot0724$; c) $0\cdot192$
9 a) $0\cdot356$; b) $0\cdot101$
10 $0\cdot323$, $0\cdot218$
11 $0\cdot3085$, $0\cdot1841$
12 $0\cdot394$, $0\cdot842$

CHAPTER 24

THE DISTRIBUTION OF SAMPLE MEANS AND THE CENTRAL LIMIT THEOREM

THE CENTRAL LIMIT THEOREM

CONFIDENCE INTERVAL FOR A POPULATION MEAN

GETTING STARTED

We have seen how a random sample taken from a population can be used to estimate unknown properties of the population. If we take all possible samples of size n from the population, then, for each sample, we can calculate its mean, variance, standard deviation, etc. So for each sample statistic (mean, variance, etc.) we have a set of values, one coming from each sample, which will itself form a distribution. For most A-level syllabuses only the sampling distribution of the means is required to be studied and we will restrict ourselves to this.

The standard deviation of the distribution of sample means is called the standard error of the mean and is often written, $\text{SE}_{\bar{x}}$.

Worked example

A population consists of 5 digits 2, 4, 6, 8, 10. Write down all possible samples of 2 different digits that can occur if random samples are taken. Find the mean of each sample and plot the distribution of the sample means. Compare the value of the mean of the sample means with the value of the population mean.

Possible samples are 2 and 4, 2 and 6, 2 and 8, 2 and 10, 4 and 6, 4 and 8, 4 and 10, 6 and 8, 6 and 10, 8 and 10. Their means respectively are 3, 4, 5, 6, 5, 6, 7, 7, 8, 9

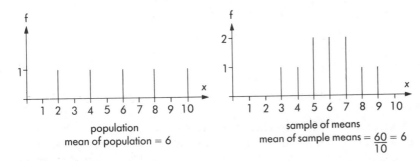

population
mean of population = 6

sample of means
mean of sample means = $\frac{60}{10}$ = 6

Fig. 24.1

Hence the mean values are equal.

In this example we have a small finite population. Consider now the case of a very large or infinite population. It is obviously not possible to take all possible samples of a given size from the population. We call the population mean μ and variance σ^2. (*Estimates* of these are usually denoted by $\hat{\mu}$ and $\hat{\sigma}^2$.)

Looking at the example above, we would expect the theoretical distribution of the means of all samples of size n taken from a population with mean μ and variance σ^2 to have a mean equal to the value μ. From common sense, we would also expect the distribution of sample means to be more closely grouped around the mean value than the population distribution is grouped around its mean value. That is, we expect $\text{SE}_{\bar{x}} < \sigma$.

ESSENTIAL PRINCIPLES

THE CENTRAL LIMIT THEOREM

The **central limit theorem** tells us more about the distribution of sample means (a proof is not required).

1   If the original population is distributed
$$X \sim N(\mu, \sigma^2)$$
then the distribution of sample means of samples of size n is distributed

$$\bar{X} \sim N\left(\mu, \frac{\sigma^2}{n}\right)$$

That is

$$SE_{\bar{x}} = \frac{\sigma}{\sqrt{n}}$$

"Properties of the central limit theorem."

2   If the original population has mean μ and variance σ^2, but is not normally distributed, then the distribution of sample means is approximately distributed

$$N\left(\mu, \frac{\sigma^2}{n}\right)$$

provided n is large enough. For most practical work $n \geqslant 30$ is acceptable. The further that the population is from normal (the more asymmetric) the larger will n have to be for us to use the approximation.

Worked example

The sugar content per litre bottle of a soft drink is known to be distributed with mean 5·8 and standard deviation 1·2. A sample of 900 bottles is taken at random and the sugar content of each bottle is measured. Estimate to 3 decimal places the probability that the mean sugar content of the 900 bottles will be less than 5·85

Since X is distributed with $\mu = 5\cdot8$, $\sigma^2 = (1\cdot2)^2$ and $n = 900$

$$\Rightarrow \bar{X} \text{ is approximately normally distributed } N\left(5\cdot8, \frac{(1\cdot2)^2}{900}\right)$$

Hence $P(\bar{X} < 5\cdot85) = \Phi \dfrac{5\cdot85 - 5\cdot8}{\dfrac{1\cdot2}{\sqrt{900}}} = \Phi(1.25) = 0\cdot8944$

$$= 0\cdot894 \text{ (to 3 decimal places)}$$

Worked example

A firm produces alternators for cars. The alternators are known to have a mean lifetime of 8 years with standard deviation 6 months. Forty samples of 144 alternators produced by the firm are tested. Estimate the number of samples which would be expected to have a mean lifetime of more than 8 years and 1 month

Let X months be the lifetime of an alternator, then X is distributed with mean 96 months and standard deviation 6 months. The sample size, n, is 144 which is large, hence $\bar{X}$ the mean sample life is approximately distributed

$$N\left(96, \frac{6^2}{144}\right)$$

That is

$$SE_{\bar{X}} = \frac{6}{\sqrt{144}} = \frac{1}{2}$$

Hence $P(\bar{X} < 97) = \Phi\left(\dfrac{1}{\frac{1}{2}}\right) = \Phi(2) = 0\cdot97725$

$\Rightarrow$ of 40 samples, $40 \times 0\cdot97725 = 39\cdot09$ would be expected to have a mean lifetime of less than 8 years and 1 month, i.e. only approximately 1 sample would be expected to have a mean lifetime of more than 8 years and 1 month.

CONFIDENCE INTERVAL FOR A POPULATION MEAN

In Chapter 22 we said that the (unknown) mean and variance of a population can be estimated from the data of a random sample, size n, taken from the population. The unbiased estimates for the population mean and variance respectively are:

$\hat{\mu} = \bar{x}$, the sample mean

$$\hat{\sigma}^2 = \frac{n}{n-1} . S^2$$

where S^2 is the sample variance.

Obviously for different random samples from the population we get different estimates for μ and σ^2. Using the central limit theorem, we try to find a symmetrical interval within which we can say with a given degree of certainty (e.g. 90%, 95%, 98%) that the population mean will lie. This is called a **confidence interval** for the population mean.

Worked example

Applicants to a large firm are given a short test. It is found over a long period that the scores for people taking this test are normally distributed with variance 12. A random sample of 15 applicants have a mean score of 22. On the basis of this result find:
i) a 98%,   ii) a 95% confidence interval for the mean score of all applicants taking this test

The population is distributed $N(\mu, 12)$

Hence the sample means are distributed $N\left(\mu, \dfrac{12}{n}\right)$ where $n = 15$

We do not know μ but we can estimate it as
$\hat{\mu} = \bar{x}$, the sample mean $= 22$
For a 98% interval we require 0·49 on each side of the mean.

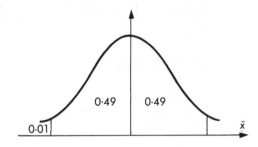

Fig. 24.2

i.e. $\Phi\left(\dfrac{A-22}{\sqrt{\left(\dfrac{12}{15}\right)}}\right) = 0\cdot99$ (i.e. $0\cdot5 + 0\cdot49$)

$$\Rightarrow A - 22 = \sqrt{\left(\frac{12}{15}\right)} \times 2\cdot327 = 2\cdot08$$

i.e. 98% confidence interval for $\mu = 22 \pm 2\cdot08 = 19\cdot92$ to $24\cdot08$
For a 95% interval we require 0·475 on each side of the mean

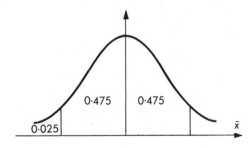

Fig. 24.3

i.e. $\Phi\left(\dfrac{A-22}{\sqrt{\left(\dfrac{12}{15}\right)}}\right) = 0\cdot975$

$$\Rightarrow A - 22 = \sqrt{\left(\frac{12}{15}\right)} \times 1\cdot960 = 1\cdot75$$

i.e. 95% confidence interval for $\mu = 22 \pm 1\cdot75 = 20\cdot25$ to $23\cdot75$.

EXERCISES 24.1

1   Estimate the percentage of random samples of size 30 taken from a continuous population whose mean is 12 with standard deviation 4 which will have means of 13 or more

2   The weight of a large collection of bags of potatoes has a mean of 25 kg and a variance of 5 kg. Estimate to 2 decimal places the probability that a random sample of 50 bags will have a mean weight of between 24·5 kg and 25·5 kg

3   An electrical firm claims that its light-bulbs have a mean life that is approximately normally distributed with mean equal to 900 hours and a standard deviation of 40 hours. Estimate the probability that a random sample of 16 bulbs will have an average life of:
a) less than 875 hours
b) between 875 hours and 925 hours

4   A random sample of 80 values of a variate X has mean 4·7 and variance 0·88. Find unbiased estimates of a) the mean, b) the variance of the population from which this sample comes. Hence find a 95% confidence interval for the mean

5   A variate X has variance 2·4. Find how large a random sample must be taken in order to:
a) ensure that the standard error is less than 0·06
b) be 95% confident that the true mean is known to within 0·25 units

NB: Examination questions on these topics can be found in Chapter 27.

ANSWERS TO EXERCISES

EXERCISES 24.1

1   8·5%
2   0·89
3   a) 0·006;   b) 0·988
4   4·7, 0·891, 4·49 to 4·91
5   a) 667;   b) 148

CHAPTER

HYPOTHESIS
TESTING

SIGNIFICANCE LEVEL

THE χ^2 (CHI-
SQUARED) TEST

GETTING STARTED

We have discussed how we can estimate the mean and variance of a population from a collected set of sample data. In practical work a problem frequently arises which is closely related to this; we wish to test whether or not a set of sample data provides us with sufficient evidence to say that the sample has come from a particular population with known or assumed parameter values (e.g. mean, variance). We set up two hypotheses about the population H_0 and H_1.

H_0, the null hypothesis, assumes that any differences between our observed sample results and the population results are due simply to sampling or chance variations. It is an assumption of no difference between the population from which the sample comes and the population with which we are concerned. Working on the assumption that H_0 is true, we find the probability that the sample results would occur. If we decide that this probability is too low (see 'significance level' a little later) then we must reject the assumption of H_0 and accept the alternative hypothesis, H_1.

H_1 may be of two kinds, one-sided or two-sided, depending on the problem we are investigating. If we are testing

H_0: $\mu = 30$ (population mean = 30)
  then H_1 could be:
  H_1: $\mu \neq 30$ (2-sided, that is the population mean is not 30)
or H_1: $\mu > 30$ (1-sided, that is the population mean is > 30)
or H_1: $\mu < 30$ (1-sided, that is the population mean is < 30)

For example, if the problem is to investigate whether certain match-boxes sold contain an average of 100 matches as stated, then we would use H_0: $\mu = 100$, H_1: $\mu < 100$, since we are only concerned if we are getting too few matches.

ESSENTIAL PRINCIPLES

We said that we calculate the probability that the sample data could have come from the assumed population by chance, and that, if this probability is too low, we reject H_0 and assume that H_1 is true. The level of probability we consider too low depends on the problem. Obviously more rigid levels must be placed on the results of a medical test than on the average number of matches in a box. Many GCE A-level questions tell you what **significance level** to take (often 5%) but it could be 10% (i.e. P = 0·10), 5% (P = 0·05), 2% (P = 0·02) or even 1% (P = 0·01). Any probability value smaller than the significance level will cause you to reject H_0. The significance level is the maximum probability at which we are willing to reject H_0 when it is in fact true.

Worked example

In a factory records over a long period have shown that the breaking strength of button thread produced there is approximately distributed N(8·20, (1·22)^2). A random sample of 50 pieces of thread is tested after the machines are serviced and these are found to have a mean breaking strength of 8·30 units. Investigate at the 5% significance level, whether or not the mean breaking strength has increased

H_0: $\mu = 8·20$, H_1: $\mu > 8·20$ (one-sided, since we are only concerned with an increase in strength)

Under H_0 (and using the central limit theorem)

$$P(x) \geqslant 8·30 = P\left(\bar{Z} \geqslant \frac{8·2 - 8·3}{\frac{1·22}{\sqrt{50}}}\right) = P(\bar{Z} \geqslant -0·5796) = 0·281$$

Since $0·281 > 0·05$ we cannot reject H_0 at the 5% level. The mean breaking strength does not appear to have increased.

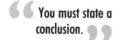

You must state a conclusion.

You must note that it is essential to finish your work by writing your conclusion in words so that it can be understood by any reader. Failure to do so will lose you marks.

SIGNIFICANCE TESTING OF A BINOMIAL PROBABILITY

Worked example

When a certain species of plant is grown commercially it is found that the probability that a seedling survives is $\frac{4}{5}$. A grower decides to add a compost to the soil and finds that of 30 seedlings planted, 28 survive. Investigate whether a claim that the survival rate has increased is justified.

$H_0: p = \frac{4}{5}$, $H_1: p > \frac{4}{5}$ (one-sided)

X, the number of seedlings surviving, $\sim$ B $(30, \frac{4}{5})$

$$P(X \geqslant 28) = \left(\frac{4}{5}\right)^{30} + 30\left(\frac{4}{5}\right)^{29}\left(\frac{1}{5}\right) + \frac{30.29}{1.2}\left(\frac{4}{5}\right)^{28}\left(\frac{1}{5}\right)^2 = 0·044 < 0·05$$

If we use a 5% significance level then $0·044 < 0·05$ and we must reject H_0 at this level and accept H_1, that the survival rate has increased. If we use a 2% level than $0·044 > 0·02$ and we would accept H_0 at this level, that the survival rate is still $\frac{4}{5}$.

SIGNIFICANCE TESTING FOR THE MEAN OF A POISSON DISTRIBUTION

Worked example

The number of accidents at a certain busy road crossing has been found to average 1·2 per week. After a new speed limit is introduced, 3 accidents happen there in the next 4 weeks. Investigate whether the mean number of accidents there has changed

$H_0: \lambda = 1·2$, $H_1: \lambda \neq 1·2$ (two-sided since we are looking for a change in λ)

Under H_0, we would expect $4 \times 1·2 = 4·8$ accidents in 4 weeks

If X is the number of accidents in 4 weeks then $X \sim$ Poi $(4·8)$

$$P(x \leqslant 3) = e^{-4\cdot8} \left[1 + \frac{4\cdot8}{1} + \frac{4\cdot8}{1.2} + \frac{4\cdot8^3}{1.2.3} \right] = 0\cdot294$$

Since we are dealing with a two-sided problem, if we are using a 5% significance level, our calculated value would need to be $<0\cdot025$ for H_0 to be rejected. Hence we must accept H_0 (at the 5% level), i.e. the mean number of accidents is unchanged.

EXERCISES 25.1

1   A machine produces nails whose lengths are normally distributed with mean 25 mm and variance $0\cdot22\,\text{mm}^2$. After adjustments to the machine a sample of 30 nails was examined and found to have a mean length of $25\cdot1\,\text{mm}$. Test at the 5% level whether the mean length of the nails produced by the machine has increased

2   A manufacturer claims that his new synthetic fishing line has a mean breaking strength of 15 kg with variance $0\cdot25\,\text{kg}^2$. A random sample of 50 lines is tested and found to have a mean breaking strength of $14\cdot8\,\text{kg}$. Investigate at the 1% significance level whether or not the mean breaking strength remains at 15 kg.

3   An electrical firm claims that its light-bulbs have a mean life that is approximately normally distributed with mean life of 800 hours and standard deviation of 40 hours. Using a 4% significance level test the hypothesis that $\mu = 800$ hours against the alternative $\mu \neq 800$ hours if a random sample of 35 bulbs has an average life of 789 hours

4   Thirty years ago 11-year-old children had an average height of 160 cm with a standard deviation of $6\cdot8\,\text{cm}$. Last year a group of 50, 11-year-old children, chosen at random were found to have an average height of $162\cdot5\,\text{cm}$. Using a 2% significance level determine whether or not there has been a change in the average height of 11-year-old children

5   The weights in grams of 10 containers of sweets chosen at random were found to be 97, 101, 102, 101, 98, 103, 98, 104, 103, 99. Find estimates of
a) the mean weight:
b) the standard deviation of the containers of sweets.
Assuming the true value of σ is equal to the estimated value and that the weights of the containers of sweets are normally distributed, test at the 1% level the manufacturer's claim that the mean weight of the container of sweets is 100 g

THE χ^2 (CHI-SQUARED) TEST

There are two ways in which you will use this test.

1 FITTING A DISTRIBUTION TO SAMPLE DATA

If we have collected in a project a set of readings for n values of a variable X with frequencies $O_1, O_2, O_3, \ldots, O_n$, and we believe that this sample can be fitted by a known distribution, we must decide if X is discrete or continuous.

For a discrete X, we might consider a binomial distribution or a Poisson distribution to be suitable; for continuous X, we might think our sample is from a normal or an exponential distribution. If the theoretical distribution is found, using the χ^2 test, to be well fitting, then the results known for the distribution can be applied to data obtained from the project.

We assume (H_0) that the chosen theoretical distribution does fit the sample data and we calculate the frequencies $E_1, E_2, E_3, \ldots, E_n$, that would be expected under this assumption.

The statistic

$$\chi^2 = \sum_{r=1}^{n} \frac{(E_r - O_r)^2}{E_r}$$

is found. We then look at the table of critical values for χ^2 (given for various significance levels). In order to use the table we need to use the "degrees of freedom, υ" of the problem. We take:

$\upsilon = n-1 - $ (the number of population parameters estimated), where n is the number of frequencies (or cells)

Worked example

The number of pages containing $0, 1, 2, 3, \ldots$ misprints in a 100-page magazine were counted with the results shown below:

No. of misprints	0	1	2	$\geqslant 3$	Total
No. of pages	63	28	8	1	100

Since the probability of a misprint is small and the number of pages large, a Poisson distribution would seem to be appropriate, i.e. $H_0 : X \sim \text{Poi}(\lambda)$

λ, the population mean, must be estimated from the sample data

i.e., $\lambda = \dfrac{0 \times 63 + 1 \times 28 + 2 \times 8 + 3 \times 1}{100} = \dfrac{47}{100}$

Under H_0, the expected frequencies are

$$E_0 = e^{-0.47} \times 100 = 62.5, \quad E_1 = 0.47 E_0 = 29.4, \quad E_2 = \frac{0.47}{2} E_1 = 6.9$$

$E_3 = 100 - (E_0 + E_1 + E_2) = 1.2$. (since the total frequency must $= 100$)

However if any value of E is <5 we should combine cells in order to avoid the small E-value distorting the χ^2 value.

We therefore write our data:

X	0	1	$\geqslant 2$	Total
O_r	63	28	9	100
E_r	62.5	29.4	8.1	100

Then $\chi^2 = \dfrac{(62.5 - 63)^2}{62.5} + \dfrac{(29.4 - 28)^2}{29.4} + \dfrac{(8.1 - 9)^2}{8.1} = 0.171$

where $\upsilon = 3 - 1 - 1 = 1$ (since the mean λ was estimated)

The table gives at $\upsilon = 1$ and $P = 0.05$ (5% level) the critical value of $\chi^2 = 3.84$.

Since 0.171 is (much) less than 3.84 we cannot reject H_0. The observed results do not differ significantly from a Poisson distribution with mean 0.47.

2 TESTING OF CONTINGENCY TABLES

Here the sample and population are recorded for more than one attribute. In the examination you will usually meet a 2×2 **contingency table**. Such an example is now shown.

Worked example

50 patients are tested for a skin disease; 30 are given the usual skin cream, 20 are given a new ointment. The results are shown below:

Treatment	Improved	Not improved	Total
Skin cream	14 (15·6)	16 (14·4)	30
New ointment	12 (10·4)	8 (9·6)	20
			50

Investigate whether these results lead you to conclude that the new ointment is more effective than the old one.

H_0: the two treatments are equally effective.

Under H_0, 26 out of 50 would improve. Thus, out of 30 treated with skin cream we would expect

$$\frac{30}{50} \times 26 = 15.6$$

to have improved and we write this in brackets besides the number 14. Using the four totals, the other three expected values 10.4, 14.4 and 9.6 can all be similarly written onto the table

$$\chi^2 = \frac{(15.6 - 14)^2}{15.6} + \frac{(10.4 - 12)^2}{10.4} + \frac{(14.4 - 16)^2}{14.4} + \frac{(9.6 - 8)^2}{9.6}$$

$$= (1 \cdot 6)^2 \left[\frac{1}{15 \cdot 6} + \frac{1}{10 \cdot 4} + \frac{1}{14 \cdot 4} + \frac{1}{9 \cdot 6} \right] = 0 \cdot 852$$

Using the table of critical values and $\upsilon = 1$, since we have needed to work out only one expected value, we find, for significance level P = 0·10 (10%) the critical χ^2 value = 2·70. Since 0·852<2·70, we cannot reject H_0. It appears therefore that the treatments are equally effective.

EXERCISES 25.2

1   A commercial traveller records over a five-year period the number of punctures that he gets each month. The results are given in the table below. Examine the results to determine at the 5% level, whether there is evidence to support the claim that the punctures occur randomly:

No. of punctures	0	1	2	3	4	5	$\geqslant 6$
Frequency	20	18	13	6	2	1	0

2   Three coins were tossed 100 times and the number of heads recorded as shown below. Determine at the 5% level of significance whether or not the coins were unbiased:

No. of heads	0	1	2	3
Frequency	7	48	30	15

3   The table below shows the number of passes and failures obtained by two driving schools, A and B. Examine the data to determine at the 2·5% level whether there is a significant difference between the two schools with respect to the number of passes:

School	Successful	Unsuccessful
A	510	25
B	177	23

4   Use the data below to test whether there is any significant difference in voting pattern between ages:

Age	Vote Conservative	Vote Labour
18–30	37	43
30–65	40	40

NB: Examination questions on these topics can be found in Chapter 27.

ANSWERS TO EXERCISES

EXERCISES 25.1

1   Cannot reject H_0
2   Mean breaking strength is less than 15 kg
3   Accept the hypothesis
4   Has been an increase in height
5   Accept claim

EXERCISES 25.2

1   Evidence is punctures occurs randomly
2   No evidence coins biased
3   There is a significant difference with respect to the number of passes ($\chi^2 = 11 \cdot 3$)
4   No significant difference at 10% level, $\chi^2 = 2 \cdot 25$

LINEAR REGRESSION

METHOD OF LEAST SQUARES

CORRELATION

RANK CORRELATION

GETTING STARTED

Our analysis of data so far has only been concerned with distributions of a single variate. However, we are often concerned with investigations which involve the measurement of more than one characteristic on each member of a sample and in particular we wish to determine whether any relationship exists between the variates. We shall concern ourselves with the investigation as to whether two quantitative variates are related. Often one of the two variates can be measured with little or no error and this is taken as the independent variable and frequently denoted by the letter x. The other variable, usually denoted by y, is the dependent variable. The values of the dependent variable are often difficult or expensive to measure directly and are subject to random error.

SCATTER DIAGRAM

Consider a random sample of size n from a population. Let the sample be represented by the set $\{(x_i, y_i), i = 1, 2, 3, \ldots, n\}$. Using the cartesian axes plot the n data points $(x_1, y_1), (x_2, y_2), (x_3, y_3), \ldots, (x_n, y_n)$ to obtain a "scatter diagram"

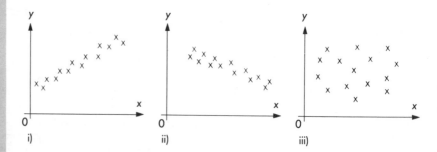

Fig. 26.1

i) Indicates a positive correlation between x and y since the plots are scattered around a straight line with positive gradient.

ii) Indicates a negative correlation between x and y since the plots are scattered around a straight line with negative gradient.

iii) Indicates no correlation between x and y since the plots provide a shapeless scatter.

GCE A-level questions will only involve bivariate data which indicate positive or negative correlation and you will be required to find the equation of the best straight line to fit the data. One can draw by eye the best straight line through the points so that there is an even spread of points above and below the line, but this is not very scientific and obviously different people will have different opinions as to which is the line of best fit. Consequently, a method known as the method of least squares is adopted.

ESSENTIAL PRINCIPLES

Consider the sample set $\{(x_i, y_i); \, i = 1, 2, 3, \ldots, n\}$ where the x_i values have little or no error and let the line of best fit have equation $y = \alpha + \beta x$ where α and β are unknown parameters. Let this line and the scatter diagram of the sample be as shown in Fig. 26.2.

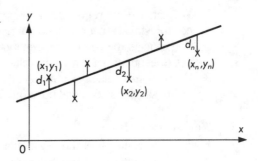

Fig. 26.2

From each plotted point draw a line parallel to the y-axis to meet the line of best fit and let d_i, $i = 1, 2, \ldots, n$ be the lengths of these lines. Clearly $d_i = y_i - (\alpha + \beta x_i)$

The **method of least squares** involves calculating α and β such that

$\sum_{i=1}^{n} d_i^2$ is a minimum.

The calculus involved in this exercise is not easy, but it gives the equation of the line of best fit as being:

$$y - \bar{y} = \beta(x - \bar{x})$$

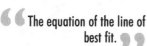
The equation of the line of best fit.

where $\bar{y} = \dfrac{1}{n}\sum_{i=1}^{n} y_i$, the mean of the y_i values,

$\bar{x} = \dfrac{1}{n}\sum_{i=1}^{n} x_i$, the mean of the x_i values,

and $\beta = \dfrac{\sum_{i=1}^{n} x_i y_i - n\bar{x}\bar{y}}{\sum_{i=1}^{n} x_i^2 - n\bar{x}^2}$

This line which minimises $\sum_{i=1}^{n} d_i^2$ (i.e. the sum of the squares of the lengths parallel to the y-axis) gives the equation of the line of regression of y on x. The gradient β of this line is called the *regression coefficient*.

An alternative form for β, the regression coefficient is

$$\beta = \dfrac{\sum_{i=1}^{n}(x_i - \bar{x})(y_i - \bar{y})}{\sum_{i=1}^{n}(x_i - \bar{x})^2}$$

Further

$$\beta = \left[\dfrac{\sum_{i=1}^{n}(x_i - \bar{x})(y_i - \bar{y})}{n}\right] \Big/ \left[\dfrac{\sum_{i=1}^{n}(x_i - \bar{x})^2}{n}\right] = \dfrac{S_{xy}}{S^2_x}$$

where $S_{xy} = \dfrac{\sum_{i=1}^{n}(x_i - \bar{x})(y_i - \bar{y})}{n}$, known as the *covariance* of x and y

$S^2_x = \dfrac{\sum_{i=1}^{n}(x_i - \bar{x})^2}{n}$, the variance of x

$\Rightarrow$ regression equation of y on x is $\quad y - \bar{y} = \dfrac{S_{xy}}{S^2_x}(x - \bar{x})$

Had the problem been changed around so that y became the independent variable, x the dependent variable and the sum of the squares of the deviations parallel to the x-axis taken, then the line of best fit would be the line of regression of x on y.

$$\Rightarrow x - \bar{x} = \frac{S_{xy}}{S^2_y} (y - \bar{y})$$

You should note:

1　You can use the regression equation of y on x to estimate y for a given value of x, but not to estimate x for a given value of y.
2　You can use the regression equation of x on y to estimate x for a given value of y, but not to estimate y for a given value of x
3　Estimation of values of x or y should be restricted to values within or close to the range of the sample since conditions outside the sample may be very different.

Worked example
Find: i) the equation of the regression line of y on x,
　　　ii) the equation of the regression line of x on y,
for the four pairs of values of the variables x and y given in the table

x	1	2	3	4
y	2	4	5	7

x	y	xy	x^2	y^2
1	2	2	1	4
2	4	8	4	16
3	5	15	9	25
4	7	28	16	49
10	18	53	30	94

$$\bar{x} = \frac{10}{4} = 2 \cdot 5 \quad \bar{y} = \frac{18}{4} = 4 \cdot 5$$

$$S_{xy} = \frac{\Sigma (x - \bar{x})(y - \bar{y})}{n} = \frac{\Sigma xy - n\bar{x}\bar{y}}{n} = \frac{53 - 45}{4} = 2$$

$$S^2_x = \frac{\Sigma (x - \bar{x})^2}{n} = \frac{\Sigma x^2 - n\bar{x}^2}{n} = \frac{30 - 25}{4} = 1 \cdot 25$$

$$\Rightarrow \text{regression line of } y \text{ on } x \text{ is } y - 4 \cdot 5 = \frac{2}{1 \cdot 25}(x - 2 \cdot 5) \quad \text{or} \quad y = 1 \cdot 6x + 0 \cdot 5$$

$$S^2_y = \frac{\Sigma (y - \bar{y})^2}{n} = \frac{\Sigma y^2 - n\bar{y}^2}{n} = \frac{94 - 81}{4} = \frac{13}{4} = 3 \cdot 25$$

$$\Rightarrow \text{regression line of } x \text{ on } y \text{ is } x - 2 \cdot 5 = \frac{2}{3 \cdot 25}(y - 4 \cdot 5) \quad \text{or} \quad x = 0 \cdot 615y - 0 \cdot 269$$

CORRELATION

In our discussion of linear regression we assumed that one of the two variates could be measured with little or no error and took this to be the independent variable. We may however have to deal with a bivariate distribution in which we are far from certain as to whether there is a relationship between the two variables and indeed if there is how close is this relationship. Can we in fact find some method of measuring the relationship, i.e. the **correlation**, between the two variates? Obviously a scatter diagram can be used to show whether any correlation exists and in this case it is immaterial as to which axis is used to represent a particular variate. Given that the scatter diagram shows signs of positive or negative correlation, i.e. that a linear relationship is apparent, then it is possible to establish a method of measuring the degree of correlation.

For variables x and y we obtained, using the method of least squares, the regression equations of y on x and x on y. They were:

$$\text{regression equation of } y \text{ on } x \Rightarrow y - \bar{y} = \frac{S_{xy}}{S^2_x}(x - \bar{x})$$

regression equation of x on $y \Rightarrow x - \bar{x} = \dfrac{S_{xy}}{S^2_y}(y - \bar{y})$

Clearly, both regression lines pass through $(\bar{x}, \bar{y})$, but have different gradients. These gradients can be used to measure the correlation.

Dividing the regression equation of y on x by S_y, the standard deviation of the

y values $\Rightarrow \dfrac{y - \bar{y}}{S_y} = \dfrac{S_{xy}}{S_x S_y}\left(\dfrac{x - \bar{x}}{S_x}\right)$

or $Y = rX$ where $Y = \dfrac{y - \bar{y}}{S_y}$, $X = \dfrac{x - \bar{x}}{S_x}$ and $r = \dfrac{S_{xy}}{S_x S_y}$

Similarly, dividing the regression equation of x on y by S_x, the standard deviation of the

x-values $\Rightarrow \dfrac{x - \bar{x}}{S_x} = \dfrac{S_{xy}}{S_x S_y}\left(\dfrac{y - \bar{y}}{S_y}\right)$

or $X = rY$

Hence, the two regression lines have equations

$Y = rX$ and $X = rY$

and

i) if $r = 1$, the two lines are coincident with positive gradient and so we have perfect positive correlation.

ii) If $r = -1$, the two lines still coincide, but with negative gradient and so we have perfect negative correlation.

iii) If $r = 0$ the two lines become $Y = 0$ and $X = 0$, two lines at right angles to each other; one parallel to the x-axis and one parallel to the y-axis, indicating zero correlation.

The quantity $r = \dfrac{S_{xy}}{S_x S_y} = \dfrac{\Sigma xy - n\bar{x}\bar{y}}{\sqrt{[(\Sigma x^2 - n\bar{x}^2)(\Sigma y^2 - n\bar{y}^2)]}}$

is called the product moment correlation coefficient.

Note:

i) r takes the sign of S_{xy},

ii) $r = 1$ indicates perfect positive correlation,

iii) $r = -1$ indicates perfect negative correlation,

iv) $r = 0$ indicates zero or no correlation.

Worked example

The data below shows the height x hundred metres above sea level (to the nearest hundred metres) of twelve towns in Austria, together with the corresponding 8·00 am air temperature (°C) on 10 June 1988. Find the product moment correlation coefficient and state any conclusion which you can draw from your result

x	4	5	5	6	6	10	11	12	13	16	17	19
y	17	17	13	17	12	4	9	12	8	3	5	7

x	y	xy	x^2	y^2
4	17	68	16	289
5	17	85	25	289
5	13	65	25	169
6	17	102	36	289
6	12	72	36	144
10	4	40	100	16
11	9	99	121	81
12	12	144	144	144
13	8	104	169	64
16	3	48	256	9
17	5	85	289	25
19	7	133	361	49
124	124	1045	1578	1568

$$\bar{x} = 10\tfrac{1}{3}$$
$$\bar{y} = 10\tfrac{1}{3}$$

$$r = \frac{\Sigma xy - n\bar{x}\bar{y}}{\sqrt{[(\Sigma x^2 - n\bar{x}^2)(\Sigma y^2 - n\bar{y}^2)]}} = \frac{1045 - 12.(10\tfrac{1}{3})^2}{\sqrt{[(1578 - 12.(10\tfrac{1}{3})^2)(1568 - 12.(10\tfrac{1}{3})^2)]}}$$

$$= \frac{1045 - 1281\cdot3}{\sqrt{[(1578 - 1281\cdot3)(1568 - 1281\cdot3)]}} = \frac{1045 - 1281\cdot3}{\sqrt{[296\cdot7 \times 286\cdot7]}} = -0\cdot81$$

The result shows good negative correlation, indicating that the higher the town is above sea level the lower the temperature is likely to be.

RANK CORRELATION

Often the variates we wish to compare are not given in the form of exact measurements, but instead are given in the form of an order or a **rank**. We may for instance be given the order in which the pupils in a class came in two examinations, say, one for physics and one for chemistry. There may of course be quite different marks obtained by a candidate who came, say, second in order in both the physics examination and the chemistry examination, but provided the actual marks do not matter, but only the order of ranking, then it is possible by using Spearman's rank correlation coefficient or Kendall's rank correlation coefficient to obtain some measure of correlation. You are not required to be able to derive either formula; you must be able to use them and know their limitations.

Consider the case of 10 students whose marks in the physics and chemistry examinations were as shown in the table.

	A	B	C	D	E	F	G	H	J	K
Physics	63	37	85	89	24	48	53	45	36	18
Chemistry	45	41	79	91	30	41	46	20	42	23

The first step is to place the marks in order of rank and we shall use 1 for the highest mark in a subject down to 10 for the lowest mark in a subject. If two or more people obtain the same mark in a subject then they are each given the mean of the ranks which they occupy. Thus B and F both obtained 41 marks in the chemistry examination; 41 occupies the 6th and 7th positions in rank and so B and F are each given a ranking of $6\tfrac{1}{2}$.

	A	B	C	D	E	F	G	H	J	K
Rank in physics	3	7	2	1	9	5	4	6	8	10
Rank in chemistry	4	$6\tfrac{1}{2}$	2	1	8	$6\tfrac{1}{2}$	3	10	5	9

SPEARMAN'S RANK CORRELATION COEFFICIENT

Spearman's **rank correlation coefficient** r_S for a sample of n pairs of ranks is defined as

$$r_S = 1 - \frac{6\sum_{r=1}^{n} d_r^{\,2}}{n(n^2 - 1)}$$

where d_r is the difference in rank value between the rth pair. (The formula can be derived directly from the product moment coefficient using the ranks as data.)

In the above example the values of d_r are

	A	B	C	D	E	F	G	H	J	K
d	-1	$\tfrac{1}{2}$	0	0	1	$-1\tfrac{1}{2}$	1	-4	3	1
d^2	1	$\tfrac{1}{4}$	0	0	1	$2\tfrac{1}{4}$	1	16	9	1

$$\Rightarrow \Sigma d^2 = 31\tfrac{1}{2}$$

$$\Rightarrow r_S = 1 - \frac{6 \times 31\cdot5}{10(10^2 - 1)} = 1 - 0\cdot1909 \approx 0\cdot81$$

Hence it would appear that there is a strong positive correlation between these students' ability in the physics and chemistry examinations.

KENDALL'S RANK CORRELATION COEFFICIENT

Kendall's formula is slightly more complicated to use, but once you have worked through an example you will soon get used to the method. The rules are as follows.

1   Place one of the samples in rank order 1, 2, 3, ..., n
2   Underneath each rank write down the rank of the corresponding member of the other sample
3   For each member of the second ranking proceed as follows. Add 1 for each rank to the right of a particular member when that rank is greater than the particular member and subtract 1 for each rank to the right of that particular member when the rank is less than the particular member
4   If S is the sum of the $+1$s and -1s for all of the members of the second ranking then **Kendall's rank correlation coefficient** r_K is defined as

$$r_K = \frac{2S}{n(n-1)}$$

Consider the previous example.

	D	C	A	G	F	H	B	J	E	K
Rank in physics	1	2	3	4	5	6	7	8	9	10
Rank in chemistry	1	2	4	3	$6\frac{1}{2}$	10	$6\frac{1}{2}$	5	8	9

For chemistry we have

$$
\begin{aligned}
1: \ &+1+1+1+1+1+1+1+1+1 = 9 \\
2: \ &+1+1+1+1+1+1+1+1 \quad = 8 \\
4: \ &-1+1+1+1+1+1+1 \quad\quad = 5 \\
3: \ &+1+1+1+1+1+1 \quad\quad\quad = 6 \\
6\tfrac{1}{2}: \ &+1+0-1+1+1 \quad\quad\quad\quad = 2 \\
10: \ &-1-1-1-1 \quad\quad\quad\quad\quad\quad = -4 \\
6\tfrac{1}{2}: \ &-1+1+1 \quad\quad\quad\quad\quad\quad\quad = 1 \\
5: \ &+1+1 \quad\quad\quad\quad\quad\quad\quad\quad = 2 \\
8: \ &+1 \quad\quad\quad\quad\quad\quad\quad\quad\quad\ = 1 \\
&\quad\quad\quad\quad\quad\quad\quad \text{Total } S = 30
\end{aligned}
$$

$$\Rightarrow r_K = \frac{2S}{n(n-1)} = \frac{60}{90} = 0.667$$

SIGNIFICANCE TESTING OF A RANK CORRELATION COEFFICIENT

If we are in doubt whether or not the value of r_S or r_K indicates a correlation (association) between the two sets of ranks, we can look up the significance of the values of r_S or of r_K in the appropriate tables. The null hypothesis will always be "no correlation exists".

i.e. $H_0: \rho_S = 0$ for Spearman
(ρ_S being the population Spearman correlation coefficient)
    $H_0: \rho_K = 0$ for Kendall
(ρ_K being the population Kendall correlation coefficient)

We look up in tables the probability that the result could have occurred for this sample simply by chance if no association exists.

For the Spearman coefficient example we have

$n = 10$, $\Sigma d^2 = 31.5$

The table gives   Prob$(\Sigma d^2 \leqslant 30)$ is 0.0029
                  Prob$(\Sigma d^2 \leqslant 36)$ is 0.0053

Interpolating we have Prob$(\Sigma d^2 \leqslant 31.5)$ is 0.0035
The probability that the results could occur by chance alone is thus only 0.35% which is very small indeed. We conclude therefore that there is strong positive correlation between the rankings of the students in physics and chemistry examinations.

Similarly for the Kendall coefficient example we have

$n = 10$, $S = 30$

The table gives $\text{Prob}(S = 29) = 0\cdot0046$
$\quad\quad\quad\quad\quad\quad\quad \text{Prob}(S = 31) = 0\cdot0023$
Interpolating
$\text{Prob}(S = 30) = 0\cdot00345$
showing that the probability the result could occur by chance alone is $0\cdot345\%$ and that therefore there is strong positive correlation between the rankings of the students in physics and chemistry examinations.

EXERCISES 26.1

1   Construct a scatter diagram to illustrate the relationship between the pairs of two variable X and Y

X	10	20	30	40	50
Y	30	25	20	15	10

2   The sag y mm at the centre of a horizontal metal bar supported at each end when a weight x kg is suspended from the mid-point is given in the table below. Assuming the x-values to be exact, use the method of least squares to express the relationship between x and y in the form $y = mx + c$

x	10	20	30	40	50	60	70
y	0·31	0·71	·078	1·11	1·15	1·40	1·71

3   The corresponding values of six pairs of the variables x and y are given in the table below

x	11	27	28	33	20	19
y	7	21	32	40	5	30

Find the equations of the regression lines of   a) y on x,   b) x on y. Calculate the product moment correlation coefficient between x and y

4   The turnover and profit for an engineering firm are given for the six consecutive years 1984–1989.

Year	1984	1985	1986	1987	1988	1989
Turnover (£10^6)	70·0	75·4	71·8	79·4	81·6	82·4
Profit (£10^6)	9·2	8·8	8·4	9·8	10·8	9·4

Calculate:
a) the product moment correlation coefficient
b) Spearman's rank correlation coefficient r_S
Using a table of critical values comment on the value of r_S obtained in b), using a 5% significance level

5   The marks obtained by 10 students in A-level examinations in   a) pure mathematics b) applied mathematics were as shown

Pure mathematics	63	79	41	34	47	43	37	39	32	28
Applied mathematics	58	56	53	24	54	55	51	26	45	48

Find the product moment correlation coefficient for the two sets of marks. Find the Spearman's rank correlation coefficient for the two examinations

6   An art master and art mistress were asked to place in order of merit the drawings of 10 children. Their decisions are shown in the table below

Pupil	A	B	C	D	E	F	G	H	J	K
Art master's rankings	1	2	3	4	5	6	7	8	9	10
Art mistress' rankings	1	1	4	1	5	8	7	6	10	9

Calculate:
a) Spearman's rank correlation coefficient
b) Kendall's rank correlation coefficient

NB: Examination questions on these topics can be found in Chapter 27.

ANSWERS TO EXERCISES

EXERCISES 26.1

2   $y = 0.02125x + 0.1743$
3   a) $y = 1.371x - 9.033$;
    b) $x = 0.424y + 13.451$; 0.763
4   a) 0.69;   b) 0.657, no correlation at 5%
    significance level
5   $0.488, 0.830$
6   $0.903, 0.711$

CHAPTER

27

BOARD EXAMINATION QUESTIONS AND ANSWERS FOR STATISTICS

GETTING STARTED

These questions have been selected from recent examination papers and relate to the statistics covered in Chapters 20–26.

EXAMINATION QUESTIONS

1 a) In a family with 3 children A, B and C, child A has an infectious disease and there is a period during which B or C (or both) may independently catch the disease from A each with probability p. At the end of the period, B and C can no longer catch the disease from A, but if B or C but not both has caught the disease, then there is another period during which the other may catch it with probability p.
   i) Find the probability that neither B nor C catches the disease.
   ii) Show that the probability that either B or C (but not both) catches the disease is $2p(1-p)^2$.
   iii) Find the probability that both B and C catch the disease.

   b) Two machines, X and Y, produce an equal number of articles. The probability of machine X producing a defective article is p and the corresponding probability for machine Y is q. If an article is drawn at random from the output of the two machines and found to be defective, find the probability that the article was produced on machine X.
   (AEB)

2 A lecturer gave a group of students an assignment consisting of two questions. The following table summarises the number of numerical errors made on each question by the group of students.

Errors on Question 2 (y)	Errors on Question 1 (x)				
	0	1	2	3	4
0				4	3
1				4	5
2			5	7	5	2
3		1	4	3	4

a) Find the product moment correlation coefficient between x and y.
b) Give a written interpretation of your answer.

The scores on each question for a random sample of 8 of the group are as shown below.

Student	1	2	3	4	5	6	7	8
Question 1	42	68	32	84	71	55	55	70
Question 2	39	75	43	79	83	65	62	68

c) Calculate the Spearman rank correlation coefficient between the scores on the two questions.
d) Give an interpretation of your result.
(AEB)

3 A certain brand of beans is sold in tins, the tins being filled and sealed by one of two machines $M1$ or $M2$. From $M1$, the mass of beans in each tin is normally distributed with mean 425 g and standard deviation 25 g and the mass of the tin is normally distributed with mean 90 g and standard deviation 10 g.

   a) Find the probability that the **total** mass of the sealed tin and its beans
      i) exceeds 550 g
      ii) lies between 466 g and 575 g

   b) Calculate an interval within which approximately 90% of the masses of the filled tins from $M1$ will lie.

   The tins from $M1$ are packed in boxes of 24, the mass of the box being normally distributed with mean 500 g and standard deviation 30 g.

   c) Find the probability that a full box weighs less than 12·75 kg.

A random sample of 10 tins was taken from the production of $M2$ and their total masses (beans and tin), measured to the nearest gram, were as follows

512, 515, 499, 528, 519, 510, 507, 522, 530, 514.

d) Find a 95% confidence interval for the mean mass of tins of beans, produced on $M2$, assuming that the masses of tins of beans from $M2$ are normally distributed with the same standard deviation as $M1$.

After a delivery of 50 boxes to a supermarket, 150 tins were found to be damaged.

e) Calculate an approximate 99% confidence interval for the proportion of damaged tins a supermarket might expect to receive. (AEB)

4 Answer the following questions using, in each case, tables of the binomial, Poisson or normal distribution according to which you think is most appropriate. In each example draw attention to any feature which either supports or casts doubt on your choice of distribution.

a) Cars pass a point on a busy city centre road at an average rate of 7 per five second interval. What is the probability that in a particular five second interval the number of cars passing will be
   i) 7 or less
   ii) exactly 7?

b) Weather records show that for a certain airport during the winter months an average of one day in 25 is foggy enough to prevent landings. What is the probability that in a period of seven winter days landings are prevented on
   i) 2 or more days
   ii) no days?

c) The working lives of a particular brand of electric light bulb are distributed with mean 1200 hours and standard deviation 200 hours. What is the probability of
   i) a bulb lasting more than 1150 hours,
   ii) the mean life of a sample of 64 bulbs exceeding 1150 hours? (AEB)

5 a) A man telephones his mother who is housebound. If her telephone is engaged he tries again until he is successful or until he has made three unsuccessful attempts. If, at each attempt, the probability of the call being successful is $0\cdot7$ and of her telephone being engaged is $0\cdot3$, what is the probability that he will be successful at
   i) the second attempt,
   ii) the third attempt?
   What is the probability that he will be successful in telephoning his mother?
   If he is successful what is the probability that it was his second attempt?

b) The man uses the same procedure for telephoning his aunt. If p is the probability of her telephone being engaged on his rth attempt, the probability of it being engaged on his $(r+1)$th attempt is $0\cdot5\,(1+p)$. If the probability of it being engaged at the first attempt is $0\cdot2$, what is the probability of it being engaged at all three attempts?

c) In the case of his aunt there is also a probability of $0\cdot4$ that at his first attempt the telephone will ring but there will be no answer. The probability of this happening at the $(r+1)$th attempt is $0\cdot8\,(1+0\cdot25q)$ where q is the probability of it happening at the rth attempt.
   i) What is the probability of it ringing but not being answered on all three attempts?
   ii) What further information would be needed in order to calculate the probability of the man making three unsuccessful attempts to telephone his aunt? (AEB)

6 Raw materials for use in an industrial process arrive by rail. A one kg sample, randomly chosen, is taken from each waggon and analysed automatically. Past experience has shown that for any waggon, the reading obtained (which is related to the percentage impurity in the raw material) is an observation from a normal distribution with standard deviation $1\cdot25$.

a) For a sample from a waggonload with a mean value 7,
   i) what is the probability of obtaining a reading of 6 or less,
   ii) what reading will be exceeded with a probability of $0\cdot995$?

b) If a waggonload with mean value 7 or more is used in the process the final product is unuseable. Current practice is to use the raw material if the reading is 6 or less, otherwise to apply a preliminary process which is certain to reduce the mean value below 7. What is the probability that a waggonload with a mean value of 5·5 will have the preliminary process applied?

c) A statistician suggests that it would be better to take five, one kg, samples at random from each waggonload and use the preliminary process if the mean of the five readings is 6 or more. If this is done what is the probability of
   i) a waggonload with mean value 7 not having the preliminary process applied,
   ii) a waggonload with mean value 5·5 having the preliminary process applied?

d) If the cost of taking each sample and obtaining a reading is £8, the cost of the preliminary process is £150, and the cost of using a waggonload with mean value 7 or more is £1500 find the expected cost for a waggonload with mean value 7 of
   i) making the decision on the basis of a single sample,
   ii) making the decision on the basis of a sample of 5.   (AEB)

7 a) In a group of 200 people, each individual is classified as either male or female and according to whether or not he or she wears glasses. The numbers falling into each category are tabulated below.

	Not wearing glasses	Wearing glasses
Male	90	24
Female	66	20

Suppose one of this group is chosen at random.
Let A be the event that the person chosen is male and B the event that the person chosen is not wearing glasses.
   i) Define the events A' and $A \cup B'$.
      (Notation: G' is the complement of G.)
   ii) Calculate the probability of occurrence of each of the events in i).
   iii) Given that the person chosen is not wearing glasses, calculate the probability that this person is male.
   iv) Use the available data to determine whether not wearing glasses is independent of sex *within the group*. Give a practical interpretation to your finding.

b) After advertising for an assistant, a manager decides to interview suitable applicants. The interview of an applicant will take place during the morning or the afternoon with probabilities 0·45 and 0·55 respectively. Each applicant is informed by telephone and in each case a message has to be left. A morning interview is wrongly transmitted to the applicant as an afternoon interview with probability 0·2, and an afternoon interview is wrongly transmitted to the applicant as a morning interview with probability 0·1.
Find the probability that an applicant arriving
   i) for a morning interview is expected for a morning interview,
   ii) for an afternoon interview is expected for an afternoon interview.   (AEB)

8 A company manufactures bars of soap. In a random sample of 70 bars, 18 were found to be mis-shaped. Calculate an approximate 99% confidence interval for the proportion of mis-shaped bars of soap.

Explain what you understand by a 99% confidence interval by considering

a) intervals in general based on the above method,
b) the interval you have calculated.

The bars of soap are either pink or white in colour and differently shaped according to colour. The masses of both types of soap are known to be normally distributed, the mean mass of the white bars being 176·2 g. The standard deviation for both bars is 6·46 g. A sample of 12 of the pink bars of soap had masses, measured to the nearest gram, as follows.
   174   164   182   169   171   187   176   177   168   171   180   175
Find a 95% confidence interval for the mean mass of pink bars of soap.
Calculate also an interval within which approximately 90% of the masses of the white bars of soap will lie.

The cost of manufacturing a pink bar of soap of mass x gm is $(15 + 0 \cdot 065x)$ p, and it is sold for 32p. If the company manufactures 9000 bars of pink soap per week, derive a 95% confidence interval for its weekly expected profit from pink bars of soap.   (AEB)

Lifetime (to nearest hour)	Number of discs
690–709	3
710–719	7
720–729	15
730–739	38
740–744	41
745–749	35
750–754	21
755–759	16
760–769	14
770–789	10

9   The table opposite shows the lifetimes of a random sample of 200 mass produced circular abrasive discs.

a) Without drawing the cumulative frequency curve, calculate estimates of the median and quartiles of these lifetimes.

b) One method of estimating the skewness of a distribution is to evaluate

$$\frac{3(\text{mean} - \text{median})}{\text{standard deviation}}$$

Carry out the evaluation for the above data and comment on your result. Use the quartiles to verify your findings.

c) The discs should have a radius of 5 cm but in practice, the values of the radius are found to be uniformly distributed over the range 4·95 to 5·15 cm. Explain why

$$\text{median of area} = \pi(\text{median of the radius})^2.$$

d) Obtain the value A_L for the area of a disc such that 25% of the discs have an area less than A_L.   (AEB)

10   i) A school video library contains 15 different films, of which 9 relate to educational matters, 4 to sport and 2 to leisure.
      Calculate the number of different selections of 4 films from the library if
      a) no restriction is placed on the selection,
      b) exactly 2 films must be educational,
      c) at least one film does not relate to education.

   ii) Alan transmits messages to Brian along an unreliable communication system. The system can only transmit strings of the two digits 0 and 1, so that both Alan and Brian have a code enabling them to translate messages into such strings. The word PROCEED is coded 00101 and the word REVERSE is coded 11001. However, independent random distortions of the digits 0 and 1 occur during transmission. The probability that a 0 is distorted to a 1 on transmission is α and the probability that a 1 is distorted to a 0 is β.
   Brian receives the message 11111, knowing it is a distorted form of either PROCEED or REVERSE. Before reading the message he believed that Alan was likely to send the message PROCEED with probability p. Having read the message, show that the probability that Alan actually sent the message PROCEED is given by

$$\frac{p\alpha}{p\alpha + (1-p)(1-\beta)}$$ (AEB)

11   In an extraction process, an ingredient can be obtained from raw material by one of two methods, P or Q. For a fixed volume of raw material, the amount X cm^3 of ingredient extracted by method P is normally distributed with mean 10 and standard deviation 2. The amount, Y cm^3, extracted from the same volume of raw material using method Q is distributed with probability density function
   $f(y) = k(y-8)$ $8 \leqslant y \leqslant 12$
   $= 0$ otherwise

a) Find the value of k.

b) Show that method Q has the greater probability of extracting more than 11 cm^3 of the ingredient.

c) Find which method extracts, on average, the greater amount of the ingredient.

The cost of applying method P is 6p per cm^3 extracted and the cost of extracting Y cm^3 using method Q is $(15 + 5Y)$p. The extracted ingredient is sold at 10p per cm^3.

d) For a fixed volume of raw material show that method P gives the higher expected profit.

e) Show that the value of v such that

$$P(X > v) = P(Y > v)$$

lies between 11·50 and 11·55.   (AEB)

12   Define a ranking scale and give an example to illustrate your definition. Explain how you would rank values of equal magnitude.

At the end of the academic year students on a particular course are given examinations in Sociology (S), Social Administration (SA) and Quantitative Methods (QM). The final grade awarded to each student is based on the total of the marks scored on the three papers. The table below shows the marks obtained by a sample of ten students who sat the three papers.

Student	Sociology (S)	Social Administration (SA)	Quantitative Methods (QM)
1	66	48	44
2	50	46	48
3	44	46	47
4	58	72	64
5	64	68	54
6	26	64	55
7	74	65	59
8	67	42	48
9	36	40	56
10	48	55	48

The following matrix of Spearman rank correlation coefficients was obtained for this sample of ten students.

	S	SA	QM	Total mark
S	1	0.24	−0.01	0.78
SA		1	x	0.77
QM			1	y
Total mark				1

Find the values of x and y.

It has been decided that in future, students should only be required to sit two papers. Use these data to decide which two examinations should be used. Give a reason for your choice.   (AEB)

13   During an epidemic of a certain disease a doctor is consulted by 110 people suffering from symptoms commonly associated with the disease. Of the 110 people, 45 are female of whom 20 actually have the disease and 25 do not. Fifteen males have the disease and the rest do not.

a) A person is selected at random. The event that this person is female is denoted by A and the event that this person is suffering from the disease is denoted by B. Evaluate.
   i) $P(A)$,
   ii) $P(A \cup B)$,
   iii) $P(A \cap B)$,
   iv) $P(A|B)$.

b) If three different people are selected at random without replacement what is the probability of
   i)   all three having the disease,
   ii) exactly one of the three having the disease,
   iii) one of the three being a female with the disease, one a male with the disease and one a female without the disease?

c) Of people with the disease 96% react positively to a test for diagnosing the disease as do 8% of people without the disease. What is the probability of a person selected at random
   i)   reacting positively,
   ii) having the disease given that he or she reacted positively?   (AEB)

14 A set of bathroom scales is known to operate with an error which is normally distributed with standard deviation 0·27 kg. Before taking a bath a man weighs himself three times with the following results (kg):

84·6, 84·8, 84·1

Find a 95% confidence interval for his mass assuming the scales are unbiased.
He wishes to obtain a more accurate estimate of his mass.

a) How many independent weighings would be necessary to obtain a 95% confidence interval width less than 0·3 kg?

b) What would be the percentage confidence associated with an interval of width 0·3 kg calculated from the data above?

c) If he decides to buy a better set of scales what must the standard deviation of the error be so that three weighings will give a 95% confidence interval of width 0·3 kg (assume normal distribution)?

Comment briefly on the advantages or disadvantages of each of these methods of attempting to improve the accuracy of the estimate.   (AEB)

15 A large civil engineering firm issues all new employees with a safety helmet. Five different sizes are available numbered 1 to 5.
A random sample of 90 employees required the following sizes:

```
2 4 2 2 2 5 4 5 4 4 4 2 4 3 4
2 3 1 5 4 3 2 3 3 3 3 3 4 3 2
4 4 3 4 4 5 3 3 3 2 4 4 2 2 3
2 3 2 3 3 5 4 2 3 4 2 4 2 2 3
2 3 4 2 3 4 5 2 3 3 2 4 3 2 2
3 3 3 2 3 4 2 3 2 4 3 3 2 2 3
```

a) Calculate an approximate 90% confidence interval for the proportion of employees requiring size 2.
b) If the proportions of employees requiring particular sizes were the same in the population of all employees as in the sample what is the probability that
i) of 16 new employees, 2 or less would require size 3,
ii) of 155 new employees, 5 or more would require size 1? (You may assume that the probability of 8 or more employees requiring size 1 is negligible.)

(Assume the new employees are a random sample from the population.)   (AEB)

16 Data, from a completed questionnaire are entered into a computer as a series of n binary digits (i.e. each digit is 0 or 1). If the probability of an error is a constant, p, for each digit entered, write down in terms of n and p the probability that the data entered will contain exactly r errors.
If $n = 7$ and $p = 0·12$ calculate the probability of

a) exactly one error,

b) two or more errors.

As a check on the accuracy of the entry the data are entered a second time, independently of the first, and the computer detects an error if the two entries are not identical. The probability of an error in the second entry is also p for each digit entered. If the first entry contains exactly one error what is the probability that no error is detected?

c) Show that the probability that the first entry contains exactly one error, and that no error is detected, is $np^2(1-p)^{2(n-1)}$.

d) Show further that the probability that the first entry contains one or more errors, and that no errors are detected, is

$$\sum_{r=1}^{n} \binom{n}{r} p^{2r} (1-p)^{2(n-r)}$$

If $n = 7$ and $p = 0·12$ calculate, correct to three decimal places, the probability of the first entry containing

e) no errors,

f) error(s) which are not detected (*only the first two terms of the summation need be evaluated*),

g) error(s) which are detected.   (AEB)

17 A sweet shop sells chocolates which appear, at first sight, to be identical. Of a random sample of 80 chocolates, 61 had hard centres and the rest soft centres. Calculate an approximate 99% confidence interval for the proportion of chocolates with hard centres.

The chocolates are in the shape of circular discs and the diameters (cm) of the 19 soft centred chocolates were:

2·79 2·63 2·84 2·77 2·81 2·69 2·66 2·71 2·62 2·75
2·77 2·72 2·81 2·74 2·79 2·77 2·67 2·69 2·75

The mean diameter of the 61 hard centred chocolates was 2.690 cm. If the diameters of both hard centred and soft centred chocolates are known to be normally distributed with standard deviation 0·042 cm, calculate a 95% confidence interval for the mean diameter of

a) the soft centred chocolates,

b) the hard centred chocolates.

Calculate also an interval within which approximately 95% of the diameters of hard centred chocolates will lie.

Discuss, briefly, how useful knowledge of the diameter of a chocolate is in determining whether it is hard or soft centred.   (AEB)

18 A meat wholesaler sells remnants of meat in 5 kg bags. The amount, in kg, of inedibles (i.e. bone and gristle) in a bag is a random variable, X, with probability density function

$$f(x) = \begin{cases} k(x-1)(3-x) & 1 \leqslant x \leqslant 3 \\ 0 & \text{otherwise.} \end{cases}$$

a) Show that $k = \frac{3}{4}$.

b) Find the mean and variance of X.

c) Find the probability that X is greater than 2·5 kg.

A butcher buys the bags and can use them in one of two processes:

i) The complete contents (including inedibles) are ground up and made into standard sausages. If the bag contains more than 2·5 kg of inedibles the resulting sausages are unsaleable and a loss of £2·20 is made on the bag, otherwise a profit of 70p is made on each bag.

ii) The inedibles are extracted and the remainder made into premium sausages. The profit on this process is $200-80X$ pence per bag.

By comparing the expected profit per bag, advise the butcher which process to use.   (AEB)

19 An unbiased cubical die has the number 1 on one face, the number 2 on two faces and the number 3 on three faces. The die is rolled twice and X is the total score. Find the probability distribution of X.

Show that $E(X) = \frac{14}{3}$ and Var $(X) = \frac{10}{9}$

A sample of 40 observations of X is taken. Find, to three decimal places, the probability that the mean of this sample exceeds $\frac{13}{3}$.   (C)

20 On average a Coastguard station receives one distress call every two days. A "bad" week is a week in which 5 or more distress calls are received. Show that the probability that a week is a bad week is 0·275, correct to three significant figures.

Find the probability that, in 8 randomly chosen weeks, at least 2 are bad weeks.

Find the probability that, in 80 randomly chosen weeks, at least 30 are bad weeks.   (C)

21   There are many complaints from passengers about the late running of trains in Ruritania. The Ruritanian Railway Board claims that the proportion of express trains that are "delayed" (i.e. at least 5 minutes late) is 43%. The Railway Passengers' Association conducts a study of a random sample of 200 trains and finds that 92 of these trains are delayed. Test, at the 8% level, whether the Railway Board is understating the proportion of trains that are delayed.

The Association also notes the time t, in seconds, by which each train in the sample is late (no trains are early). The results are summarised by $\Sigma(t-300) = 2012$, $\Sigma(t-300)^2 = 525\,262$. Find a symmetric 95% confidence interval for the mean time by which a train is late.   (C)

22   In a sales campaign, a petrol company gives each motorist who buys their petrol a card with a picture of a film star on it. There are 10 different pictures, one each of 10 different film stars, and any motorist who collects a complete set of all 10 pictures gets a free gift. On any occasion when a motorist buys petrol, the card received is equally likely to carry any one of the 10 pictures in the set.
   i) Find the probability that the first four cards the motorist receives all carry different pictures.
   ii) Find the probability that the first four cards received result in the motorist having exactly three different pictures.
   iii) Two of the ten film stars in the set are X and Y. Find the probability that the first four cards received result in the motorist having a picture of X or of Y (or both).
   iv) At a certain stage the motorist has collected nine of the ten pictures. Find the least value of n such that
   P(at most n more cards are needed to complete the set) $> 0 \cdot 99$.   (C)

23   The continuous random variable X has cumulative distribution function given by
$$F(x) = \begin{cases} 0, & \text{for } x < 0, \\ 2x - x^2, & \text{for } 0 \leqslant x \leqslant 1, \\ 1, & \text{for } x > 1. \end{cases}$$
   i) Find $P(X > \frac{1}{2})$.
   ii) Find the value of q such that $P(X < q) = \frac{1}{4}$.
   iii) Find the probability density function of X, and sketch its graph.
   iv) Find $E(X)$ and $E(\sqrt{X})$.   (C)

24   At an early stage in analysing the marks scored by the large number of candidates in an examination paper, the Examining Board takes a random sample of 250 candidates and finds that the marks, x, of these candidates give $\Sigma x = 11\,872$ and $\Sigma x^2 = 646\,193$. Calculate a 90% confidence interval for the population mean mark (μ) for this paper.

Using the figures obtained in this sample, the null hypothesis $\mu = 49 \cdot 5$ is tested against the alternative hypothesis $\mu < 49 \cdot 5$ at the α% significance level. Determine the set of values of α for which the null hypothesis is rejected in favour of the alternative hypothesis.

It is subsequently found that the population mean and standard deviation for the paper are $45 \cdot 292$ and $18 \cdot 761$ respectively. Find the probability of a random sample of size 250 giving a sample mean at least as high as the one found in the sample above.   (C)

25   In each month, the proportion of "Prize" bonds that win a prize is 1 in 11 000. There is a large number of prizes and all bonds are equally likely to win each prize. Show that, for a given month, the probability that a bondholder with 5000 bonds wins at least one prize is $0 \cdot 365$, correct to three significant figures.

For a given month find

   i) the probability that, in a group of 10 bondholders each holding 5000 bonds, four or more win at least one prize,
   ii) the probability that, in a group of 100 bondholders each holding 5000 bonds, 40 or more win at least one prize.

Find the expected number of prizes for a bondholder holding 550 bonds for 24 months.   (C)

26   A fruit grower uses a machine to sort apples into various grades. Grade C apples have weights uniformly distributed in the interval 100 to 110 grams. Find the variance of the weight of a grade C apple.

Ten randomly chosen grade C apples are packed in a bag. Using the central limit theorem, find an approximate value for the probability that the weight of the ten apples in the bag exceeds 1030 grams.

The grower suspects that the machine is not working correctly and that the mean weight, μ grams, of a grade C apple may be less than 105 grams. Devise a test, at the 10% level of significance, based on the weight of the apples in five randomly chosen bags, each containing ten apples, of the null hypothesis $\mu = 105$, with alternative hypothesis $\mu < 105$. (C)

27 In a public opinion poll, 1000 randomly chosen electors were asked whether they would vote for the "Purple Party" at the next election and 357 replied "Yes". Find a 95% confidence interval for the proportion p of the population who would answer "Yes" to the same question.

Twenty similar polls are taken and the 95% confidence interval is determined for each poll. State the expected number of these intervals which will enclose the true value of p.

The leader of the "Purple Party" believes that the true value of p is 0·4. Test, at the 8% level, whether he is overestimating his support. (C)

28 An electronic device is advertised as being able to retain information stored in it "for 70 to 90 hours" after power has been switched off. In experiments carried out to test this claim, the retention time in hours, X, was measured on 250 occasions, and the data obtained is summarised by $\Sigma(x-76) = 683$ and $\Sigma(x-76)^2 = 26\,132$. The population mean and variance of X are denoted by μ and σ^2 respectively.

i) Show that, correct to one decimal place, an unbiased estimate of σ^2 is 97·5.
ii) Test the hypothesis that $\mu = 80$ against the alternative hypothesis that $\mu < 80$, using a 5% significance level.
iii) Calculate a symmetric 95% confidence interval for μ. (C)

29 A bag contains 5 white balls and 3 red balls. Two players, A and B, take turns at drawing one ball from the bag at random, and balls drawn are not replaced. The player who first gets two red balls is the winner, and the drawing stops as soon as either player has drawn two red balls. Player A draws first. Find the probability

i) that player A is the winner on his second draw,
ii) that player A is the winner, given that the winning player wins on his second draw,
iii) that neither player has won after two draws each, given that A draws a red ball on his first draw. (C)

30 The continuous random variable X has probability density function given by

$$f(x) = \begin{cases} \dfrac{k}{x} & \text{for } 1 \leq x \leq 9, \\ 0 & \text{otherwise,} \end{cases}$$

where k is a constant. Giving your answers correct to three significant figures where appropriate,

i) find the value of k, and find also the median value of X,
ii) find the mean and variance of X,
iii) find the cumulative distribution function, F, of X, and sketch the graph of $y = F(x)$. (C)

31 The random variable X is normally distributed with unknown mean μ cm and standard deviation 2 cm.
i) A random sample of 16 observations of X had values which summed to 118·4 cm. Calculate a 95% confidence interval for the value of μ.
ii) Find the smallest sample size, n, of observations of X that should be taken for the width of the 95% confidence interval for μ calculated from the sample values to be less than 1 cm. (WJEC)

32 In an investigation of the relationship $y = \alpha + \beta x$ connecting the two variables x and y, five experiments were conducted with x having the values 20, 30, 40, 50 and 60, respectively, and the corresponding values of y were measured. The least-squares estimate of the relationship was calculated from the results to be
$$y = 3 \cdot 4 - 0 \cdot 65x.$$
Assuming that the errors in the y-measurements are independent and normally distributed with mean zero and standard deviation $0 \cdot 2$, calculate 90% confidence limits for the values of
i) α,
ii) β. (WJEC)

33 The continuous random variable X has probability density function f, where
$$f(x) = \frac{25}{12(x+1)^3} \quad \text{for } 0 \leq x \leq 4,$$
$$f(x) = 0, \quad \text{otherwise.}$$
i) Evaluate $E[X+1]$. Hence, or otherwise, find the mean of X.
ii) Find the value of $c > 0$ for which $P(X \leq c) = c$.
iii) Find the cumulative distribution function of
$$Y = (X+1)^{-2}.$$
Hence, or otherwise, find the probability density function of Y. (WJEC)

34 a) Five independent measurements of the diameter of a ball bearing were made using a certain instrument. The results obtained, in millimetres, were:
8·9, 9·1, 9·1, 8·9, 8·9.
i) Given that the true diameter of the ball bearing is 9 mm, calculate unbiased estimates of the mean and the variance of the measurement error of the instrument.
ii) Assuming that the measurement errors are independent and normally distributed, calculate 90% confidence limits for the mean measurement error.
b) The weekly wages received by a random sample of 80 personnel employed in factory A had a mean of £86·40 and a standard deviation of £3·80. Use this information to calculate an approximate 99% confidence interval for the mean weekly wage of all personnel employed in factory A.

The weekly wages received by a random sample of 100 personnel employed in factory B had a mean of £87·60 and a standard deviation of £5·20. Calculate an approximate 95% confidence interval for the difference between the mean weekly wages in the two factories. State, with your reason, whether or not your interval discredits the claim that the mean weekly wages in the two factories are equal. (WJEC)

35 The continuous random variable X had the probability density function f, where
$$f(x) = \tfrac{1}{2}(x-2), \text{ for } 2 \leq x \leq 4,$$
$$f(x) = 0, \quad \text{otherwise.}$$

By first expanding $(X-c)^2$, or otherwise, find two values of c such that $E[(X-c)^2] = \tfrac{2}{3}$.

36 Of the ten cards in a pack, five are numbered 1, three are numbered 2, and the remaining two cards are numbered 3. Three cards are selected at random without replacement from the pack. Calculate the probabilities that

i) all three selected cards have the same number,
ii) the numbers on the three selected cards are 1, 2 and 3 (in any order),
iii) the largest (or equal largest) of the numbers on the selected cards is 2.

Let X denote the sum of the numbers on the three selected cards.

iv) Find the probability distribution of X.
v) Deduce the most probable value of the sum of the numbers on the seven cards remaining in the pack. (WJEC)

		x		
		0	1	2
	1	0	0.05	0
y	2	0.2	0	0.15
	3	0	0.6	0

37 The two discrete random variables X and Y have the joint probability distribution displayed in the table opposite.

i) Show that X and Y are not independent.

ii) Evaluate $E(XY)$.

iii) Derive the joint probability distribution of $U = Y + X$ and $W = Y - X$. Show that U and W are independent. Express $Y - 3X$ in terms of U and W and hence evaluate $\mathrm{Var}(Y - 3X)$. (WJEC)

38 X and Y are independent random variables having Poisson distributions with means 2 and 4, respectively. Find, to four decimal places in each case, the probabilities that a random observation of X and a random observation of Y will be such that

i) each has the value 2,

ii) at least one of them has a value of 2 or more,

iii) one of them has the value 2 and the other has a value less than 2.

Evaluate iv) $E(2^X)$;

v) $E(2^{X-Y})$. (WJEC)

39 The time, in hours, required to roast a chicken of weight w kg is a normally distributed random variable having mean $(0 \cdot 4w + 2 \cdot 2)$ and standard deviation $0 \cdot 05w$.

i) Find, to three decimal places, the probability that at least $3\frac{1}{4}$ hours will be required to roast a chicken weighing 2 kg.

ii) Find the weight in kg, to three decimal places, of a chicken for which there is a probability of $0 \cdot 95$ that it will require less than 4 hours to roast it.

iii) Find, to three decimal places, the probability that a chicken weighing 4 kg will require at least half an hour longer to roast than a chicken weighing 2 kg.

Three chickens, two weighing 2 kg and one weighing 4 kg, are to be roasted successively in random order. Find, to three decimal places, the probabilities that

iv) the first chicken roasted will require at least $3\frac{1}{4}$ hours,

v) the total time required to roast all three chickens will be less than $9\frac{1}{2}$ hours. (WJEC)

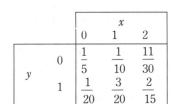

40 The two discrete random variables X and Y have the joint probability distribution displayed in the table opposite.

i) Determine whether or not X and Y are independent.

ii) Show that $E[XY] = E[X]E[Y]$. (WJEC)

41 The amount of a certain chemical in a type A cell is normally distributed with mean 10 and standard deviation 1, while that in a type B cell is normally distributed with mean 14 and standard deviation 2. To determine whether a cell is of type A or of type B, the amount of the chemical in the cell is measured and the cell is classified as being of type A if the amount is less than a specified value c, and as being of type B otherwise.

i) If $c = 12$ calculate, correct to three decimal places, the probability that a type A cell will be misclassified and the probability that a type B cell will be misclassified.

ii) Find the value of c for which the two probabilities of misclassification will be equal.

iii) Calculate, correct to three decimal places, the probability that the sum of the amounts of the chemical in one type A cell and two type B cells will be greater than 41.

iv) One cell is chosen at random from a collection of cells of which 70% are of type A and 30% are of type B. Calculate, correct to three decimal places, the probability that the amount of the chemical in the chosen cell will be at least 12. (WJEC)

42 An analysis of the membership of a large organisation shows that 60% of the members are over 50 years old and $2 \cdot 5\%$ are under 20 years old.

i) Using appropriate distributional approximations *when necessary*, calculate the probabilities, correct to three decimal places, that a random sample of 100 members will include

a) 60 or more members over 50 years old,

b) 3 or fewer members under 20 years old,

c) exactly 40 members aged from 20 years to 50 years, inclusive.

ii) Use a normal approximation to find the smallest number of members that should be sampled at random in order that the probability is at least 0·9 that 55% or more of the sampled members will be over 50 years old.   (WJEC)

43 The score X that may be obtained in one play of a certain game is a discrete random variable whose probability distribution is shown in the following table.

x	0	1	2	3
$P(X = x)$	0·4	0·2	0·2	0·2

i) Find the mean and the variance of X.
ii) Let $\bar{X}$ denote the mean of the scores in 136 independent plays of the game. Write down the mean value of $\bar{X}$ and show that its standard deviation is 0·1. Using a normal approximation to the sampling distribution of $\bar{X}$, find, correct to three decimal places, the probability that $\bar{X}$ will be less than unity.   (WJEC)

44 It is known that the two variables u and y are related in the form

$$y = \alpha + \frac{\beta}{u}$$

for values of u in the interval $0 < u \leqslant 1$. An experiment was conducted in which the value of y was observed for each of four specific values of u. The values of u and the corresponding observed values of y are given in the following table.

u	0·1	0·2	0·5	1·0
y	4·5	9·6	14·2	15·1

i) By introducing a new variable instead of u, show that the least-squares estimate of β is equal to $-1\cdot2$, and find the least-squares estimate of α.
ii) The errors in the observed values of y are independent and normally distributed with mean zero and standard deviation 0·7. Calculate 95% confidence limits for
a) the true value of y when $u = 0\cdot4$,
b) the true increase in the value of y when the value of u is increased from 0·5 to 1·0.   (ODLE)

45 Index numbers are a frequently encountered use of statistics. Describe what you understand by an index. (You may refer to any familiar index to illuminate your answer.)
A statistician is compiling an index of fish prices. Show, using the information in the table given, that the weighted mean unit price for 1981 is £274. Calculate the equivalent 1985 figure and from this find the price index for 1985 using 1981 as base year. What price change is reflected in this index?

	1981		1985	
	Unit Price £	Weight	Unit Price £	Weight
Cod	555	164	749	121
Haddock	380	256	510	247
Saithe	269	25	265	25
Whiting	305	108	422	73
Herring	130	99	121	180
Mackerel	96	348	106	354

(ODLE)

46 During the course of one year a tutor marked 111 assignments. The grades he awarded and the comparable national proportions are given in the table:

Grade	A	B	C	D
Number he awarded	86	18	6	1
National proportion	71%	16%	7%	6%

Calculate the expected numbers (to 1 decimal place) based on the national proportions. The χ^2 goodness of fit test requires the summation of terms of the form $(O-E)^2/E$

where O and E are observed and expected frequencies. Suggest reasons why

a) the difference between O and E is used
b) this difference is squared, and
c) the squared difference is divided by E.

Test, at the 5% level, whether there is any difference between the tutor's and the national awarding of grades. State your conclusions clearly. (ODLE)

47 Under what circumstances would you reasonably expect to be able to use the Binomial Distribution to model a probability distribution? When may a Binomial Distribution be approximated by a Normal Distribution?
Recent astronomical observations indicate that, of the 16 stars closest to our Sun, about half are accompanied by an orbiting planet at least the size of Jupiter.

a) Assume that the proportion of such stars in the Galaxy is 50%. Using standard cumulative probability tables, write down the probability that, in a group of 16 stars,
   i) exactly 8
   ii) at least 8
   have such a planetary system. Explain, carefully, how it is possible to state that there is a 5% chance of there being more than 12 or less than 5 in such a group.
b) The Pleiades are a cluster of some 500 stars. Use an appropriate approximation to determine the probability that there are between 230 and 270 (inclusive) stars in the Pleiades with accompanying planets at least the size of Jupiter. (ODLE)

48 Close encounters involving scheduled aircraft in the air space over Britain are recorded by the Civil Aviation Authority as near misses. In 1977 there were 45 near misses. Write down the mean monthly number of such incidences.
Explain briefly why it may be appropriate to model the occurrence of these near misses using the Poisson Distribution. Under what circumstances may a Poisson Distribution be approximated using a Normal Distribution?
Calculate the probability that, in June 1977, there were

a) exactly three
b) at least three near misses. (ODLE)

Obtain approximate 99% confidence limits for the mean number of near misses **per year** carefully explaining the steps you take to arrive at your answer. Why is the word approximate used here?
In 1986, there were 16 near misses; is there sufficient evidence to suggest that air safety has improved? How would you modify your answer in the light of the additional knowledge that there were over 40% more scheduled aircraft movements in 1986 compared with 1977?

49 State conditions under which the Poisson distribution is a suitable model to use in statistical work. Describe briefly how a Poisson distribution was used, or could have been used, in a project.
The number, X, of breakdowns per day of the lifts in a large block of flats has a Poisson distribution with mean $0 \cdot 2$. Find, to 3 decimal places, the probability that on a particular day

a) there will be at least one breakdown,
b) there will be at most two breakdowns.

Find, to 3 decimal places, the probability that, during a 20 day period, there will be no lift breakdowns.
The maintenance contract for the lifts is given to a new company. With this company it is found that there are 2 breakdowns over a period of 30 days. Perform a significance test at the 5% level to decide whether or not the number of breakdowns has decreased. (L)

50 The random variables X_1 and X_2 are both normally distributed such that $X_1 \sim N(\mu_1, \sigma_1^2)$ and $X_2 \sim N(\mu_2, \sigma_2^2)$. Given that $\mu_1 < \mu_2$ and $\sigma_1^2 < \sigma_2^2$, sketch both distributions on the same diagram.

State the "2σ rule" for a normal random variable. Explain how you used, or could have used, a normal distribution in a project.

The weights of vegetable marrows supplied to retailers by a wholesaler have a normal distribution with mean 1·5 kg and standard deviation 0·6 kg. The wholesaler supplies 3 sizes of marrow:

Size 1, under 0·9 kg,
Size 2, from 0·9 kg to 2·4 kg,
Size 3, over 2·4 kg.

Find, to 3 decimal places, the proportions of marrows in the three sizes.

Find, in kg to one decimal place, the weight exceeded on average by 5 marrows in every 200 supplied.

The prices of the marrows are 16p for Size 1, 40p for Size 2 and 60p for Size 3. Calculate the expected total cost of 100 marrows chosen at random from those supplied.   (L)

51   State the conditions under which the binomial distribution is a suitable model to use in statistical work. Describe briefly how you used, or could have used, a binomial distribution in a project, giving the parameters of your distribution.

A large store sells a certain size of nail either in a small packet at 50p per packet, or loose at £3 per kg. On any shopping day the number, X, of packets sold is a random variable where $X \sim B(8, 0·6)$, and the weight, Y kg, of nails sold loose is a continuous random variable with probability density function f given by

$$f(y) = \frac{2(y-1)}{25}, 1 \leqslant y \leqslant 6,$$

$$f(y) = 0, \qquad \text{otherwise}$$

Find, to 3 decimal places, the probability that, on any shopping day, the number of packets sold will be

a) more than one,
b) seven or fewer,

Find the probability that

c) the weight of nails sold loose on any shopping day will be between 4 kg and 5 kg,
d) on any one shopping day the shop will sell exactly 2 packets of nails and less than 2 kg of nails sold loose, giving your answer to 2 significant figures.
e) Calculate the expected money received on any shopping day from the sale of this size of nail in this store.   (L)

52   i) Six fuses, of which two are defective and four are good, are to be tested one after another in random order until both defective fuses are identified. Find the probability that the number of fuses that will be tested is
a) three
b) four or fewer.

ii) A random variable R takes the integer value r with probability p(r) where
$$p(r) = kr^3, \quad r = 1, 2, 3, 4$$
$$p(r) = 0, \qquad \text{otherwise.}$$

Find
a) the value of k, and display the distribution on graph paper
b) the mean and the variance of the distribution,
c) the mean and the variance of $5R - 3$.   (L)

53   Describe briefly how the Central Limit Theorem may be demonstrated.

The distance driven by a long distance lorry driver in a week is a normally distributed variable having mean 1130 km and standard deviation 106 km. Find, to 3 decimal places, the probability that in a given week he will drive less than 1000 km.

Find, to 3 decimal places, the probability that in 20 weeks his average distance driven per week is more than 1200 km.

New driving regulations are introduced and, in the first 20 weeks after their introduction, he drives a total of 21 900 km. Assuming that the standard deviation of the weekly distances he drives is unchanged, test, at the 10% level of significance, whether his mean weekly driving distance has been reduced. State clearly your null and alternative hypotheses.   (L)

54   Explain how you used, or could have used, a correlation coefficient to analyse the results of an experiment. State briefly when it is appropriate to use a rank correlation coefficient rather than a product-moment correlation coefficient.

Seven rock samples taken from a particular locality were analysed. The percentages, C and M, of two oxides contained in each sample were recorded. The results are shown in the table.

Sample	1	2	3	4	5	6	7
C	0·60	0·42	0·51	0·56	0·31	1·04	0·80
M	1·06	0·72	0·94	1·04	0·84	1·16	1·24

Given that
$$\Sigma CM = 4 \cdot 459, \quad \Sigma C^2 = 2 \cdot 9278, \quad \Sigma M^2 = 7 \cdot 196,$$
find, to 3 decimal places, the product-moment correlation coefficient of the percentages of the two oxides.

Calculate also, to 3 decimal places, a rank correlation coefficient.

Using the tables provided state any conclusions which you draw from the value of your rank correlation coefficient. State clearly the null hypothesis being tested.   (L)

55   Define in words and in symbols the meaning of the terms 'mutually exclusive' and 'independent' applied to two simple events E and F.

Given that the events E and F are independent, show that the events E and F' are also independent.

A research study indicates that 16% of the total population are aged 60 years or more and that 18% of the total population have a measurable hearing defect. Also, 65% of those aged 60 or more have a measurable hearing defect. Given that a person is chosen at random from the population find, to 3 decimal places, the probability that the chosen person is

a) less than 60 years old and has a measurable hearing defect,
b) aged 60 years or over and has no measurable hearing defect,
c) aged 60 years or over, given that the person has no measurable hearing defect,
d) either aged 60 years or over or has a measurable hearing defect or both.

Find, to 3 decimal places, the probability that, if two persons are chosen at random from the population, at least one of them will be aged 60 years or over and at least one of them will have a measurable hearing defect.   (L)

56   i) A factory uses four raw materials A, B, C and D to produce motor-car tyres. The masses of the materials used in the production are in the ratios $1 : 2 : 5 : 3$ respectively. The prices of the materials, to the nearest pound sterling per tonne, in the years 1984 and 1987 are shown in the table.

	A	B	C	D
1984	60	42	31	12
1987	82	54	52	21

Using 1984 as the base year, calculate an index number for the total cost of the materials used in the manufacture of tyres in 1987.

ii) The queuing time, X minutes, of a traveller at the ticket office of a large railway station has probability density function, f, defined by
$$f(x) = kx(100 - x^2), \quad 0 \leqslant x \leqslant 10,$$
$$f(x) = 0, \qquad\qquad \text{otherwise.}$$
Find

a) the value of k,
b) the mean of the distribution,
c) the standard deviation of the distribution to 2 decimal places,
d) the probability that a traveller at the ticket office will have to queue for more than 2 minutes.

Given that 3 travellers go independently to the booking office, find, to 2 significant figures, the probability that one has to queue for less than one minute, one has to queue for between one and two minutes and one has to queue for more than two minutes.   (L)

57 Give an example of data that you have met in which linear regression was used.
The amount of tetraethyl lead present (Lppm) in grass alongside a motorway was measured at varying distances (Dm) from the outer edge of the hard shoulder of the motorway. A summary of the results, in terms of $D' = \lg D$ and $L' = \lg L$, is as follows:

$$\bar{D}' = 2\cdot034, \qquad \bar{L}' = 2\cdot016,$$
$$\Sigma(D'-\bar{D}')^2 = 0\cdot7310,$$
$$\Sigma(L'-\bar{L}')^2 = 0\cdot3844,$$
$$\Sigma(D'-\bar{D}')(L'-\bar{L}') = -0\cdot5190.$$

Find the equation of the line of regression of $\lg L$ on $\lg D$ in the form $\lg L = a + b\lg D$, giving the values of a and b to 2 decimal places.
Explain why, in this problem, you would expect the value of b to be negative.
Find the product moment correlation coefficient between $\lg L$ and $\lg D$, and interpret your result in words.
Express L in terms of D and estimate, to 2 decimal places, the amount of lead present at 150 m from the outer edge of the hard shoulder.   (L)

58 i) Cotton cloth produced by a factory is tested for shrinkage by immersion in water at 10°C for one hour. It is then removed and the shrinkage is measured. Three tests are carried out. In the first, second and third tests 50, 25 and 35 strips of cloth respectively are put into the water. A summary of the results obtained is shown in the table:

Test	Size of sample	Mean (%)	SD (%)
Shrinkage of cotton cloth strips			
1	50	2·3	0·25
2	25	2·2	0·30
3	35	2·0	0·10

Find, to 2 decimal places, the mean and the standard deviation of the percentage shrinkages of the 110 strips of cloth.

ii) The weighted mean cost in 1984 of 5 raw materials, A, B, C, D, E, used in the production of a car component was £1·00. From 1984 to 1985 the costs of A, B, C, D and E rose by 12%, 10%, 32%, 10% and 20% respectively. Assuming that the costs of the 5 materials were equal in 1984 and that the quantities used in the production are in the ratios $2:1:3:3:1$ respectively, find the index for 1985 relative to 1984 as a base.   (L)

59 i) Write down the mean and the variance of the distribution of the means of all possible samples of size n taken from an infinite population having mean μ and variance σ^2.
Describe the form of this distribution of sample means when
a) n is large,
b) the distribution of the population is normal.
Explain briefly how you acquired empirical evidence for the Central Limit theorem.

ii) The standard deviation of all the till receipts of a supermarket during 1984 was £4·25.
a) Given that the mean of a random sample of 100 of the till receipts is £18.50, obtain an approximate 95% confidence interval for the mean of all the till receipts during 1984.
b) Find the size of sample that should be taken so that the management can be 95% confident that the sample mean will not differ from the true mean by more than 50p.
c) The mean of all the till receipts of the supermarket during 1983 was £19·40. Using a 5% significance level, investigate whether the sample in a) above provides sufficient evidence to conclude that the mean of all the 1984 till receipts is different from that in 1983.   (L)

60 i) Explain the difference between a discrete variable and a continuous variable, giving an example of each. Describe a suitable diagram which you have used, or

could have used, in a project to illustrate the distribution of a sample of
a) discrete data,
b) continuous data.
The distribution by age, in completed years, of 2000 patients attending a clinic for
rheumatic diseases during 1985 is shown in the table:

Ages of patients attending the clinic in 1985								
Age (years)	0–4	5–14	15–19	20–34	35–49	50–64	65–74	75–89
Frequency	10	32	30	270	300	405	338	615

Draw a histogram to illustrate these data, and estimate the mean age of the
patients to the nearest month.

ii) In a sixth form, the numbers of days of absence through illness of 10 pupils in a
school year were
  10, 4, 2, 0, 42, 6, 4, 12, 1, 8.
Find the mean, the mode, the geometric mean and the median for these data.
State, giving your reasons, the value which you consider gives the best measure
of the typical number of days absence through illness of the pupils during that year.
(L)

61 i) Explain briefly, referring to your projects if possible, the conditions under which
you would measure association using a rank correlation coefficient.
Nine applicants for places at a particular college were interviewed by 2 tutors.
Each tutor ranked the applicants in order of merit. The rankings of the applicants
by each tutor are shown in the table:

Rankings of students									
Applicant	A	B	C	D	E	F	G	H	J
Tutor 1	1	2	3	4	5	6	7	8	9
Tutor 2	1	3	5	4	2	7	9	8	6

Investigate the extent of the agreement between the rankings of the two tutors.

ii) In an experiment the temperature of a metal rod was raised from 300 K. The
extensions E mm of the rod at selected temperatures T K are shown in the table:

Extension of a metal rod								
T	300	350	400	450	500	550	600	650	700
E	0	0·38	0·80	1·22	1·60	2·00	2·42	2·80	3·18

Draw a scatter diagram of the data and mark on your diagram the point
representing the means of E and T.
Find the equation of the regression line of E on T and draw this line on your
diagram.
Estimate the extension of the rod at 430 K. (L)

62 Express $P(E \cup F)$ in terms of $P(E)$ and $P(F)$ when the events E and F are a) mutually
exclusive, b) independent.
State in words the meaning of $P(G|H)$ for two events G and H.
It is known that 0·3% of the population suffer from a certain kidney disease. A urine
test for detecting the presence of the disease will show a positive reaction for 94% of
people suffering from the disease, and for 1% of people who do not suffer from the
disease. Find, to 3 significant figures, the probability that a randomly selected person
who has the test will show a positive reaction.
Given that a randomly selected person who takes the urine test shows a negative
reaction, find, to 2 significant figures, the probability that this person does, in fact,
have the disease.
Use an appropriate Poisson approximation to find, to 3 significant figures, the
probability that not more than 3 out of 1000 randomly selected persons have the
disease. (L)

63   i)   Fifteen professional and ten amateur gymnasts took part in a trampoline contest. Each gymnast was awarded an integer score from 0 to 20 inclusive. A summary of the results is shown below. Calculate the mean and, to 3 decimal places, the standard deviation of the 25 scores.

	Number	Mean Score (marks)	Standard Deviation (marks)
Professionals	15	14.6	2.1
Amateurs	10	10.1	3.6

One of the gymnasts fell during the competition, injuring an ankle, with the result that he scored only 2 marks. Calculate the mean and, to 3 decimal places, the standard deviation of the other 24 scores.

ii)   The weights, to the nearest 0·5 kg, of a random sample of 500 army recruits at a large camp were measured. The results are shown in the table:

Weight of army recruits			
Weight (kg) 59·5–63·0	63·5–67·0	67·5–71·0	71·5–75·0
Frequency 18	45	88	109

Weight (kg) 75·5–79·0	79·5–83·0	83·5–87·0	87·5–91·0
Frequency 80	63	68	29

Construct a cumulative frequency table.
Draw on graph paper the cumulative relative frequency polygon.
Estimate, to the nearest half kg, the median and the interquartile range of the weights of this sample of recruits.   (L)

64   Explain, briefly, the roles of a null hypothesis, an alternative hypothesis and a level of significance in a statistical test, referring to your projects where possible.
A shopkeeper complains that the average weight of chocolate bars of a certain type that he is buying from a wholesaler is less than the stated value of 8·50 g. The shopkeeper weighed 100 bars from a large delivery and found that their weights had a mean of 8·36 g and a standard deviation of 0·72 g. Using a 5% significance level, determine whether or not the shopkeeper is justified in his complaint. State clearly the null and alternative hypotheses that you are using, and express your conclusion in words.
Obtain, to 2 decimal places, the limits of a 98% confidence interval for the mean weight of the chocolate bars in the shopkeeper's delivery.   (L)

65   i)   Define, in words or in symbols, the meaning of each of the following statements.
a) Two events E and F are independent.
b) Two events G and H are mutually exclusive.
Three events A, B and C are defined in the same sample space. The events A and C are mutually exclusive. The events A and B are independent. Given that
$$P(A) = \frac{1}{3}, \ P(C) = \frac{1}{5}, \ P(A \cup B) = \frac{2}{3},$$
find
$$P(A \cup C), \ P(B), \ P(A \cap B).$$
Given also that $P(B \cup C) = \frac{3}{5}$, determine whether or not B and C are independent.

ii)   Jane and Mary play five games. Independently for each game, Jane's probability of winning is $\frac{3}{4}$ in each of the first two games, and $\frac{2}{3}$ in each of the remaining three games. Find the probability that Jane will win exactly three of the five games.   (L)

66   Referring to your projects if possible, give an example of a graphical representation of
a) a discrete frequency distribution,
b) a grouped frequency distribution.

Given the frequency distribution

x	1	2	3	4	5	6	7	8	9
f	1	3	7	9	13	9	5	2	1

find the median and the semi-interquartile range when
c) x is a discrete variable,
d) x is a continuous variable whose values were recorded to the nearest integer.
Calculate also, to 2 decimal places, the mean and the variance of the above distribution. (L)

67 i) Given that X is the number showing when a fair die is thrown, name the distribution of X, and write down its probability function.
Determine the values of $E(X)$ and $Var(X)$.

ii) A child is asked to place 10 objects in order and she gives the ordering
     A C H F B D G E J I.
The correct ordering is
     A B C D E F G H I J.
Find a coefficient of rank correlation between the child's ordering and the correct ordering.
Using a 5% significance level state the conclusion which you draw. (L)

68 A continuous random variable X has probability density function

$$f(x) = k(2-x) \quad 0 \leqslant x \leqslant 2$$
$$f(x) = 0 \quad\quad\quad\ \text{otherwise}$$

where k is a constant.
Obtain an expression for $P(X \leqslant t)$, the probability that X is less than or equal to t, in terms of k and t, where t is a value such that $0 \leqslant t \leqslant 2$. By substituting $t = 2$, verify that $k = \dfrac{1}{2}$.
Using the expression for $P(X \leqslant t)$ above, or otherwise, determine the probability

$$P\left(\frac{1}{2} \leqslant X \leqslant \frac{3}{2}\right).$$ (NI)

69 A class of 9 children has mean weight 35 kg. When a boy B joins the class the mean weight of the class increases by 1 kg. Determine the weight of B.
When a girl G now joins the class of 10, the mean weight of the class is unchanged. What is G's weight?
Determine the ratio of the standard deviation of the weights of the class of 10 without G to the standard deviation of the weights of the class of 11 with G. (NI)

70 a) In a large town with two weekly newspapers, 10% of people read newspaper I, 15% read newspaper II and 3% read both newspapers. What is the probability that a randomly selected person
     i) reads at least one of these newspapers,
     ii) reads newspaper II given that they do not read newspaper I?

b) A test used to detect a certain disease is 90% effective i.e. in 10% of cases in which a person actually has the disease, the test does not detect the disease. When used with persons free of the disease, the test indicates 3% to be affected. What is the probability that the test will indicate the presence of the disease in a person chosen at random from a large population in which 1% have the disease?

c) Four girls A, B, C, D have identical school bags. On the way out of a particular class each girl randomly selects a bag. What is the probability that
     i) A gets her own bag,
     ii) both A and B get their own bags,
     iii) at least one of A and B gets her own bag? (NI)

ANSWERS TO EXAMINATION QUESTIONS

1 a) i) $(1-p)^2$; iii) b) $p^2(3-2p)$; $\dfrac{p}{p+q}$

2 a) -0.0589; c) 0.923

3 a) i) 0.0968; ii) 0.953; b) 470.7–559.3;
   c) 0.209; d) 498.9–532.3;
   e) 0.100–0.150

4 a) i) 0.5987; ii) 0.1490; b) i) 0.0294;
   ii) 0.7514; c) i) 0.5987; ii) 0.97725

5 a) i) 0.21; ii) 0.063, 0.973,
   0.216; b) 0.096; c) 0.344

6 a) i) 0.212, ii) 3.78 b) 0.345;
   c) i) 0.0368; ii) 0.1857; d) i) £444;
   ii) £239.7

7 a) ii) $\dfrac{43}{100}, \dfrac{67}{100}$; iii) $\dfrac{45}{78}$; b) i) 0.867;
   ii) 0.846

8 0.122–0.392; 170.84–178.16;
   165.6–186.8; £487.79 to £530.56

9 a) $744, 736, 752$; b) $744.24, 14.898$,
   0.048 $(0.04$–$0.05)$; d) 25π

10 i) a) 1365; b) 540; c) 1239

11 a) $\dfrac{1}{8}$; c) $E(Y) = 10.67$, method Q

12 $x = 0.4818$, $y = 0.3576$

13) a) i) $\dfrac{9}{22}$; ii) $\dfrac{6}{11}$; iii) $\dfrac{2}{11}$; iv) $\dfrac{4}{7}$;
   b) i) 0.0303; ii) 0.450; iii) 0.0348;
   c) i) 0.36; ii) 0.848

14 84.19–84.81 kg; a) 13; b) 66.4%;
   c) 0.133 kg

15 a) 0.231–0.391;
   b) i) 0.0410; ii) 0.0306

16 a) 0.3901; b) 0.201; c) 0.4087;
   b) 0.02295; g) 0.568

17 0.640–0.885; a) 2.717–2.755;
   b) 2.6795–2.7005; 2.608–2.775

18 b) $\mu = 2$, $\sigma^2 = 0.2$; c) 0.15625;
   Expected profit I) 24.7 p, II) 40 p

19 0.977

20 0.691; 0.0296

21 Not understating; $303.08 < \mu < 317.04$

22 0.504; 0.432; 0.5904; 22

23 i) 0.25; ii) 0.134; iii) $f(x) = 2 - 2x$,
   $0 \leqslant x \leqslant 1$; $f(x) = 0$ elsewhere; iv) $\dfrac{1}{3}, \dfrac{8}{15}$

24 $45.60 < \mu < 49.38$; $> 5\%$; 0.0322

25 i) 0.5274; ii) 0.2684, 1.2

26 $8\dfrac{1}{3}$; 0.9857

27 $0.327 < p < 0.387$; 19; overestimating
   his support

28 ii) $\mu < 80$; iii) $77.51 < \mu < 79.96$

29 i) $\dfrac{3}{28}$; ii) $\dfrac{1}{2}$; iii) $\dfrac{2}{3}$

30 i) $0.455, 3.00$; ii) $3.64, 4.95$;
   ii) $F(x) = 0$, $x < 1$; $F(x) = 0.4551 \ln x$,
   $1 \leqslant x \leqslant 9$; $F(x) = 1$, $x > 9$

31 i) 7.4 ± 0.98; ii) 62

32 i) 3.4 ± 0.44; ii) -0.65 ± 0.01

33 i) $1\dfrac{2}{3}, \dfrac{2}{3}$; ii) $c = \dfrac{2}{3}$

34 a) i) $-0.02, 0.012$;
   ii) -0.02 ± 0.104;
   b) 86.40 ± 1.09; -1.2 ± 1.32

35 $2\dfrac{2}{3}$, 4

36 i) $\dfrac{11}{120}$; ii) $\dfrac{1}{4}$; iii) $\dfrac{23}{60}$;
   iv)

x	3	4	5	6	7	8
$120p(x)$	10	30	35	31	11	3

;
   v) 12

37 ii) 2.45; iii) 3.31

38 i) 0.0397; ii) 0.9628; iii) 0.0843;
   iv) $e^2 = 7.389$; v) 1

39 i) 0.006; ii) 3.733 kg; iii) 0.910;
   iv) 0.336; v) 0.110

40 $E(XY) = \dfrac{5}{12}$

41 i) $0.023, 0.159$; ii) $11\dfrac{1}{3}$; iii) 0.159;
   iv) 0.268

42 i) a) 0.543; b) 0.758; c) 0.072;
   ii) 158

43 i) $1.2, 1.36$; ii) $1.2, 0.023$

44 i) 16.25; ii) a) $(12.46, 14.04)$;
   b) $(1.00, 1.40)$

45 £313; 114; 14% increase

46 No difference

47 a) i) 0.1964; ii) 0.5982; b) 0.9334

48 3.75; a) 0.2067; b) 0.5162;
   45 ± 26.1; Yes

49 a) 0.181; b) 0.999; 0.018; Not
   decreased

50 0.159; 0.7745; 0.975; 2.7 kg;
   £37.53

51 a) 0.991; b) 0.983; c) $\dfrac{7}{25}$; d) 0.0017
   e) £15.40

52 i) a) $\dfrac{2}{15}$; b) $\dfrac{7}{15}$; ii) a) $\dfrac{1}{100}$;
   b) $3.54, 0.47$; c) $14.7, 11.71$

53 $0.110, 0.002$; mean weekly distance
   reduced

54 0.825; $r_s = 0.929$

55 a) 0.076; b) 0.056; c) 0.068;
   d) 0.236; 0.206

56 i) 1.531; ii) a) $\dfrac{1}{2500}$; b) $\dfrac{16}{3}$;
   c) 2.21; d) $0.922, 0.0064$

57 $\lg L = 3.46 - 0.71 \lg D$; -0.98;
   $L = 2884 D^{-0.71}$; 82.22 ppm

58 i) $2.18, 0.26$; ii) £1.18

59 ii) a) £$17\dfrac{2}{3}$–£$19\dfrac{1}{3}$; b) 278; c) Different

60 i) 59 years 4 months;
   ii) $8.9, 4, 0, 5$; Median

61 i) considerable agreement;
   ii) $125E = T - 300$, 1.04 mm

62 0.0128; 0.00018; 0.647

63 i) $12.8, 3.562, 13.25$,
   2.856; ii) 95 kg; 11.5 kg

64 Complaint justified; 8.19–8.53

65   i) $\dfrac{8}{15}, \dfrac{1}{6}$; Independent;. ii) $\dfrac{67}{216}$

66   c) 5, 1;   d) 4·9, 1.15; 4·86, 2·84

67   i) $3\dfrac{1}{2}, \dfrac{35}{12}$;   ii) 0·673; acquired some
measure of learning order of objects

68   $\dfrac{Kt}{2}(4-t)$; $\dfrac{1}{2}$

69   45 kg, 36 kg; $\dfrac{\sqrt{11}}{\sqrt{10}}$

70   a) 0·22, $\dfrac{2}{15}$;   b) 0·0387;   c) $\dfrac{1}{4}, \dfrac{1}{12}, \dfrac{5}{12}$

INDEX